CLOSED-FORM SOLUTIONS FOR DRUG TRANSPORT THROUGH CONTROLLED-RELEASE DEVICES IN TWO AND THREE DIMENSIONS

CLOSED-FORM SOLUTIONS FOR DRUG TRANSPORT THROUGH CONTROLLED-RELEASE DEVICES IN TWO AND THREE DIMENSIONS

LAURENT SIMON, PH.D.
New Jersey Institute of Technology,
Otto H. York Department of Chemical, Biological and Pharmaceutical Engineering,
Newark, NJ 07102

JUAN OSPINA
EAFIT University
Logic and Computation Group, Physics Engineering Program,
School of Sciences and Humanities
Medellin, Colombia

WILEY

Published by John Wiley & Sons, Inc., Hoboken, New Jersey
Published simultaneously in Canada

Library of Congress Cataloging-in-Publication Data:

Simon, Laurent, 1968- , author.
Closed-form solutions for drug transport through controlled-release devices in two and three dimensions / Laurent Simon and Juan Ospina.
p. ; cm.
Includes index.
ISBN 978-1-118-94725-8 (cloth)
I. Ospina, Juan, 1965- , author. II. Title.
[DNLM: 1. Technology, Pharmaceutical–methods. 2. Drug Delivery Systems. 3. Models, Theoretical. QV 778]
RS420
615.1′9–dc23

2014047534

Printed in the United States of America

10 9 8 7 6 5 4 3 2 1

1 2016

CONTENTS

PREFACE

This book covers various classical and analytical techniques to solve boundary-value problems involving two- and three-dimensional partial differential equations (PDEs): separation of variables, Fourier series, Laplace transform, Bromwich integral, residue theorem, Bessel functions, Legendre polynomials, Spherical harmonics, Fourier-Bessel series and Legendre series. These methods are applied to study drug-transport mechanisms in 2-D and 3-D coordinate systems and result in a detailed picture of the evolution of active pharmaceutical ingredients (APIs) through a controlled-released (CR) device or membrane. CR drug-delivery systems are designed to administer an exact dosage of an API to a target site during a treatment period. Contrary to sustained release, which simply prolongs the drug action, CR technologies involve regulating the therapeutic agent release rate while it is being delivered to a particular location. The kinetics, e.g., Higuchi, zero-order or first-order models, is relevant to the development of such devices. Mathematical modeling platforms, that can represent the transport mechanisms adequately, are important assets in the fabrication of these products, as well. This contribution shows how analytical tools, routinely used by physicists, mathematicians and engineers, can be implemented to guide the design of CR devices. A host of diverse real-world applications are taken from the literature to help illustrate the methods in Cartesian, cylindrical and spherical coordinate systems.

Solution strategies for transient problems include the use of separation of variables and Laplace transforms. In some examples, the original PDE, with initial condition, is conveniently written in the Laplace domain. Separation of variables is then used to solve the resulting PDE. After expanding the transformed variables in Fourier series, boundary conditions are applied to calculate the coefficient of the series using orthogonality properties of trigonometric or special functions such as Bessel functions and Legendre polynomials. The inverse Laplace transform of the resulting Fourier series is expressed as a Bromwich integral that can be evaluated using the residue theorem. Green's Function in 3-D Spherical Coordinates and Weber's integral theorem are also exemplified in the text.

Mathematical techniques are implemented using Computer Algebra Software (CAS): Maple, Mathematica and Maxima. In Maple, for example, a subset of packages used includes

VectorCalculus (vector calculus in a number of coordinate systems), *inttrans* (computation of integral transforms such as Fourier, Laplace and Hankel transforms) and *PDETools* (manipulation od PDEs). Some symbolic computations performed in Maple are reproduced in Mathematica and Maxima (a free software).

An expected outcome, of the proposed perspective, is an increase in the understanding of CR devices, which will give students broader career choices and a competitive edge in the job market. For researchers in the field, the integration of modeling and simulations at an early design stage is crucial in the development of new technologies. The materials covered in the book will help provide a good foundation for anyone who wishes to be involved in cutting-edge drug-delivery research.

The textbook is presented in a manner to help the reader apply the theory to their problems. Step-by-step solutions are accompanied by practical case studies. Chapter 1 analyzes the steady-state behavior of a two-dimensional model for percutaneous drug transport. Separation of variables and the principle of linear superposition were applied to solve the Laplace's equation and derive the spatial distribution of drug across the skin. In Chapter 2, multiplicative separation of variables leads to the solution of the Laplace's equation in cylindrical coordinates. The concentration profile inside the walls of a hollow cylinder was developed in terms of Bessel functions. Separation of variables and Bessel functions are applied in Chapter 3 to evaluate drug transport through a double-layered scaffold system. In Chapter 4, Legendre series are needed to develop expressions for the drug concentration inside spherical matrices. Chapter 5 covers the steady-state three-dimensional diffusion of a drug through a membrane and from a distributed source. The reader is introduced to double Fourier series. Chapter 6 deals with the study of Laplace's equation in 3-D cylindrical coordinates. An analytical solution, which involves the Bessel function, was obtained for the drug concentration in the walls of a hollow cylinder with finite length. In Chapter 7, the Green's function was used to derive the solutions for problems of drug diffusion in spherical matrices. A method that combines Laplace transforms and Fourier-Bessel series was applied in Chapter 8 to investigate 2-D drug transport through cylindrical devices. The case study consists of the transient transport behavior of growth factor through double-layered scaffolds. In Chapter 9, Laplace transforms were used, with spherical harmonics, to solve problems of the transient diffusion of drugs inside devices with 3-D spherical geometry. Steady-state and transient diffusion of a ligand in a cylinder were studied in Chapter 10. A disk, covered with receptors, was positioned at the center of the cylinder's bottom. Both the stationary and dynamic solutions were generated via Weber's integral and separation of variables. In Chapter 11, two-dimensional analysis of a cylindrical matrix device was performed using Laplace transforms and Bessel functions. The cumulative percentages of drug released were derived for the two systems considered. In Chapter 12, Laplace transforms and double Fourier series are applied to obtain the average dimensionless concentration and flux for a 3-D Cartesian model. A method for computing the effective time constant for CR devices was provided in Chapter 13. This technique was used to estimate the time it takes to reach an average drug concentration in a cylindrical vitreous body model and to study the release of a medication through a small area of a rectangular-shaped matrix. Optimal drug diffusion coefficients were computed in Chapter 14 using regression algorithms available in Mathematica and analytical solutions of two and three-dimensional transport models. Finally, the optimization of two- and three-dimensional CR systems was examined in Chapter 15. These approaches are based on solution procedures discussed in previous chapters. In the illustrations provided, the polymeric devices were designed with a pre-determined time constant for releasing the drug.

ACKNOWLEDGEMENTS

We are grateful to our families for their love, support and patience. A special thanks goes to Lisa Simon, for her guidance and constructive suggestions and to Francesca Simon for her patience.

1

STEADY-STATE ANALYSIS OF A TWO-DIMENSIONAL MODEL FOR PERCUTANEOUS DRUG TRANSPORT

1.1 SEPARATION OF VARIABLES IN 2-D CARTESIAN COORDINATES

The Laplace's equation in two-dimensional Cartesian coordinates takes the form

$$\frac{\partial^2 c(x,y)}{\partial x^2} + \frac{\partial^2 c(x,y)}{\partial y^2} = 0 \tag{1.1}$$

which is solved to give

$$c(x,y) = F_1(y - xi) + F_2(y + xi) \tag{1.2}$$

where F_1 and F_2 are arbitrary functions of $y - xi$ and $y + xi$, respectively, and $i^2 = -1$. This solution can be obtained in Maple using the command *pdsolve*. However, Eq. (1.2) is rarely used, in practice. Instead, the method of **separation of variables** is adopted. The goal of this technique is to reduce the original problem into a system of ordinary differential equations in one variable (Rice & Do, 1995).

A solution of Eq. (1.1) is $c(x,y) = f(x) + g(y)$. These two functions satisfy the equations

$$\frac{d^2 f(x)}{dx^2} = -c_1 \tag{1.3}$$

Closed-form Solutions for Drug Transport through Controlled-Release Devices in Two and Three Dimensions, First Edition. Laurent Simon and Juan Ospina.

and

$$\frac{d^2g(y)}{dy^2} = c_1 \tag{1.4}$$

where c_1 represents an arbitrary constant. After solving Eqs. (1.3) and (1.4) for $f(x)$ and $g(y)$, we obtain

$$c(x,y) = -\frac{1}{2}c_1x^2 + C_1x + C_2 + \frac{1}{2}c_1y^2 + C_3y \tag{1.5}$$

The solution (1.5) is expressed as an **additive separation of variables** (Cherniavsky, 2010). Another method for solving Eq. (1.1) is the use of a **multiplicative separation of variables** such that $c(x, y) = f(x)g(y)$. In this case, $f(x)$ and $g(y)$ satisfy the following ordinary differential equations:

$$\frac{d^2f(x)}{dx^2} = c_2f(x) \tag{1.6}$$

and

$$\frac{d^2g(y)}{dy^2} = -c_2g(y) \tag{1.7}$$

After solving Eqs. (1.6) and (1.7), the solution is

$$c(x,y) = (D_1e^{x\sqrt{c_2}} + D_2e^{-x\sqrt{c_2}})(D_3\sin(y\sqrt{c_2}) + D_4\cos(y\sqrt{c_2})) \tag{1.8}$$

Given that c_2 is an arbitrary constant, it is possible to apply the **principle of superposition** (Farlow, 1993) to get

$$c(x,y) = \int [(D_1(\eta)e^{x\sqrt{\eta}} + D_2(\eta)e^{-x\sqrt{\eta}})(D_3(\eta)\sin(y\sqrt{\eta}) + D_4(\eta)\cos(y\sqrt{\eta}))]d\eta \tag{1.9}$$

The discrete form of Eq. (1.9) is

$$c(x,y) = \sum_{\eta} [(D_{1\eta}e^{x\sqrt{\eta}} + D_{2\eta}e^{-x\sqrt{\eta}})(D_{3\eta}\sin(y\sqrt{\eta}) + D_{4\eta}\cos(y\sqrt{\eta}))] \tag{1.10}$$

The types of boundary conditions determine the choice of Eq. (1.9) or (1.10). In cases where Eqs. (1.5) and (1.9) are both solutions, their sum is also a solution:

$$\begin{aligned} c(x,y) = {} & -\frac{1}{2}c_1x^2 + C_1x + C_2 + \frac{1}{2}c_1y^2 + C_3y \\ & + \int [(D_1(\eta)e^{x\sqrt{\eta}} + D_2(\eta)e^{-x\sqrt{\eta}})(D_3(\eta)\sin(y\sqrt{\eta}) + D_4(\eta)\cos(y\sqrt{\eta}))]d\eta \end{aligned} \tag{1.11}$$

or

$$\begin{aligned} c(x,y) = {} & -\frac{1}{2}c_1x^2 + C_1x + C_2 + \frac{1}{2}c_1y^2 + C_3y \\ & + \sum_{\eta} [(D_{1\eta}e^{x\sqrt{\eta}} + D_{2\eta}e^{-x\sqrt{\eta}})(D_{3\eta}\sin(y\sqrt{\eta}) + D_{4\eta}\cos(y\sqrt{\eta}))] \end{aligned} \tag{1.12}$$

after using the discretized form of Eq. (1.9).

1.2 MODEL FOR DRUG TRANSPORT ACROSS THE SKIN

The steady-state drug transport across the skin is described by Laplace's equation (1.1). The drug is contained in a patch of length h_c (Fig. 1.1). During treatment, the drug concentration in the reservoir remains unchanged (Simon & Ospina, 2013). Two segments perpendicular to the skin surface, h_u and h_d, are chosen in this application. There is no exchange of material with the environment except at the skin/capillary boundary. A first-order elimination kinetics is observed at the interface. After using the dimensionless variables and constants,

$$x = \frac{x_1}{l_s}, y = \frac{x_2}{l_s}, c = \frac{C}{C_b}, w = \frac{l_s K_{cl}}{D}, L_d = \frac{h_d}{l_s}$$
$$L_c = \frac{h_c}{l_s}, L_u = \frac{h_c + h_u}{l_s} \tag{1.13}$$

the boundary conditions are

$$\alpha(y)c(0,y) + \beta(y)\left.\frac{\partial c(x,y)}{\partial x}\right|_{x=0} = \delta(y) \tag{1.14}$$

$$\left.\frac{\partial c(x,y)}{\partial y}\right|_{y=-L_d} = 0, 0 \le x \le 1 \tag{1.15}$$

$$\left.\frac{\partial c(x,y)}{\partial y}\right|_{y=L_u} = 0, 0 \le x \le 1 \tag{1.16}$$

$$\left.\frac{\partial c(x,y)}{\partial x}\right|_{x=1} + wc(1,y,\tau) = 0, -L_d \le y \le L_u \tag{1.17}$$

where

$$\alpha(y) = \delta(y) = \text{Heaviside}(y) - \text{Heaviside}(y - L_c) \tag{1.18}$$

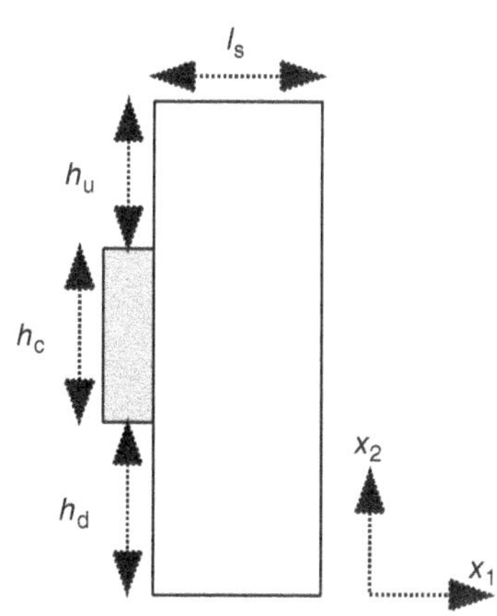

Figure 1.1 Diagram of the drug absorption model.

and "Heaviside$(y-a)$" is the step function defined as

$$\text{Heaviside}(y-a) = \begin{cases} 1 & a \le y \\ 0 & \text{otherwise} \end{cases} \tag{1.19}$$

The coefficients D and K_{cl} are the drug diffusivity and clearance at the skin/capillary boundary; C_b and C are the concentrations in the reservoir and in the skin, respectively. Also,

$$\beta(y) = \alpha(y) - 1 \tag{1.20}$$

Note that Eq. (1.14) is equivalent to the following three conditions:

$$\frac{\partial c(0,y)}{\partial x} = 0, -L_d \le y < 0 \tag{1.21}$$

$$c(0,y) = 1, 0 \le y \le L_c \tag{1.22}$$

and

$$\frac{\partial c(0,y)}{\partial x} = 0, L_c < y \le L_u \tag{1.23}$$

1.3 ANALYTICAL SOLUTION OF THE DIFFUSION MODEL IN 2-D CARTESIAN SYSTEMS

We look for a solution to Eq. (1.1) of the form $c(x,y) = f(x)g(y)$ (i.e., multiplicative separation of variables). Eq (1.8) is used in this case. Condition (1.15) leads to

$$D_4 = -\frac{D_3 \cos(L_d\sqrt{c_2})}{\sin(L_d\sqrt{c_2})} \tag{1.24}$$

Replacing Eq. (1.24) in Eq. (1.8) and applying condition (1.16) yield

$$\sin((L_u + L_d)\sqrt{c_2}) = 0 \tag{1.25}$$

leading to

$$c_2 = \frac{n^2\pi^2}{(L_u + L_d)^2} \tag{1.26}$$

Equation (1.8) becomes

$$c(x,y) = -\frac{\cos\left(\frac{n\pi(y+L_d)}{L_u+L_d}\right)\left(D_1 e^{\frac{n\pi x}{L_u+L_d}} + D_2 e^{-\frac{n\pi x}{L_u+L_d}}\right)}{\sin\left(\frac{n\pi L_d}{L_u+L_d}\right)} \tag{1.27}$$

Given that there is a solution for every value of $n = 0, 1, 2, \cdots$, we write

$$c_n(x,y) = -\frac{\cos\left(\frac{n\pi(y+L_d)}{L_u+L_d}\right)\left(A_n e^{\frac{n\pi x}{L_u+L_d}} + B_n e^{-\frac{n\pi x}{L_u+L_d}}\right)}{\sin\left(\frac{n\pi L_d}{L_u+L_d}\right)} \tag{1.28}$$

Applying the principle of superposition for linear equations, we have

$$c(x,y) = \sum_{n=0}^{\infty}\left[-\frac{\cos\left(\frac{n\pi(y+L_d)}{L_u+L_d}\right)\left(A_n e^{\frac{n\pi x}{L_u+L_d}} + B_n e^{-\frac{n\pi x}{L_u+L_d}}\right)}{\sin\left(\frac{n\pi L_d}{L_u+L_d}\right)}\right] \tag{1.29}$$

The use of condition (1.17) in Eq. (1.29) results in

$$B_n = \frac{A_n e^{\left(\frac{2n\pi}{L_u+L_d}\right)}(n\pi + wL_u + wL_d)}{n\pi - wL_u - wL_d} \tag{1.30}$$

Therefore, the concentration is

$$c(x,y) = -\frac{2A_0(wx-1-w)}{L_d w} + \sum_{n=1}^{\infty}\left[-\frac{\cos\left(\frac{n\pi(y+L_d)}{L_u+L_d}\right)\left(e^{\frac{n\pi x}{L_u+L_d}} + \frac{e^{\left(\frac{2n\pi}{L_u+L_d}\right)}(n\pi+wL_u+wL_d)}{n\pi-wL_u-wL_d}e^{-\frac{n\pi x}{L_u+L_d}}\right)A_n}{\sin\left(\frac{n\pi L_d}{L_u+L_d}\right)}\right] \tag{1.31}$$

Finally, after applying Eq. (1.14) to Eq. (1.31), we have

$$\alpha(y)\left[\frac{2A_0(1+w)}{L_d w} + \sum_{n=1}^{\infty}\left[-\frac{\cos\left(\frac{n\pi(y+L_d)}{L_u+L_d}\right)\left(1 + \frac{e^{\left(\frac{2n\pi}{L_u+L_d}\right)}(n\pi+wL_u+wL_d)}{n\pi-wL_u-wL_d}\right)A_n}{\sin\left(\frac{n\pi L_d}{L_u+L_d}\right)}\right]\right]$$
$$-\beta(y)\left[\frac{2A_0}{L_d} + \sum_{n=1}^{\infty}\left[\frac{n\pi}{L_u+L_d}\frac{\cos\left(\frac{n\pi(y+L_d)}{L_u+L_d}\right)}{\sin\left(\frac{n\pi L_d}{L_u+L_d}\right)}\times\left(e^{\frac{n\pi x}{L_u+L_d}} - \frac{e^{\left(\frac{2n\pi}{L_u+L_d}\right)}(n\pi+wL_u+wL_d)}{n\pi-wL_u-wL_d}e^{-\frac{n\pi x}{L_u+L_d}}\right)A_n\right]\right] = \delta(y) \tag{1.32}$$

The solution to the problem, defined by Eqs. (1.1), (1.14)–(1.17), is given by Eqs. (1.31) and (1.32). Using Eq. (1.32), it is possible to develop successive approximations. For example, a zero-order solution can be obtained by setting $A_n = 0$ with $n = 1, 2, \ldots, \infty$. In this case, Eq. (1.31) reduces to

$$c(x,y) = -\frac{2A_0(wx-1-w)}{L_d w} \tag{1.33}$$

The coefficient A_0 is calculated from Eq. (1.32) resulting in the following "zero-order" approximation of the concentration (Fig. 1.2):

$$c(x,y) = -\frac{L_c(wx-1-w)}{L_c + wL_u + wL_d} \tag{1.34}$$

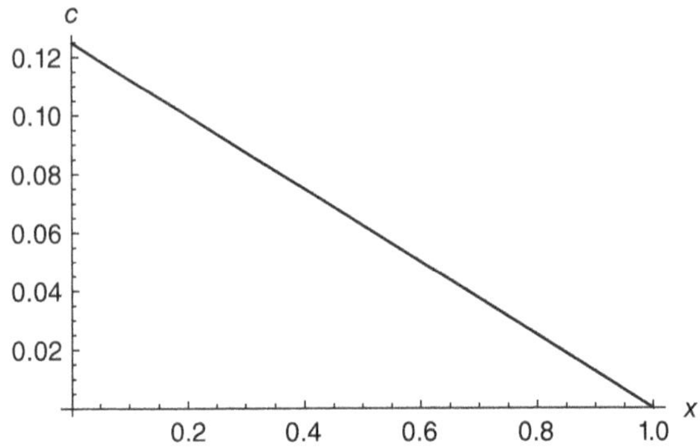

Figure 1.2 Normalized drug concentration in the skin: $L_c = 0.5$, $L_d = 2.5$, $L_u = 3.0$, and $w = 54825$.

1.4 SUMMARY

The method of separation of variables was applied to solve Laplace's equation in two dimensions. In this technique, the partial differential equation is reduced to ordinary differential equations in one variable. The principle of linear superposition was implemented to add the solutions of the subproblems and generate the solution of the initial PDE model. This procedure helps derive the spatial distribution of drug across the skin.

1.5 APPENDIX: MAPLE, MATHEMATICA, AND MAXIMA CODE LISTINGS

1.5.1 Maple Code: steadytwo.mws

```
> restart:with(VectorCalculus):with(PDETools);
[CanonicalCoordinates, ChangeSymmetry, CharacteristicQ,
   CharacteristicQInvariants, ConservedCurrentTest, ConservedCurrents,
   ConsistencyTest, D_Dx, DeterminingPDE, Eta_k, Euler, FromJet,
   FunctionFieldSolutions, InfinitesimalGenerator, Infinitesimals,
   IntegratingFactorTest, IntegratingFactors, InvariantEquation,
   InvariantSolutions, InvariantTransformation, Invariants, Laplace,
   Library, PDEplot, PolynomialSolutions, ReducedForm,
   SimilaritySolutions, SimilarityTransformation, Solve,
   SymmetryCommutator, SymmetryGauge, SymmetrySolutions, SymmetryTest,
   SymmetryTransformation, TWSolutions, ToJet, build, casesplit,
   charstrip, dchange, dcoeffs, declare, diff_table, difforder,
   dplolyform, dsubs,mapde, separability, splitstrip, splitsys,
   undeclare]

> eq:=Laplacian(c(x,y),cartesian[x,y])=0;
```

$$eq := \frac{\partial^2}{\partial x^2} c(x, y) + \frac{\partial^2}{\partial y^2} c(x, y) = 0$$

```
> eq1:=alpha(y)*c(0,y)+beta(y)*Eval(diff(c(x,y),x),x=0)=delta(y):
> eq2:=Eval(diff(c(x,y),y),y=-L[d])=0:
```

```
> eq3:=Eval(diff(c(x,y),y),y=L[u])=0:
> eq4:=Eval(diff(c(x,y),x),x=1)+w*c(1,y)=0:
> eq5:=pdsolve(eq,HINT=f(x)*g(y)):
> eq6:=factor(build(eq5)):
> eq7:=eval(diff(rhs(eq6),y),y=-L[d])=0:
>
> eq8:=isolate(eq7,_C4):
> eq9:=factor(subs(eq8,eq6)):
> eq10:=subs(_C3=1,eq9):
> eq11:=factor(combine((eq10),sin)):
> eq12:=eval(diff(rhs(eq11),y),y=L[u])=0:
> eq13:=sin(_c[1]^(1/2)*(L[u]+L[d]))=0:
>
> eq14:=_c[1]^(1/2)*(L[u]+L[d])=n*Pi:
> eq15:=isolate(eq14,_c[1]):
> eq16:=simplify(subs(eq15,eq11),power,symbolic):
> eq17:=subs(_C1=A[n],_C2=B[n],c=c[n],eq16):
> eq18:=c(x,y)=Sum(rhs(eq17),n=0..infinity);
```

$$eq18 := c(x, y) = \sum_{n=0}^{\infty} \left[-\frac{\cos\left(\frac{n\pi\ (y+L_d)}{L_u+L_d}\right)\left(A_n e^{\frac{n\pi x}{L_u+L_d}} + B_n e^{-\frac{n\pi x}{L_u+L_d}}\right)}{\sin\left(\frac{\pi n L_d}{L_u+L_d}\right)} \right]$$

```
>
> eq22:=subs(c=c[n],eq4):
>
> eq23:=subs(x=1,eq17):
> eq24:=factor(eval(subs(Eval=eval,subs(eq23,subs(eq17,eq22))))):
> eq25:=simplify(simplify(factor(isolate(eq24,B[n])),power,symbolic),
exp):
> eq26:=collect(subs(eq25,eq17),A[n]):
> eq26A:=subs(n=0,simplify(series(rhs(eq26),n=0,4))):
> eq27:=c(x,y)=eq26A+Sum(rhs(eq26),n=1..infinity):
>
> eq19:=eq1:
>
> eq19A:=eval(subs(x=0,eq27)):
> eq20:=factor(eval(subs(Eval=eval,subs(eq19A,subs(eq27,eq19))))):
>
============= Zero-order approximation=====================
> eq28:=factor(eval(subs(Sum=sum,subs(A[n]=0,eq20)))):
> eq29:=simplify(subs(y=z,Int(lhs(eq28),y=-L[d]..L[u])=Int(rhs(eq28),
y=-L[d]..L[u]))):
> eq30:=isolate(eq29,A[0]):
> eq31:=eval(subs(Sum=sum,subs(A[n]=0,eq27))):
> eq32:=subs(eq30,eq31):
> ñas:=alpha(y)=Heaviside(y)-Heaviside(y-L[c]):
```

```
> ñasA:=beta(y)=alpha(y)-1:
> ñasB:=delta(y)=alpha(y):
> eq32A:=subs(y=z,ñas):
> eq32B:=subs(y=z,ñasA):
> eq32C:=subs(y=z,ñasB):
> eq32D:=eval(subs(Int=int,subs(eq32A,subs(eq32C,eq32B,eq32))))
assuming L[c]>0 and L[d]>0 and L[u]>0 and L[u]>L[c];
```

eq32D := $c(x, y) = -\frac{L_c(wx-w-1)}{wL_d+wL_u+L_c}$

```
>
```

1.5.2 wxMaxima Code: steadytwo.wxm

```
(%i1) eq: diff(C(x,y),x,2)+diff(C(x,y),y,2)=0;
```

(%o1) $\frac{d^2}{d\ y^2}C(x,\ y)+\frac{d^2}{d\ x^2}C(x,\ y)=0$

```
(%i2) eq1: C(x,y)=f(x)*g(y)$
(%i3) eq2: (subst(eq1,eq))$
(%i4) eq3: eq2,simp, diff$
(%i5) eq4: eq3/rhs(eq1)$
(%i6) eq5: eq4, expand$
(%i7) eq6: f(x)=f$
(%i8) eq7: subst(eq6,lhs(eq5))= -alpha $
(%i9) eq8: eq7, simp, diff $
(%i10) eq9: g(y) =g $
(%i11) eq10: subst(eq9,lhs(eq5))=alpha $
(%i12) eq11: eq10, simp, diff $
(%i13) eq12: desolve(diff(g(y),y,2)=-alpha*g(y),g(y)) $
Is α positive, negative or zero? positive;
(%i14) eq13: desolve(diff(f(x),x,2)=alpha*f(x),f(x))$
Is α positive, negative or zero? positive;
(%i15) eq14: subst([g(0)=A,(at('diff(g(y),y,1),y=0))=B*sqrt(alpha)],
eq12)$
(%i16) eq14A: subst([f(0)=C,(at('diff(f(x),x,1),x=0))=D*sqrt(alpha)],
eq13)$
(%i17) eq15: C(x,y)=rhs(eq14A)*rhs(eq14)$
(%i18) eq16: subst(y=-L[d],diff(rhs(eq15),y))=0 $
(%i19) eq17: solve(eq16, A) $
(%i20) eq18: subst(eq17,eq15) $
(%i21) eq19: eq18, factor $
(%i22) eq20: subst(B=sin(sqrt(alpha)*L[d]),eq19)$
(%i23) eq21: sin(sqrt(alpha)*L[d])*sin(sqrt(alpha)*y)-cos(sqrt(alpha)*
L[d])*cos(sqrt(alpha)*y)=-cos(sqrt(alpha)*(y+L[d]))$
(%i24) eq22: subst(eq21,eq20)$
(%i25) eq23: subst(y=L[u],diff(rhs(eq22),y))=0$
(%i26) eq24: sin(sqrt(alpha)*(L[u]+L[d]))=0 $
(%i27) eq25: sqrt(alpha)*(L[u]+L[d])= n*%pi $
(%i28) eq26: solve(eq25,alpha) $
```

Is $n(L_u L_d)$ positive, negative or zero? positive;

```
(%i29) eq26A: factor(eq26) $
(%i30) eq27: subst(eq26A,eq22) $
(%i31) assume_pos: n $
(%i32) eq28: eq27, simp $
(%i33) eq29: C = A[n] $
(%i34) eq30: D = B[n] $
(%i35) eq31: C[n](x,y)=subst(eq30,subst(eq29,rhs(eq28))) $
(%i36) eq32: C(x,y)=sum (rhs(eq31), n, 0, inf);
```

(%o36) $C(x, y) = -\sum_{n=0}^{\infty}\left(B_n \sinh\left(\frac{\pi n x}{L_u + L_d}\right) + A_n \cosh\left(\frac{\pi n x}{L_u + L_d}\right)\right) \cos\left(\frac{\pi n (y + L_d)}{L_u + L_d}\right)$

```
(%i37) eq33: subst(x=1,diff(rhs(eq31),x))+w*subst(x=1,rhs(eq31))=0 $
(%i38) eq34: factor(solve(eq33,B[n])) $
(%i39) eq35: factor(subst(eq34,eq31)) $
(%i40) eq36: taylor(rhs(eq35),n,0,1), factor $
(%i41) eq37: C(x,y)=sum(rhs(eq35),n,1,inf)+eq36 $
(%i42) eq38: alpha(y)*subst(x=0,rhs(eq37))+b(y)*subst(x=0,diff(rhs
(eq37),x))=delta(y) $
Zero order approximation:
(%i43) eq39: subst(A[n]=0,eq38) $
(%i44) eq40: subst(y=z,eq39) $
(%i45) eq41: factor(integrate(lhs(eq40),z,-L[d],L[u]))=integrate(rhs
(eq40),z,-L[d],L[u]) $
(%i46) eq42: solve(eq41,A[0]) $
(%i47) eq43: subst(A[n]=0,eq37) $
(%i48) eq44: subst(eq42,eq43);
```

(%o48) $C(x, y) = \frac{(wx - w - 1)\int_{-L_d}^{L_u} \delta(z)dz}{\int_{-L_d}^{L_u} w\ b(z) - w\ \alpha(z) - \alpha(z)dz}$

1.5.3 Mathematica Code: steadytwo.nb

$eq = \partial_{x,x}C[x,y] + \partial_{y,y}C[x,y]$

$C^{(0,2)}[x,y] + C^{(2,0)}[x,y]$

$eq2 = eq/.C \to Function[\{x,y\}, X[x]Y[y]];$

$eq3 = Expand[eq2/(X[x] * Y[y])] == 0;$

$eq4 = \frac{X''[x]}{X[x]} == \alpha;$

$eq4A = \frac{Y''[y]}{Y[y]} == -\alpha;$

$eq5 = DSolve[eq4, X, x];$

$eq5A = (DSolve[eq4A, Y, y]/.C[1] -> A[1])/.C[2] \to B[1];$

$eq6 = (e^{x\sqrt{\alpha}}C[1] + e^{-x\sqrt{\alpha}}C[2]) * (A[1]Cos[y\sqrt{\alpha}] + B[1]Sin[y\sqrt{\alpha}]);$

$eq7 = ((\partial_y eq6)/.y \to -L_d) == 0;$

$eq8 = Solve[eq7, A[1]];$

$eq9 = eq6/.A[1] \to -B[1]Cot[\sqrt{\alpha}L_d];$

$eq10 = eq9/.B[1] \to 1;$

$eq11 = Simplify[eq10];$

$eq12 = ((\partial_y eq11)/.y \to L_u) == 0;$

```
eq13 = Sin[√α(Ld + Lu)] == 0;
eq14 = √α(Ld + Lu) == nπ;
eq15 = Solve[eq14, α];
eq16 = eq11/.α → n^2 π^2/(Ld+Lu)^2;
eq17 = Simplify[Simplify[eq16, n > 0], Ld + Lu > 0];
eq18 = (eq17/.C[1] → An)/.C[2] → Bn;
eq19 = C[x, y] == Sum[-e^(-nπx/(Ld+Lu)) Cos[nπ(y+Ld)/(Ld+Lu)] Csc[nπLd/(Ld+Lu)] (e^(2nπx/(Ld+Lu)) An + Bn), {n, 0, inf}];
eq20 = ((∂x eq18)/.x → 1) + (w(eq18/.x → 1)) == 0;
eq21 = Solve[eq20, Bn];
eq22 = eq19/.Bn → e^(2nπ/(Ld+Lu)) An(nπ+wLd+wLu)/(nπ-wLd-wLu);
eq23 = eq18/.Bn → e^(2nπ/(Ld+Lu)) An(nπ+wLd+wLu)/(nπ-wLd-wLu);
eq24 = Simplify[Series[Expand[eq23An], {n, 0, 0}]];
eq25 =
C[x, y] == A0 Factor[(2+2w-2wx)/(wLd)] +
Sum[-e^(-nπx/(Ld+Lu)) Cos[nπ(y+Ld)/(Ld+Lu)] Csc[nπLd/(Ld+Lu)] (e^(2nπx/(Ld+Lu)) An + e^(2nπ/(Ld+Lu)) An(nπ+wLd+wLu)/(nπ-wLd-wLu)), {n, 0, inf}];
eq26 = A0 Factor[(2+2w-2wx)/(wLd)] +
Sum[-e^(-nπx/(Ld+Lu)) Cos[nπ(y+Ld)/(Ld+Lu)] Csc[nπLd/(Ld+Lu)] (e^(2nπx/(Ld+Lu)) An + e^(2nπ/(Ld+Lu)) An(nπ+wLd+wLu)/(nπ-wLd-wLu)), {n, 0, inf}];
eq27 = α[y](eq26/.x → 0) + b[y]((∂x eq26)/.x → 0) == d[y];
(* Approximation of zero order *)
eq28 = α[y]((-2(-1-w+wx)A0/(wLd))/.x → 0) + b[y]((∂x(-2(-1-w+wx)A0/(wLd)))/.x → 0) == d[y];
eq29 = eq28/.y → z;
eq30 = A0(Integrate[-2b[z]/Ld - 2(-1-w)α[z]/(wLd), {z, -Ld, Lu}]) == Integrate[d[z], {z, -Ld, Lu}];
eq31 = Simplify[Solve[eq30, A0]];
eq32 = C[x, y] == (A0 Factor[(2+2w-2wx)/(wLd)])/.A0 → ∫_{-Ld}^{Lu} d[z] dz / ∫_{-Ld}^{Lu} 2(-wb[z]+(1+w)α[z])/(wLd) dz;
eq33 = Apart[eq32]
C[x, y] == -2(-1-w+wx) ∫_{-Ld}^{Lu} d[z] dz / (w (∫_{-Ld}^{Lu} 2(-wb[z]+(1+w)α[z])/(wLd) dz) Ld)
```

PROBLEMS

1.1. Show that for a zero-order approximation, the concentration is

$$c(x,y) = \frac{(wx - 1 - w)\int_{-L_d}^{L_u} \delta(z)dz}{\int_{-L_d}^{L_u} (-\alpha(z) - w\alpha(z) + w\beta(z))dz} \tag{1.1}$$

1.2. Consider the case when

$$\alpha(y) = \delta(y) = \text{Heaviside}(y) - \text{Heaviside}(y - L_c)$$
$$\beta(y) = \alpha(y) - 1 \quad (1.1)$$

Show that the zero-order approximation of the concentration is

$$c(x,y) = -\frac{L_c(wx - 1 - w)}{L_c + wL_u + wL_d} \quad (1.2)$$

1.3. For

$$\alpha(y) = \delta(y) = \text{Heaviside}(y) - \text{Heaviside}(y - L_c)$$
$$\beta(y) = \alpha(y) - 1 \quad (1.1)$$

with $L_c = 1$, $L_d = 2$, and $L_u = 2$, show that the first-order approximation of the concentration is

$$c(x,y) = -\frac{2A_0(wx - 1 - w)}{L_d w}$$
$$-\frac{\cos\left(\frac{\pi(y+L_d)}{L_u+L_d}\right)\left(e^{\frac{\pi x}{L_u+L_d}} + \frac{e^{\left(\frac{2\pi}{L_u+L_d}\right)}(\pi+wL_u+wL_d)}{\pi-wL_u-wL_d}e^{-\frac{\pi x}{L_u+L_d}}\right)A_1}{\sin\left(\frac{\pi L_d}{L_u+L_d}\right)} \quad (1.2)$$

where

$$A_0 = \frac{w(2872.4w + 1514.5)}{11357.5w^2 + 9319.2w + 1514.5}$$
$$A_1 = -\frac{6.3w(75.0w - 58.9)}{11357.5w^2 + 9319.2w + 1514.5} \quad (1.3)$$

1.4. For

$$\alpha(y) = \delta(y) = \text{Heaviside}(y) - \text{Heaviside}(y - L_c)$$
$$\beta(y) = \alpha(y) - 1 \quad (1.1)$$

with $L_c = 2$, $L_d = 3$, and $L_u = 4$, show that the second-order approximation of the concentration is

$$c(x,y) = -\frac{2}{3}\frac{A_0(wx - 1 - w)}{w} - \frac{\cos\left(\frac{\pi(y+3)}{7}\right)\left(e^{\frac{\pi x}{7}} + \frac{e^{\frac{2\pi}{7}}(\pi+7w)e^{-\frac{\pi x}{7}}}{\pi-7w}\right)A_1}{\sin\left(\frac{3\pi}{7}\right)}$$
$$-\frac{\cos\left(\frac{2\pi(y+3)}{7}\right)\left(e^{\frac{2\pi x}{7}} + \frac{e^{\frac{4\pi}{7}}(2\pi+7w)e^{-\frac{2\pi x}{7}}}{2\pi-7w}\right)A_2}{\sin\left(\frac{\pi}{7}\right)} \quad (1.2)$$

with

$$A_0 = \frac{0.46w(w+0.66)(w+0.22)}{(w+0.94)(w+0.29)(w+0.16)}$$

$$A_1 = -\frac{0.08w(w+0.71)(w-0.45)}{(w+0.94)(w+0.29)(w+0.16)}$$

$$A_2 = -\frac{0.03w(w+0.22)(w-0.90)}{(w+0.94)(w+0.29)(w+0.16)} \tag{1.3}$$

1.5. Consider the case when

$$\alpha(y) = \delta(y) = \text{Heaviside}(y) - \text{Heaviside}(y - L_c)$$

$$\beta(y) = \alpha(y) - 1 \tag{1.1}$$

Derive an expression for the zero-order approximation of the flux defined by

$$J = -\int_{-L_d}^{L_u} \frac{\partial}{\partial x} c\,(x,y)\bigg|_{x=1} dy \tag{1.2}$$

REFERENCES

Cherniavsky, Y. (2010). A note on separation of variables. *Int J Math Educ Sci Technol*, 42(1), 129–131.

Farlow, S. J. (1993). *Partial differential equations for scientists and engineers*. New York: Dover Publications.

Rice, R. G., & Do, Duong D. (1995). *Applied mathematics and modeling for chemical engineers*. New York: Wiley.

Simon, L., & Ospina, J. (2013). Two-dimensional transport analysis of transdermal drug absorption with a non-perfect sink boundary condition at the interface. *Math Biosci*, 244(1), 58–67.

2

CONSTANT DRUG RELEASE FROM A HOLLOW CYLINDER OF FINITE LENGTH IN TWO DIMENSIONS

2.1 SEPARATION OF VARIABLES IN 2-D CYLINDRICAL COORDINATES

The Laplace's equation in three-dimensional cylindrical coordinates takes the form

$$\frac{\frac{\partial}{\partial r}c(r,\theta,z)}{r} + \left(\frac{\partial^2}{\partial r^2}c(r,\theta,z)\right) + \frac{\frac{\partial^2}{\partial \theta^2}c(r,\theta,z)}{r^2} + \left(\frac{\partial^2}{\partial z^2}c(r,\theta,z)\right) = 0 \quad (2.1)$$

For a problem with perfect axial symmetry, the solution does not depend on the variable θ. Equation (2.1) becomes

$$\frac{\frac{\partial}{\partial r}c(r,z) + r\left(\frac{\partial^2}{\partial r^2}c(r,z)\right) + r\left(\frac{\partial^2}{\partial z^2}c(r,z)\right)}{r} = 0 \quad (2.2)$$

Based on the mechanism of **additive separation of variables** (Cherniavsky, 2010), we seek a solution of the form $c(r,z) = f(r) + g(z)$ where the functions f and g satisfy the system:

$$\frac{d^2 f(r)}{dr^2} = -c_1 - \frac{1}{r}\frac{df(r)}{dr} \quad (2.3)$$

Closed-form Solutions for Drug Transport through Controlled-Release Devices in Two and Three Dimensions, First Edition. Laurent Simon and Juan Ospina.

and

$$\frac{d^2g(z)}{dz^2} = c_1 \tag{2.4}$$

where c_1 is an arbitrary constant. Solving Eqs. (2.3) and (2.4) for $f(r)$ and $g(z)$ yields

$$c(r,z) = -\frac{1}{4}c_1r^2 + C_1 \ln(r) + C_2 + \frac{1}{2}c_1z^2 + C_3z \tag{2.5}$$

The technique of **multiplicative separation of variables** can be used to solve Eq. (2.2). In this respect, $c(r,z) = f(r)g(z)$ where $f(r)$ and $g(z)$ satisfy the two ODEs:

$$\frac{d^2f(r)}{dr^2} = c_2f(r) - \frac{1}{r}\frac{df(r)}{dr} \tag{2.6}$$

and

$$\frac{d^2g(z)}{dz^2} = -c_2g(z) \tag{2.7}$$

After solving Eqs. (2.6) and (2.7), $c(r,z)$ becomes

$$c(r,z) = (A_1J_0(r\sqrt{-c_2}) + A_2Y_0(r\sqrt{-c_2}))(B_1 \sin(z\sqrt{c_2}) + B_2 \cos(z\sqrt{c_2})) \tag{2.8}$$

where J_0 is the zero-order Bessel function of the first kind and Y_0 is the zero-order Bessel function of the second kind. Application of the **principle of superposition** (Tyn Myint & Debnath, 2007; Zill et al., 2013) leads to

$$c(r,z) = \int \begin{bmatrix} (A_1(\eta)J_0(r\sqrt{-\eta}) + A_2(\eta)Y_0(r\sqrt{-\eta}))\times \\ (B_1(\eta)\sin(z\sqrt{\eta}) + B_2(\eta)\cos(z\sqrt{\eta})) \end{bmatrix} d\eta \tag{2.9}$$

The discrete version of Eq. (2.9) is

$$c(r,z) = \sum_n \begin{bmatrix} (A_{1,n}(\eta)J_0(r\sqrt{-\eta}) + A_{2,n}(\eta)Y_0(r\sqrt{-\eta}))\times \\ (B_{1,n}(\eta)\sin(z\sqrt{\eta}) + B_{2,n}(\eta)\cos(z\sqrt{\eta})) \end{bmatrix} \tag{2.10}$$

If Eqs. (2.5) and (2.9) are both solutions, the concentration $c(r,z)$ is

$$\begin{aligned} c(r,z) &= -\frac{1}{4}c_1r^2 + C_1 \ln(r) + C_2 + \frac{1}{2}c_1z^2 + C_3z \\ &+\int \begin{bmatrix} (A_1(\eta)J_0(r\sqrt{-\eta}) + A_2(\eta)Y_0(r\sqrt{-\eta}))\times \\ (B_1(\eta)\sin(z\sqrt{\eta}) + B_2(\eta)\cos(z\sqrt{\eta})) \end{bmatrix} d\eta \end{aligned} \tag{2.11}$$

or

$$\begin{aligned} c(r,z) &= -\frac{1}{4}c_1r^2 + C_1 \ln(r) + C_2 + \frac{1}{2}c_1z^2 + C_3z \\ &+\sum_n \begin{bmatrix} (A_{1,n}(\eta)J_0(r\sqrt{-\eta}) + A_{2,n}(\eta)Y_0(r\sqrt{-\eta}))\times \\ (B_{1,n}(\eta)\sin(z\sqrt{\eta}) + B_{2,n}(\eta)\cos(z\sqrt{\eta})) \end{bmatrix} \end{aligned} \tag{2.12}$$

when the discretized form of Eq. (2.10) is applied.

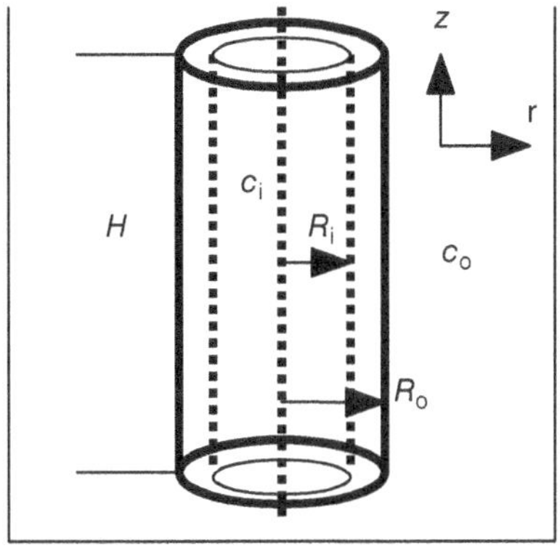

Figure 2.1 Diffusion through the walls of the hollow cylinder.

2.2 MODEL FOR DRUG RELEASE FROM A HOLLOW CYLINDER

Drug-release experiments are often performed by placing dialysis tubes, containing a drug solution, in a buffer medium (Gao et al., 2007; Sanson et al., 2010). These investigations can be conducted to assess formulations of drug-loaded microspheres. Several factors, such as a suitable mathematical model, the sampling method, and maintaining a perfect sink condition, are to be considered when analyzing *in-vitro* release data. In a recent study, several dialysis equilibrium techniques were evaluated for the study of topical liposomal gel formulations to deliver hydrophobic drugs (Hua, 2014). It was stated that the kinetics of drug release was essential to a rational design of delivery systems. Therefore, it is important to study how drug molecules diffuse through the walls of these semipermeable membranes.

The dialysis bag is modeled as a hollow cylindrical device in two dimensions with inner radius R_i and outer radius R_o (Fig. 2.1). The tube has a finite length H and the steady-state transport through the walls is described by Fick's law of diffusion:

$$\frac{1}{r}\frac{\partial}{\partial r}\left(r\frac{\partial c}{\partial r}\right)+\frac{\partial^2 c}{\partial z^2}=0 \qquad R_i \le r \le R_o,\ \ 0 \le z \le H \tag{2.13}$$

The concentration in the release medium is kept at c_0 (i.e., sink condition):

$$c(r,0)=c(r,H)=c(R_o,z)=c_o \tag{2.14}$$

In the laboratory, the sink condition is maintained by introducing a volume of a fresh medium equal to the volume collected at each sampling time. The value of c_0 is usually zero. The drug concentration in the bag is c_i:

$$c(R_i,z)=c_i \tag{2.15}$$

2.3 ANALYTICAL SOLUTION OF THE TRANSPORT MODEL IN 2-D CYLINDRICAL COORDINATES

The concentration $c(r,z)$ is first written as $c_o + C(r,z)$, where $C(r,z) = f(r)g(z)$. Then, Eq. (2.14) is converted into **homogeneous boundary conditions** before applying the method of separation of variables. As described in Section 2.1, f and g satisfy Eqs. (2.6) and (2.7). The function $C(r, z)$ is

$$C(r,z) = (C_1 J_0(r\sqrt{-c_2}) + C_2 Y_0(r\sqrt{-c_2}))(C_3 \sin(z\sqrt{c_2}) + C_4 \cos(z\sqrt{c_2})) \tag{2.16}$$

Applying the boundary condition (2.14) at $r = R_o$ to Eq. (2.16) yields

$$(C_1 J_0(R_o\sqrt{-c_2}) + C_2 Y_0(R_o\sqrt{-c_2}))(C_3 \sin(z\sqrt{c_2}) + C_4 \cos(z\sqrt{c_2})) = 0 \tag{2.17}$$

since $C(R_o, z) = c(R_o, z) - c_o$. From Eq. (2.17), we calculate C_2 as

$$C_2 = -\frac{C_1 J_0(R_o\sqrt{-c_2})}{Y_0(R_o\sqrt{-c_2})} \tag{2.18}$$

because the term $C_3 \sin(z\sqrt{c_2}) + C_4 \cos(z\sqrt{c_2})$ is not zero for all values of z. In addition, the radial solution has to be zero at the outer radius. After replacing C_2 in Eq. (2.16), we obtain

$$C(r,z) = \left(C_1 J_0\left(r\sqrt{-c_2}\right) - \frac{C_1 J_0(R_o\sqrt{-c_2})}{Y_0(R_o\sqrt{-c_2})} Y_0(r\sqrt{-c_2}) \right) (C_3 \sin(z\sqrt{c_2}) + C_4 \cos(z\sqrt{c_2})) \tag{2.19}$$

or

$$C(r,z) = \left(\frac{C_1 J_0\left(r\sqrt{-c_2}\right) Y_0(R_o\sqrt{-c_2}) - C_1 J_0(R_o\sqrt{-c_2}) Y_0(r\sqrt{-c_2})}{Y_0(R_o\sqrt{-c_2})} \right) \times$$
$$(C_3 \sin(z\sqrt{c_2}) + C_4 \cos(z\sqrt{c_2})) \tag{2.20}$$

Without loss of generality, C_1 is set equal to 1:

$$C(r,z) = \left(\frac{J_0\left(r\sqrt{-c_2}\right) Y_0(R_o\sqrt{-c_2}) - J_0(R_o\sqrt{-c_2}) Y_0(r\sqrt{-c_2})}{Y_0(R_o\sqrt{-c_2})} \right) \times$$
$$(C_3 \sin(z\sqrt{c_2}) + C_4 \cos(z\sqrt{c_2})) \tag{2.21}$$

The boundary condition (2.14) at $z = 0$ gives

$$0 = \left(\frac{J_0\left(r\sqrt{-c_2}\right) Y_0(R_o\sqrt{-c_2}) - J_0(R_o\sqrt{-c_2}) Y_0(r\sqrt{-c_2})}{Y_0(R_o\sqrt{-c_2})} \right) \times$$
$$(C_3 \sin(0\sqrt{c_2}) + C_4 \cos(0\sqrt{c_2})) \tag{2.22}$$

or C_4, resulting in

$$C(r,z) = C_3 \sin(z\sqrt{c_2}) \left(\frac{J_0\left(r\sqrt{-c_2}\right) Y_0(R_o\sqrt{-c_2}) - J_0(R_o\sqrt{-c_2}) Y_0(r\sqrt{-c_2})}{Y_0(R_o\sqrt{-c_2})} \right) \tag{2.23}$$

At $z = H$, Eq. (2.23) becomes

$$C_3 \sin(H\sqrt{c_2}) \left(\frac{J_0\left(r\sqrt{-c_2}\right) Y_0(R_o\sqrt{-c_2}) - J_0(R_o\sqrt{-c_2}) Y_0(r\sqrt{-c_2})}{Y_0(R_o\sqrt{-c_2})} \right) = 0 \tag{2.24}$$

after using Eq. (2.14). The value of c_2 is derived from Eq. (2.24) as

$$c_2 = \frac{n^2\pi^2}{H^2} \tag{2.25}$$

Hence,

$$C(r,z) = C_3 \sin\left(\frac{n\pi z}{H}\right) \left(\frac{I_0\left(\frac{n\pi r}{H}\right) Y_0\left(\frac{n\pi R_o}{H} i\right) - I_0\left(\frac{n\pi R_o}{H}\right) Y_0\left(\frac{n\pi r}{H} i\right)}{Y_0\left(\frac{n\pi R_o}{H} i\right)} \right) \tag{2.26}$$

Because there is a solution for every value of $n = 1, 2, \cdots$, Eq. (2.26) can be written as

$$C_n(r,z) = A_n \sin\left(\frac{n\pi z}{H}\right) \left(\frac{I_0\left(\frac{n\pi r}{H}\right) Y_0\left(\frac{n\pi R_o}{H} i\right) - I_0\left(\frac{n\pi R_o}{H}\right) Y_0\left(\frac{n\pi r}{H} i\right)}{Y_0\left(\frac{n\pi R_o}{H} i\right)} \right) \tag{2.27}$$

According to the principle of superposition, we have

$$c(r,z) = c_o + \sum_{n=1}^{\infty} A_n \sin\left(\frac{n\pi z}{H}\right) \left(\frac{I_0\left(\frac{n\pi r}{H}\right) Y_0\left(\frac{n\pi R_o}{H} i\right) - I_0\left(\frac{n\pi R_o}{H}\right) Y_0\left(\frac{n\pi r}{H} i\right)}{Y_0\left(\frac{n\pi R_o}{H} i\right)} \right) \tag{2.28}$$

The sink condition (2.14) is applied to Eq. (2.28):

$$c_o + \sum_{n=1}^{\infty} A_n \sin\left(\frac{n\pi z}{H}\right) \left(\frac{I_0\left(\frac{n\pi R_i}{H}\right) Y_0\left(\frac{n\pi R_o}{H} i\right) - I_0\left(\frac{n\pi R_o}{H}\right) Y_0\left(\frac{n\pi R_i}{H} i\right)}{Y_0\left(\frac{n\pi R_o}{H} i\right)} \right) = c_i \tag{2.29}$$

or

$$\sum_{n=1}^{\infty} A_n \sin\left(\frac{n\pi z}{H}\right) \left(\frac{I_0\left(\frac{n\pi R_i}{H}\right) Y_0\left(\frac{n\pi R_o}{H} i\right) - I_0\left(\frac{n\pi R_o}{H}\right) Y_0\left(\frac{n\pi R_i}{H} i\right)}{Y_0\left(\frac{n\pi R_o}{H} i\right)} \right) = c_i - c_o \tag{2.30}$$

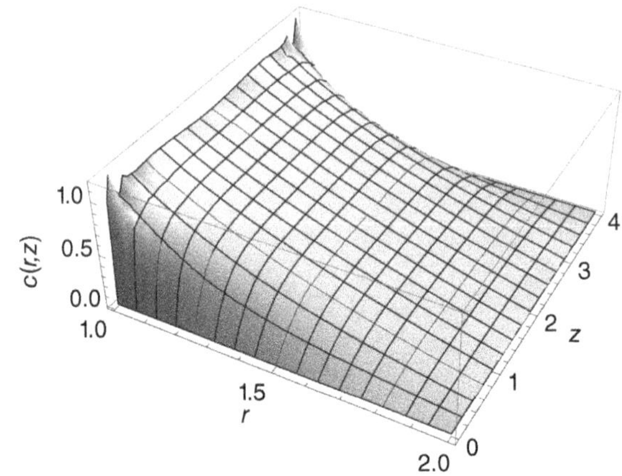

Figure 2.2 Concentration profile when $R_i = 1$, $R_o = 2$, $H = 4$, $c_o = 0$ and $c_i = 1$. Fifteen terms are used in the series.

We multiply both sides of Eq. (2.30) by $\sin\left(\frac{n\pi z}{H}\right)$ and integrate from 0 to H to get

$$-\frac{1}{2}\frac{HA_n\left(-\mathrm{I}_0\left(\frac{\pi nR_i}{H}\right)\mathrm{Y}_0\left(\frac{\pi nR_0 I}{H}\right)+\mathrm{I}_0\left(\frac{\pi nR_0}{H}\right)\mathrm{Y}_0\left(\frac{\pi nR_i I}{H}\right)\right)}{\mathrm{Y}_0\left(\frac{\pi nR_0 I}{H}\right)} = \frac{H(-c_i+c_0)(-1+(-1)^n)}{n\pi} \tag{2.31}$$

Therefore,

$$A_n = \frac{2(c_i-c_0)(-1+(-1)^n)\mathrm{Y}_0\left(\frac{\pi nR_0 I}{H}\right)}{n\pi\left(-\mathrm{I}_0\left(\frac{\pi nR_i}{H}\right)\mathrm{Y}_0\left(\frac{\pi nR_0 I}{H}\right)+\mathrm{I}_0\left(\frac{\pi nR_0}{H}\right)\mathrm{Y}_0\left(\frac{\pi nR_i I}{H}\right)\right)} \tag{2.32}$$

The final solution is given by Eq. (2.28) with A_n defined by Eq. (2.32). We can calculate the amount of drug inside the walls of the hollow cylinder by using the following formula:

$$M = \int_0^H \int_{R_i}^{R_0} rc(r,z)\ \mathrm{d}r\ \mathrm{d}z \tag{2.33}$$

Figure 2.2 shows the concentration profile for the following values: $R_i = 1$, $R_0 = 2$, $H = 4$, $c_0 = 0$ and $c_i = 1$. The amount M is 1.997 when 15 terms (i.e., $N = 15$) are used in the series. When $N = 50$, we obtain $M = 1.998$.

2.4 SUMMARY

The method of multiplicative separation of variables was applied to solve Laplace's equation in cylindrical coordinates. As some boundary conditions were not homogeneous, it was necessary to use an appropriate change of variables before implementing the separation process. A profile for the concentration inside the walls of a hollow cylinder was developed in terms of Bessel functions. The method exhibited superior convergence properties. The results can help interpret drug-release kinetics by dialysis-tubing techniques.

2.5 APPENDIX: MAPLE CODE LISTINGS

2.5.1 Maple Code: steadycyltwo.mws

```
> restart:
> with(VectorCalculus):with(PDETools);
[CanonicalCoordinates, ChangeSymmetry, CharacteristicQ,
   CharacteristicQInvariants, ConservedCurrentTest, ConservedCurrents,
   ConsistencyTest, D_Dx, DeterminingPDE, Eta_k, Euler, FromJet,
   FunctionFieldSolutions, InfinitesimalGenerator, Infinitesimals,
   IntegratingFactorTest, IntegratingFactors, InvariantEquation,
   InvariantSolutions, InvariantTransformation, Invariants, Laplace,
   Library, PDEplot, PolynomialSolutions, ReducedForm,
   SimilaritySolutions, SimilarityTransformation, Solve,
   SymmetryCommutator, SymmetryGauge, SymmetrySolutions, SymmetryTest,
   SymmetryTransformation, TWSolutions, ToJet, build, casesplit,
   charstrip, dchange, dcoeffs, declare, diff_table, difforder,
   dplolyform, dsubs,mapde, separability, splitstrip, splitsys,
   undeclare]
> eq0:=expand(Laplacian(C(r,theta,z),cylindrical[r,theta,z]))=0;
```

$$eq0 := \frac{\frac{\partial}{\partial r} C(r,\theta,z)}{r} + \frac{\partial^2}{\partial r^2} C(r,\theta,z) + \frac{\frac{\partial^2}{\partial \theta^2} C(r,\theta,z)}{r^2} + \frac{\partial^2}{\partial z^2} C(r,\theta,z) = 0$$

```
> eq0A:=expand(Laplacian(C(r,z),cylindrical[r,theta,z]))=0:
> eq0B:=pdsolve(eq0A,HINT=f(r)+g(z)):
> eq0C:=build(eq0B):
> eq:=Laplacian(C(r,z),cylindrical[r,theta,z])=0:
> eq1:=pdsolve(eq,HINT=f(r)*g(z)):
> eq2:=factor(build(eq1)):
>
eq2A := C(r,z) = subs({_C1 = A[1],_C2 = A[2],_C3 = B[1],
_C4 = B[2],_c[1] = eta},rhs(eq2)) :
eq2B := C(r,z) = subs({_C1 = A[1](eta),_C2 = A[2](eta),_C3 = B[1](eta),
_C4 = B[2](eta),_c[1] = eta},rhs(eq2)) :
> eq2C:=C(r,z)=Int(rhs(eq2B),eta):
>
eq2D := C(r,z) = subs({_C1 = A[1,eta],_C2 = A[2,eta],_C3 = B[1,eta],_
C4 = B[2,eta],_c[1] = eta},rhs(eq2)) :
> eq2E:=C(r,s)=Sum(rhs(eq2D),eta):
> eq2F:=C(r,z)=rhs(eq0C)+rhs(eq2C):
> eq2G:=C(r,z)=rhs(eq0C)+rhs(eq2E):
```

```
> eq3:=subs(r=R[0],(rhs(eq2)))=0:
> eq4:=isolate(eq3,_C2):
> eq5:=factor(subs(eq4,eq2)):
> eq6:=subs(_C1=1,eq5):
> eq7:=eval(subs(z=0,rhs(eq6)))=0:
> eq8:=isolate(eq7,_C4):
> eq9:=subs(eq8,eq6):
> eq10:=subs(z=H,rhs(eq9))=0:
> eq11:=sin(_c[1]^(1/2)*H)=0:
> eq12:=_c[1]^(1/2)*H=n*Pi:
> eq13:=isolate(eq12,_c[1]):
> eq14:=simplify(subs(eq13,eq9),power,symbolic):
> eq15:=subs(C=C[n],_C3=A[n],eq14):
> eq16:=c(r,z)=c[0]+Sum(C[n](r,z),n=1..infinity):
> eq17:=subs(eq15,eq16):
> eq18:=subs(r=R[i],rhs(eq17))=c[i]:
> eq19:=int(subs(r=R[i],rhs(eq15))*sin(n*Pi*z/H),z=0..H)=int((c[i]-
c[0])*sin(n*Pi/H*z),z=0..H) assuming n::integer:
> eq20:=isolate(eq19,A[n]):
> eq20A:=subs(eq20,eq17);
```

$$\text{eq20A} := c(r, z) = c_0 + \sum_{n=1}^{\infty} \left(2\left(\text{BesselI}\left(0, \frac{\pi n R_0}{H}\right) \text{BesselY}\left(0, \frac{I\pi n r}{H}\right) - \text{BesselY}\left(0, \frac{I\pi n R_0}{H}\right) \text{BesselI}\left(0, \frac{\pi n r}{H}\right)\right) (-c_i + c_0)\,(-1+(-1)^n) \sin\left(\frac{n\pi z}{H}\right)\right) \Big/ \left(\pi\, n \left(\text{BesselI}\left(0, \frac{\pi n R_0}{H}\right) \text{BesselY}\left(0, \frac{I\pi n R_i}{H}\right) - \text{BesselY}\left(0, \frac{I\pi n R_0}{H}\right) \text{BesselY}\left(0, \frac{\pi n R_i}{H}\right)\right)\right)$$

\>

PROBLEMS

2.1. Develop an equation for the concentration profile when $c_o = 0$.

2.2. Derive the expression for A_n in Eq. (2.32) using the properties of Fourier series.

2.3. Derive an expression for the total amount of drug in the walls of the hollow cylinder (M). Verify that for the following values $R_i = 1$, $R_o = 2$, $H = 4$, $c_o = 0$ and $c_i = 1$, the amount of drug M is 2.0.

2.4. Develop an expression for the total flux through the cylinder.

2.5. Plot the concentration profile for the following values $R_i = 0.5$, $R_o = 1$, $H = 5$, $c_o = 0$ and $c_i = 1$. Calculate the amount of drug.

REFERENCES

Cherniavsky, Y. (2010). A note on separation of variables. *Int J Math Educ Sci Technol*, 42(1), 129–131.

Gao, P., Xu, H., Ding, P., Gao, Q., Sun, J., & Chen, D. (2007). Controlled release of huperzine A from biodegradable microspheres: In vitro and in vivo studies. *Int J Pharm*, 330(1–2), 1–5.

Hua, S. (2014). Comparison of in vitro dialysis release methods of loperamide-encapsulated liposomal gel for topical drug delivery. *Int J Nanomedicine*, 9, 735–744.

Sanson, C., Schatz, C., Le Meins, J. F., Soum, A., Thevenot, J., Garanger, E., & Lecommandoux, S. (2010). A simple method to achieve high doxorubicin loading in biodegradable polymersomes. *J Control Release*, 147(3), 428–435.

Tyn Myint, U., & Debnath, L. (2007). *Linear partial differential equations for scientists and engineers* (4th ed.). Boston: Birkhäuser.

Zill, D. G., Wright, W. S., & Cullen, M. R. (2013). *Differential equations with boundary-value problems* (8th ed.). Boston, MA: Brooks/Cole, Cengage Learning.

3

ANALYSIS OF STEADY-STATE GROWTH FACTOR TRANSPORT THROUGH DOUBLE-LAYERED SCAFFOLDS

3.1 GOVERNING STEADY-STATE TRANSPORT EQUATIONS

Poly (lactide-co-glycolide) (PLG) scaffold systems have been used to facilitate localized and sustained delivery of growth factors that contribute to tissue regeneration. The development of new blood vessels, known as angiogenesis, is critical for wound repair and the maturation of viable and healthy offspring (Grazul-Bilska et al., 2011; Reynolds & Redmer, 2001). This mechanism is initiated by pro-angiogenic factors such as the vascular endothelial growth factor (VEGF) (Chen et al., 2007b). Subsequent vessel stabilization is promoted by the platelet-derived growth factor (PDGF).

Consider a scaffold with two connected compartments, layers 1 and 2, packed on top of each other (Fig. 3.1). This system was adapted for sequential delivery of VEGF and PDGF yielding spatially organized tissues (Chen et al., 2007b). The device is fabricated in such a way that the active agent remains confined in each cell prior to the release experiment. A mathematical model, consisting of two reaction–diffusion equations, is built to describe transport across the device (Chen et al., 2007a, b):

$$\eta\left(\frac{1}{r}\frac{\partial}{\partial r}\left(r\frac{\partial c_1}{\partial r}\right)+\frac{\partial^2 c_1}{\partial z^2}\right)+k_2c_{10}-k_cc_1=0 \tag{3.1}$$

$$\eta\left(\frac{1}{r}\frac{\partial}{\partial r}\left(r\frac{\partial c_2}{\partial r}\right)+\frac{\partial^2 c_2}{\partial z^2}\right)+k_2c_{20}-k_cc_2=0 \tag{3.2}$$

where $c_1(r,z)$ is the concentration of the medication (VEGF or PDGF) in the cylindrical region defined by $0<r<R$ and $0<z<b$ (i.e., layer 1); $c_2(r,z)$ is its concentration in the space

---

Closed-form Solutions for Drug Transport through Controlled-Release Devices in Two and Three Dimensions, First Edition. Laurent Simon and Juan Ospina.

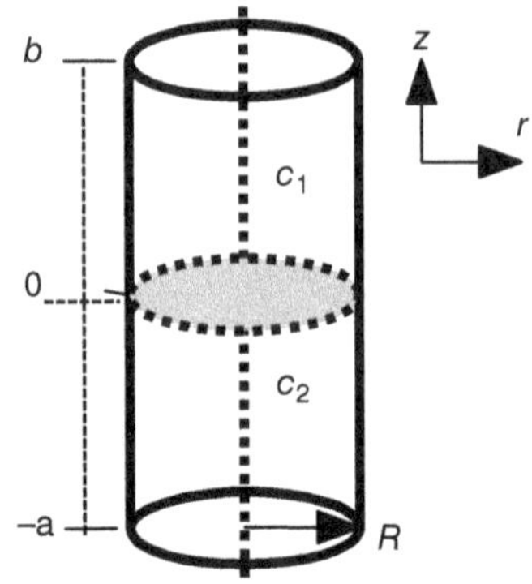

Figure 3.1 Schematic of the double-layered scaffold system.

defined by $0 < r < R$ and $-a < z < 0$ (i.e., layer 2); η is the diffusivity in the polymeric matrix; k_2 and k_c are release and degradation rates, respectively; c_{10} and c_{20} are the initial drug concentrations. The steady-state system, governed by diffusion, generation, and degradation, includes the following boundary conditions:

$$\left.\frac{\partial c_1(r,z)}{\partial r}\right|_{r=0} = 0 \tag{3.3}$$

$$\left.\frac{\partial c_2(r,z)}{\partial r}\right|_{r=0} = 0 \tag{3.4}$$

$$c_1(R,z) = 0 \tag{3.5}$$

$$c_2(R,z) = 0 \tag{3.6}$$

$$\left.\frac{\partial c_1(r,z)}{\partial z}\right|_{z=0} = \left.\frac{\partial c_2(r,z)}{\partial z}\right|_{z=0} \tag{3.7}$$

$$c_1(r,0) = c_2(r,0) \tag{3.8}$$

$$c_1(r,b) = 0 \tag{3.9}$$

$$c_2(r,-a) = 0 \tag{3.10}$$

Equations (3.3) and (3.4) are the symmetry boundary conditions at the center of the device. Perfect-sink conditions, represented by Eqs. (3.5), (3.6), (3.9), and (3.10), are applied at the surface; Eq. (3.7) is the flux continuity at the interface between the layers; Eq. (3.8) is a local equilibrium partition condition.

3.2 SOLUTION PROCEDURE FOR TRANSPORT THROUGH A TWO-LAYERED SCAFFOLD

The system of Eqs. (3.1)–(3.10) is solved using **separation of variables**. We look for solutions of the forms

$$c_1(r,z) = G_1(r) + F_1(r,z) \tag{3.11}$$

and

$$c_2(r,z) = G_2(r) + F_2(r,z) \tag{3.12}$$

where $G_1(r)$ and $F_1(r,z)$ satisfy Eqs. (3.3) and (3.5); $G_2(r)$ and $F_2(r,z)$ are constrained by Eqs. (3.4) and (3.6); $F_1(r,z)$ and $F_2(r,z)$ satisfy Eq. (3.7).

- Determination of $G_1(r)$ and $G_2(r)$

The first step in the procedure is to solve for $G_1(r)$. This function is such that

$$\eta\left(\frac{1}{r}\frac{\partial}{\partial r}\left(r\frac{\partial G_1}{\partial r}\right)\right) + k_2c_{10} - k_cG_1 = 0 \tag{3.13}$$

The general solution of Eq. (3.13) is

$$G_1(r) = C_1\mathrm{Y}_0\left(\sqrt{-\frac{k_C}{\eta}}r\right) + C_2\mathrm{J}_0\left(\sqrt{-\frac{k_C}{\eta}}r\right) + \frac{k_2c_{10}}{k_C} \tag{3.14}$$

where C_1 and C_2 are integration constants; J_0 and Y_0 are Bessel functions of the first and second kind, respectively. Applying Eqs. (3.3) and (3.5) to Eq. (3.14) leads to

$$C_1 = 0 \tag{3.15}$$

and

$$C_2 = -\frac{k_2c_{10}}{k_C\mathrm{J}_0\left(R\sqrt{-\frac{k_C}{\eta}}\right)} \tag{3.16}$$

Finally,

$$G_1(r) = -\frac{k_2c_{10}}{k_C\mathrm{J}_0\left(R\sqrt{-\frac{k_C}{\eta}}\right)}\mathrm{J}_0\left(\sqrt{-\frac{k_C}{\eta}}r\right) + \frac{k_2c_{10}}{k_C} \tag{3.17}$$

Following a method similar to the one described above, it can be shown that

$$G_2(r) = -\frac{\mathrm{J}_0\left(r\sqrt{-\frac{k_C}{\eta}}\right)k_2c_{20}}{k_C\mathrm{J}_0\left(R\sqrt{-\frac{k_C}{\eta}}\right)} + \frac{k_2c_{20}}{k_C} \tag{3.18}$$

- Determination of the general solution $F_1(r, z)$

From Eq. (3.11), $c_1(r, z)$ takes the form

$$c_1(r,z) = -\frac{\mathrm{J}_0\left(r\sqrt{-\frac{k_C}{\eta}}\right) k_2 c_{10}}{k_C \mathrm{J}_0\left(R\sqrt{-\frac{k_C}{\eta}}\right)} + \frac{k_2 c_{10}}{k_C} + F_1(r,z) \tag{3.19}$$

After replacing Eq. (3.19) into Eq. (3.1) and simplifying further, the following expression is obtained:

$$-\frac{\left(\frac{\partial}{\partial r}F_1(r,z)\right)\eta}{r} - \left(\frac{\partial^2}{\partial r^2}F_1(r,z)\right)\eta - \left(\frac{\partial^2}{\partial z^2}F_1(r,z)\right)\eta + k_C F_1(r,z) = 0 \tag{3.20}$$

The solution of Eq. (3.20) gives

$$F_1(r,z) = f_1(r) g_1(z) \tag{3.21}$$

where

$$\frac{\mathrm{d}^2 f_1(r)}{\mathrm{d}r^2} = C_3 f_1(r) - \frac{1}{r}\frac{\mathrm{d}f_1(r)}{\mathrm{d}r} \tag{3.22}$$

and

$$\frac{\mathrm{d}^2 g_1(z)}{\mathrm{d}z^2} = -C_3 g_1(z) + \frac{k_C g_1(z)}{\eta} \tag{3.23}$$

Note that C_3 is a constant from the separation-of-variable method. As a result, $F_1(r, z)$ becomes

$$F_1(r,z) = \left(C_4 \sin\left(\frac{z\sqrt{C_3\eta - k_C}}{\sqrt{\eta}}\right) + C_5 \cos\left(\frac{z\sqrt{C_3\eta - k_C}}{\sqrt{\eta}}\right)\right) \times (C_6 \mathrm{J}_0(r\sqrt{-C_3}) + C_7 \mathrm{Y}_0(r\sqrt{-C_3})) \tag{3.24}$$

after solving Eqs. (3.22) and (3.23). After applying condition (3.3) to Eq. (3.24) and setting $C_6 = 1$, we have

$$F_1(r,z) = \mathrm{J}_0(r\sqrt{-C_3})\left(C_4 \sin\left(\frac{z\sqrt{C_3\eta - k_C}}{\sqrt{\eta}}\right) + C_5 \cos\left(\frac{z\sqrt{C_3\eta - k_C}}{\sqrt{\eta}}\right)\right) \tag{3.25}$$

- Determination of the general solution $F_2(r, z)$

Following a technique similar to the one that yielded Eq. (3.19), $c_2(r, z)$ can be expressed as

$$c_2(r,z) = -\frac{\mathrm{J}_0\left(r\sqrt{-\frac{k_C}{\eta}}\right)k_2c_{20}}{k_C\mathrm{J}_0\left(R\sqrt{-\frac{k_C}{\eta}}\right)} + \frac{k_2c_{20}}{k_C} + F_2(r,z) \tag{3.26}$$

and, by analogy to Eq. (3.20), $F_2(r, z)$ is given as

$$-\frac{\left(\frac{\partial}{\partial r}F_2(r,z)\right)\eta}{r} - \left(\frac{\partial^2}{\partial r^2}F_2(r,z)\right)\eta - \left(\frac{\partial^2}{\partial z^2}F_2(r,z)\right)\eta + k_CF_2(r,z) = 0 \tag{3.27}$$

As a result, $F_2(r, z)$ is

$$F_2(r,z) = \mathrm{J}_0(r\sqrt{-C_3})\left(C_8\sin\left(\frac{z\sqrt{C_3\eta - k_C}}{\sqrt{\eta}}\right) + C_9\cos\left(\frac{z\sqrt{C_3\eta - k_C}}{\sqrt{\eta}}\right)\right) \tag{3.28}$$

Note that the same constant C_3 appears in $F_1(r, z)$ and $F_2(r, z)$ because the diffusion equations contain the same model parameters and Eqs. (3.20) and (3.27) satisfy identical boundary conditions, that is, Eqs. (3.5) and (3.6). Condition (3.7) yields

$$\frac{C_4\sqrt{C_3\eta - k_C}\mathrm{J}_0(\sqrt{-C_3}r)}{\sqrt{\eta}} = \frac{C_8\sqrt{C_3\eta - k_C}\mathrm{J}_0(\sqrt{-C_3}r)}{\sqrt{\eta}} \tag{3.29}$$

Therefore, $C_4 = C_8$, as well. The function $F_2(r, z)$ becomes

$$F_2(r,z) = \mathrm{J}_0(r\sqrt{-C_3})\left(C_4\sin\left(\frac{z\sqrt{C_3\eta - k_C}}{\sqrt{\eta}}\right) + C_9\cos\left(\frac{z\sqrt{C_3\eta - k_C}}{\sqrt{\eta}}\right)\right) \tag{3.30}$$

- Determination of $c_1(r, z)$ and $c_2(r, z)$

To solve for the integration constants in Eqs. (3.25) and (3.30), we first apply Eq (3.5) to Eq. (3.25) to give

$$\mathrm{J}_0(R\sqrt{-C_3})\left(C_4\sin\left(\frac{z\sqrt{C_3\eta - k_C}}{\sqrt{\eta}}\right) + C_5\cos\left(\frac{z\sqrt{C_3\eta - k_C}}{\sqrt{\eta}}\right)\right) = 0 \tag{3.31}$$

As a result,

$$R\sqrt{-C_3} = \alpha_{0,n} \tag{3.32}$$

or

$$C_3 = -\frac{\alpha_{0,n}^2}{R^2} \tag{3.33}$$

where $\alpha_{0,n}$ are the zeros of $J_0(x)$, that is to say, $J_0(\alpha_{0,n}) = 0$. Replacing Eq. (3.33) into Eq. (3.25) yields

$$F_1(r,z) = \left(C_4 \sinh\left(\frac{z\sqrt{\alpha_{0,n}^2 \eta + k_C R^2}}{R\sqrt{\eta}} \right) i + C_5 \cosh\left(\frac{z\sqrt{\alpha_{0,n}^2 \eta + k_C R^2}}{R\sqrt{\eta}} \right) \right) J_0\left(\frac{\alpha_{0,n} r}{R} \right) \tag{3.34}$$

Similarly, $F_2(r,z)$ becomes

$$F_2(r,z) = \left(C_4 \sinh\left(\frac{z\sqrt{\alpha_{0,n}^2 \eta + k_C R^2}}{R\sqrt{\eta}} \right) i + C_9 \cosh\left(\frac{z\sqrt{\alpha_{0,n}^2 \eta + k_C R^2}}{R\sqrt{\eta}} \right) \right) J_0\left(\frac{\alpha_{0,n} r}{R} \right) \tag{3.35}$$

Using the **principle of superposition**, we seek solutions of the forms

$$\begin{aligned} c_1(r,z) = &- \frac{J_0\left(r\sqrt{-\frac{k_C}{\eta}} \right) k_2 c_{10}}{k_C J_0\left(R\sqrt{-\frac{k_C}{\eta}} \right)} + \frac{k_2 c_{10}}{k_C} \\ &+ \sum_{n=1}^{\infty} \left(A_n \sinh\left(\frac{z\sqrt{\alpha_{0,n}^2 \eta + k_C R^2}}{R\sqrt{\eta}} \right) i + E_n \cosh\left(\frac{z\sqrt{\alpha_{0,n}^2 \eta + k_C R^2}}{R\sqrt{\eta}} \right) \right) J_0\left(\frac{\alpha_{0,n} r}{R} \right) \end{aligned} \tag{3.36}$$

and

$$\begin{aligned} c_2(r,z) = &- \frac{J_0\left(r\sqrt{-\frac{k_C}{\eta}} \right) k_2 c_{20}}{k_C J_0\left(R\sqrt{-\frac{k_C}{\eta}} \right)} + \frac{k_2 c_{20}}{k_C} \\ &+ \sum_{n=1}^{\infty} \left(A_n \sinh\left(\frac{z\sqrt{\alpha_{0,n}^2 \eta + k_C R^2}}{R\sqrt{\eta}} \right) i + B_n \cosh\left(\frac{z\sqrt{\alpha_{0,n}^2 \eta + k_C R^2}}{R\sqrt{\eta}} \right) \right) J_0\left(\frac{\alpha_{0,n} r}{R} \right) \end{aligned} \tag{3.37}$$

The boundary condition (3.8) yields

$$\begin{aligned} &- \frac{J_0\left(r\sqrt{-\frac{k_C}{\eta}} \right) k_2 c_{10}}{k_C J_0\left(R\sqrt{-\frac{k_C}{\eta}} \right)} + \frac{k_2 c_{10}}{k_C} + \sum_{n=1}^{\infty} E_n J_0\left(\frac{\alpha_{0,n} r}{R} \right) \\ &= - \frac{J_0\left(r\sqrt{-\frac{k_C}{\eta}} \right) k_2 c_{20}}{k_C J_0\left(R\sqrt{-\frac{k_C}{\eta}} \right)} + \frac{k_2 c_{20}}{k_C} + \sum_{n=1}^{\infty} B_n J_0\left(\frac{\alpha_{0,n} r}{R} \right) \end{aligned} \tag{3.38}$$

Multiplying both sides of Eq. (3.38) by $r\mathrm{J}_0\left(\frac{\alpha_{0,m}r}{R}\right)$ and integrating from 0 to R lead to

$$\left\{\begin{array}{l}\dfrac{k_2c_{10}R^4\mathrm{J}_1\left(\alpha_{0,m}\right)}{(k_CR^2+\alpha_{0,m}^2\eta)\alpha_{0,m}}\\ +\displaystyle\sum_{n=1}^{\infty}E_n\left(\int_0^R \mathrm{J}_0\left(\frac{\alpha_{0,m}r}{R}\right)\mathrm{J}_0\left(\frac{\alpha_{0,n}r}{R}\right)dr\right)\end{array}\right. = \left\{\begin{array}{l}\frac{k_2c_{20}R^4\mathrm{J}_1\left(\alpha_{0,m}\right)}{(k_CR^2+\alpha_{0,m}^2\eta)\alpha_{0,m}}\\ +\displaystyle\sum_{n=1}^{\infty}B_n\left(\int_0^R \mathrm{J}_0\left(\frac{\alpha_{0,m}r}{R}\right)\mathrm{J}_0\left(\frac{\alpha_{0,n}r}{R}\right)dr\right)\end{array}\right. \tag{3.39}$$

It can be shown that

$$\int_0^R r\mathrm{J}_0\left(\frac{\alpha_{0,m}r}{R}\right)\mathrm{J}_0\left(\frac{\alpha_{0,n}r}{R}\right)\,\mathrm{d}r = \frac{R^2(-\mathrm{J}_0(\alpha_{0,m})\mathrm{J}_1(\alpha_{0,n})\alpha_{0,n}+\mathrm{J}_1(\alpha_{0,m})\alpha_{0,m}\mathrm{J}_0(\alpha_{0,n}))}{-\alpha_{0,n}^2+\alpha_{0,m}^2} \tag{3.40}$$

Therefore,

$$\int_0^R r\mathrm{J}_0\left(\frac{\alpha_{0,m}r}{R}\right)\mathrm{J}_0\left(\frac{\alpha_{0,n}r}{R}\right)\,\mathrm{d}r = 0;\quad m\neq n \tag{3.41}$$

and

$$\int_0^R r\mathrm{J}_0\left(\frac{\alpha_{0,n}r}{R}\right)^2\,\mathrm{d}r = \frac{1}{2}\frac{R^2(\sqrt{\pi}\alpha_{0,n}\mathrm{J}_0(\alpha_{0,n})^2+\sqrt{\pi}\alpha_{0,n}\mathrm{J}_1(\alpha_{0,n})^2)}{\sqrt{\pi}\alpha_{0,n}};\quad m=n \tag{3.42}$$

Since $\mathrm{J}_0(\alpha_{0,n})=0$ (see Eqs. (3.31) and (3.32)),

$$\int_0^R r\mathrm{J}_0\left(\frac{\alpha_{0,n}r}{R}\right)^2\,\mathrm{d}r = \frac{1}{2}R^2\mathrm{J}_1(\alpha_{0,n})^2;\quad m=n \tag{3.43}$$

In summary,

$$\int_0^R r\mathrm{J}_0\left(\frac{\alpha_{0,m}r}{R}\right)\mathrm{J}_0\left(\frac{\alpha_{0,n}r}{R}\right)\,\mathrm{d}r = \frac{1}{2}R^2\mathrm{J}_1(\alpha_{0,n})^2\delta_{n,m} \tag{3.44}$$

After substituting Eq. (3.44) into Eq. (3.39), we have

$$\frac{k_2c_{10}R^4\mathrm{J}_1(\alpha_{0,n})}{(k_CR^2+\alpha_{0,n}^2\eta)\alpha_{0,n}}+\frac{1}{2}E_nR^2\mathrm{J}_1(\alpha_{0,n})^2 = \frac{k_2c_{20}R^4\mathrm{J}_1(\alpha_{0,n})}{(k_CR^2+\alpha_{0,n}^2\eta)\alpha_{0,n}}+\frac{1}{2}B_nR^2\mathrm{J}_1(\alpha_{0,n})^2 \tag{3.45}$$

Applying the boundary condition (3.10) to (3.37), multiplying both sides of the resulting equation by $r\mathrm{J}_0\left(\frac{\alpha_{0,m}r}{R}\right)$ and integrating from 0 to R result in

$$\begin{aligned}&\frac{k_2c_{20}R^4\mathrm{J}_1(\alpha_{0,m})}{(k_CR^2+\alpha_{0,m}^2\eta)\alpha_{0,m}}\\ &+\sum_{n=1}^{\infty}\left(-A_n\sinh\left(\frac{a\sqrt{\alpha_{0,n}^2\eta+k_CR^2}}{R\sqrt{\eta}}\right)i+B_n\cosh\left(\frac{a\sqrt{\alpha_{0,n}^2\eta+k_CR^2}}{R\sqrt{\eta}}\right)\right)\times\\ &\left(\int_0^R r\mathrm{J}_0\left(\frac{\alpha_{0,m}r}{R}\right)\mathrm{J}_0\left(\frac{\alpha_{0,n}r}{R}\right)\mathrm{d}r\right)=0\end{aligned} \tag{3.46}$$

Using Eq. (3.44), Eq. (3.46) becomes

$$\frac{k_2 c_{20} R^4 \mathrm{J}_1(\alpha_{0,m})}{(k_C R^2 + \alpha_{0,m}^2 \eta)\alpha_{0,m}} + \sum_{n=1}^{\infty} \frac{1}{2}\left(-A_n \sinh\left(\frac{a\sqrt{\alpha_{0,n}^2 \eta + k_C R^2}}{R\sqrt{\eta}}\right) i + B_n \cosh\left(\frac{a\sqrt{\alpha_{0,n}^2 \eta + k_C R^2}}{R\sqrt{\eta}}\right)\right) \times R^2 \mathrm{J}_1(\alpha_{0,n})^2 \delta_{n,m} = 0 \quad (3.47)$$

which simplifies to

$$\frac{k_2 c_{20} R^4 \mathrm{J}_1(\alpha_{0,n})}{(k_C R^2 + \alpha_{0,n}^2 \eta)\alpha_{0,n}} + \left(-\frac{1}{2} A_n \sinh\left(\frac{a\sqrt{\alpha_{0,n}^2 \eta + k_C R^2}}{R\sqrt{\eta}}\right) i + \frac{1}{2} B_n \cosh\left(\frac{a\sqrt{\alpha_{0,n}^2 \eta + k_C R^2}}{R\sqrt{\eta}}\right)\right) R^2 \mathrm{J}_1(\alpha_{0,n})^2 = 0 \quad (3.48)$$

Similarly, applying the boundary condition (3.9) to (3.36), multiplying both sides of the resulting equation by $r\mathrm{J}_0\left(\frac{\alpha_{0,m} r}{R}\right)$, and integrating from 0 to R yield an equation analogous to Eq. (3.48)

$$\frac{k_2 c_{10} R^4 \mathrm{J}_1(\alpha_{0,n})}{(k_C R^2 + \alpha_{0,n}^2 \eta)\alpha_{0,n}} + \left(\frac{1}{2} A_n \sinh\left(\frac{b\sqrt{\alpha_{0,n}^2 \eta + k_C R^2}}{R\sqrt{\eta}}\right) i + \frac{1}{2} E_n \cosh\left(\frac{b\sqrt{\alpha_{0,n}^2 \eta + k_C R^2}}{R\sqrt{\eta}}\right)\right) R^2 \mathrm{J}_1(\alpha_{0,n})^2 = 0 \quad (3.49)$$

Solving Eqs. (3.45), (3.48), and (3.49) for A_n, B_n, and E_n and replacing the results in Eqs. (3.36) and (3.37) give the concentrations in the Appendix.

The fluxes can be obtained by the following expressions:

$$J_1 = -2\eta\pi R\left(\int_0^b \left.\frac{\partial c_1(r,z)}{\partial r}\right|_{r=R} dz\right) \quad (3.50)$$

$$J_2 = -2\eta\pi R\left(\int_{-a}^0 \left.\frac{\partial c_2(r,z)}{\partial r}\right|_{r=R} dz\right) \quad (3.51)$$

$$J_3 = 2\eta\pi\left(\int_0^R \left(\left.\frac{\partial c_2(r,z)}{\partial z}\right|_{z=-a}\right) r dr\right) \quad (3.52)$$

and

$$J_4 = -2\eta\pi \left(\int_0^R \left(\frac{\partial c_1(r,z)}{\partial z}\bigg|_{z=b} \right) rdr \right) \tag{3.53}$$

Finally, the total flux is

$$J = J_1 + J_2 + J_3 + J_4 \tag{3.54}$$

3.3 CONCENTRATION PROFILE OF VASCULAR ENDOTHELIAL GROWTH FACTOR IN TWO LAYERS

When describing the VEGF profile, researchers used the following model parameters (Chen et al., 2007b):

$\eta = 0.06048$ cm^2/day,
$R = 0.235$ cm,
$a = b = 0.15$ cm,
$k_2 = 0.040608$ (1/day),
$k_c = 19.9584$ (1/day),
$c_{10} = 57638.7$ ng/ml,
$c_{20} = 115277$ ng/ml

The contour and 3-D plots are shown in Figures 3.2 and 3.3 with only ten terms in the series. Based on the expressions for $c_1(r, z)$ and $c_2(r, z)$, it is possible to draw the steady-state concentration profile at a specific location in the scaffold system (Fig. 3.4). Information about

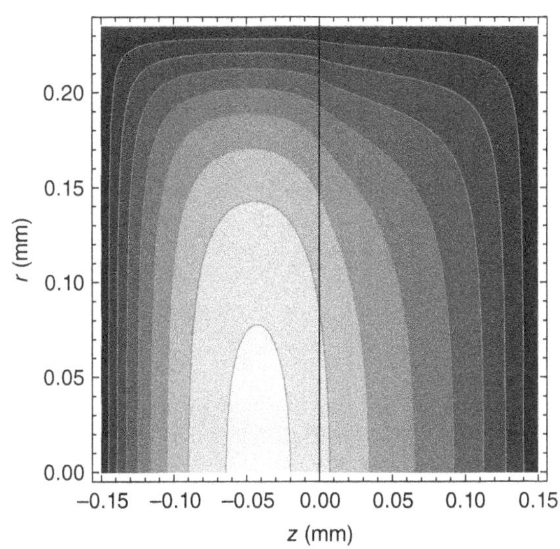

Figure 3.2 Contour plot of the steady-state VEGF concentration in the bilayered scaffold.

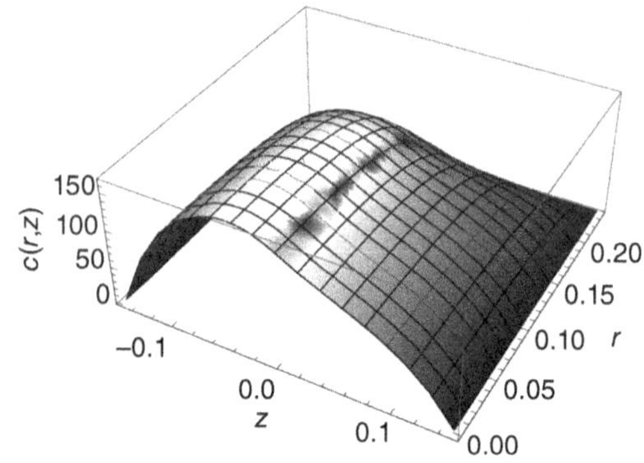

Figure 3.3 Three-dimensional steady-state VEGF concentration profile in the bilayered scaffold.

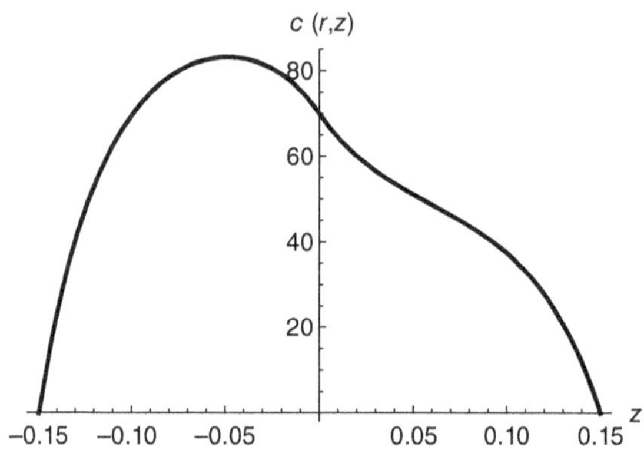

Figure 3.4 Steady-state VEGF concentration profile at $r = 0.2$ mm.

the growth-factor distribution in the device and the flux can be used by researchers and help in the rational design of bilayered porous scaffolds for tissue regeneration.

3.4 SUMMARY

A scaffold with two connected layers stacked on top of each other was analyzed. This system would deliver VEGF and PDGF sequentially to create spatially organized tissues. Two reaction–diffusion equations were written in cylindrical coordinates to model transport through the device. An analytical solution, which included the Bessel function, was derived using the method of separation of variables. We were able to link the growth-factor distribution in the device and the flux to specific design properties. This approach may help the development of bilayered porous scaffolds for tissue regeneration.

3.5 APPENDIX: MAPLE CODE LISTINGS

3.5.1 Expressions for the Concentration Profiles

$$
\begin{aligned}
c_1(r,z) = &-\frac{\mathrm{J}_0\left(\sqrt{-\frac{k_C}{\eta}}r\right)k_2c_{10}}{k_C\mathrm{J}_0\left(\sqrt{-\frac{k_C}{\eta}}R\right)}+\frac{k_2c_{10}}{k_C}+\\
&\sum_{n=1}^{\infty}\Bigg\{\Bigg[\Bigg(2c_{20}\mathrm{e}^{\frac{\sqrt{\alpha_{0,n}^2\eta+k_CR^2}(a+2b)}{\sqrt{\eta}R}}+c_{10}\mathrm{e}^{\frac{2\sqrt{\alpha_{0,n}^2\eta+k_CR^2}a}{\sqrt{\eta}R}}-c_{20}\mathrm{e}^{\frac{2\sqrt{\alpha_{0,n}^2\eta+k_CR^2}a}{\sqrt{\eta}R}}-c_{20}+\mathrm{e}^{\frac{2\sqrt{\alpha_{0,n}^2\eta+k_CR^2}(b+a)}{\sqrt{\eta}R}}c_{10}\\
&+\mathrm{e}^{\frac{2\sqrt{\alpha_{0,n}^2\eta+k_CR^2}b}{\sqrt{\eta}R}}c_{10}+2c_{20}\mathrm{e}^{\frac{\sqrt{\alpha_{0,n}^2\eta+k_CR^2}a}{\sqrt{\eta}R}}-2c_{10}\mathrm{e}^{\frac{\sqrt{\alpha_{0,n}^2\eta+k_CR^2}(b+2a)}{\sqrt{\eta}R}}-2c_{10}\mathrm{e}^{\frac{\sqrt{\alpha_{0,n}^2\eta+k_CR^2}b}{\sqrt{\eta}R}}\\
&+c_{10}-\mathrm{e}^{\frac{2\sqrt{\alpha_{0,n}^2\eta+k_CR^2}(b+a)}{\sqrt{\eta}R}}c_{20}-\mathrm{e}^{\frac{2\sqrt{\alpha_{0,n}^2\eta+k_CR^2}b}{\sqrt{\eta}R}}c_{20}\Bigg)k_2R^2\sinh\left(\frac{\sqrt{\alpha_{0,n}^2\eta+k_CR^2}z}{\sqrt{\eta}R}\right)\Bigg]\Bigg/\\
&\left[\mathrm{J}_1(\alpha_{0,n})\alpha_{0,n}\left(\alpha_{0,n}^2\eta\mathrm{e}^{\frac{2\sqrt{\alpha_{0,n}^2\eta+k_CR^2}(b+a)}{\sqrt{\eta}R}}-\alpha_{0,n}^2\eta+k_CR^2\mathrm{e}^{\frac{2\sqrt{\alpha_{0,n}^2\eta+k_CR^2}(b+a)}{\sqrt{\eta}R}}-k_CR^2\right)\right]+\\
&\Bigg[k_2R^2\Bigg(-2c_{10}\mathrm{e}^{\frac{\sqrt{\alpha_{0,n}^2\eta+k_CR^2}(b+2a)}{\sqrt{\eta}R}}-\mathrm{e}^{\frac{2\sqrt{\alpha_{0,n}^2\eta+k_CR^2}(b+a)}{\sqrt{\eta}R}}c_{10}-c_{20}\mathrm{e}^{\frac{2\sqrt{\alpha_{0,n}^2\eta+k_CR^2}a}{\sqrt{\eta}R}}+c_{10}\mathrm{e}^{\frac{2\sqrt{\alpha_{0,n}^2\eta+k_CR^2}a}{\sqrt{\eta}R}}\\
&+\mathrm{e}^{\frac{2\sqrt{\alpha_{0,n}^2\eta+k_CR^2}(b+a)}{\sqrt{\eta}R}}c_{20}+2c_{20}\mathrm{e}^{\frac{\sqrt{\alpha_{0,n}^2\eta+k_CR^2}a}{\sqrt{\eta}R}}-2c_{20}\mathrm{e}^{\frac{\sqrt{\alpha_{0,n}^2\eta+k_CR^2}(a+2b)}{\sqrt{\eta}R}}+2c_{10}\mathrm{e}^{\frac{\sqrt{\alpha_{0,n}^2\eta+k_CR^2}b}{\sqrt{\eta}R}}\\
&-\mathrm{e}^{\frac{2\sqrt{\alpha_{0,n}^2\eta+k_CR^2}b}{\sqrt{\eta}R}}c_{10}-c_{20}+c_{10}+\mathrm{e}^{\frac{2\sqrt{\alpha_{0,n}^2\eta+k_CR^2}b}{\sqrt{\eta}R}}c_{20}\Bigg)\cosh\left(\frac{\sqrt{\alpha_{0,n}^2\eta+k_CR^2}z}{\sqrt{\eta}R}\right)\Bigg]\Bigg/\\
&\left[\left(\alpha_{0,n}^2\eta+k_CR^2\right)(\mathrm{e}^{\frac{2\sqrt{\alpha_{0,n}^2\eta+k_CR^2}(b+a)}{\sqrt{\eta}R}}-1)\mathrm{J}_1(\alpha_{0,n})\alpha_{0,n}\right]\Bigg\}\mathrm{J}_0\left(\frac{\alpha_{0,n}r}{R}\right)
\end{aligned}
\tag{A.3.1}
$$

and

$$
\begin{aligned}
c_2(r,z) = &-\frac{\mathrm{J}_0\left(\sqrt{-\frac{k_C}{\eta}}r\right)k_2c_{20}}{k_C\mathrm{J}_0\left(\sqrt{-\frac{k_C}{\eta}}R\right)}+\frac{k_2c_{20}}{k_C}+\\
&\sum_{n=1}^{\infty}\Bigg\{\Bigg[\Bigg(2c_{20}\mathrm{e}^{\frac{\sqrt{\alpha_{0,n}^2\eta+k_CR^2}(a+2b)}{\sqrt{\eta}R}}+c_{10}\mathrm{e}^{\frac{2\sqrt{\alpha_{0,n}^2\eta+k_CR^2}a}{\sqrt{\eta}R}}-c_{20}\mathrm{e}^{\frac{2\sqrt{\alpha_{0,n}^2\eta+k_CR^2}a}{\sqrt{\eta}R}}-c_{20}+\mathrm{e}^{\frac{2\sqrt{\alpha_{0,n}^2\eta+k_CR^2}(b+a)}{\sqrt{\eta}R}}c_{10}
\end{aligned}
$$

$$
+ e^{\frac{2\sqrt{\alpha_{0,n}^2\eta+k_C R^2}\,b}{\sqrt{\eta}R}} c_{10} + 2c_{20} e^{\frac{\sqrt{\alpha_{0,n}^2\eta+k_C R^2}\,a}{\sqrt{\eta}R}} - 2c_{10} e^{\frac{\sqrt{\alpha_{0,n}^2\eta+k_C R^2}\,(b+2a)}{\sqrt{\eta}R}} - 2c_{10} e^{\frac{\sqrt{\alpha_{0,n}^2\eta+k_C R^2}\,b}{\sqrt{\eta}R}}
$$

$$
\left. + c_{10} - e^{\frac{2\sqrt{\alpha_{0,n}^2\eta+k_C R^2}\,(b+a)}{\sqrt{\eta}R}} c_{20} - e^{\frac{2\sqrt{\alpha_{0,n}^2\eta+k_C R^2}\,b}{\sqrt{\eta}R}} c_{20} \right) k_2 R^2 \sinh\left(\frac{\sqrt{\alpha_{0,n}^2\eta + k_C R^2}\,z}{\sqrt{\eta}R} \right) \Bigg] \Bigg/
$$

$$
\left[J_1(\alpha_{0,n})\,\alpha_{0,n} \left(\alpha_{0,n}^2 \eta e^{\frac{2\sqrt{\alpha_{0,n}^2\eta+k_C R^2}\,(b+a)}{\sqrt{\eta}R}} - \alpha_{0,n}^2\eta + k_C R^2 e^{\frac{2\sqrt{\alpha_{0,n}^2\eta+k_C R^2}\,(b+a)}{\sqrt{\eta}R}} - k_C R^2 \right) \right] +
$$

$$
\Bigg[k_2 R^2 \Bigg(-2c_{10} e^{\frac{\sqrt{\alpha_{0,n}^2\eta+k_C R^2}\,(b+2a)}{\sqrt{\eta}R}} + e^{\frac{2\sqrt{\alpha_{0,n}^2\eta+k_C R^2}\,(b+a)}{\sqrt{\eta}R}} c_{10} - c_{20} e^{\frac{2\sqrt{\alpha_{0,n}^2\eta+k_C R^2}\,a}{\sqrt{\eta}R}} + c_{10} e^{\frac{2\sqrt{\alpha_{0,n}^2\eta+k_C R^2}\,a}{\sqrt{\eta}R}}
$$

$$
- e^{\frac{2\sqrt{\alpha_{0,n}^2\eta+k_C R^2}\,(b+a)}{\sqrt{\eta}R}} c_{20} + 2c_{20} e^{\frac{\sqrt{\alpha_{0,n}^2\eta+k_C R^2}\,a}{\sqrt{\eta}R}} - 2c_{20} e^{\frac{\sqrt{\alpha_{0,n}^2\eta+k_C R^2}\,(a+2b)}{\sqrt{\eta}R}} + 2c_{10} e^{\frac{\sqrt{\alpha_{0,n}^2\eta+k_C R^2}\,b}{\sqrt{\eta}R}}
$$

$$
\left. - e^{\frac{2\sqrt{\alpha_{0,n}^2\eta+k_C R^2}\,b}{\sqrt{\eta}R}} c_{10} + c_{20} - c_{10} + e^{\frac{2\sqrt{\alpha_{0,n}^2\eta+k_C R^2}\,b}{\sqrt{\eta}R}} c_{20} \right) \cosh\left(\frac{\sqrt{\alpha_{0,n}^2\eta + k_C R^2}\,z}{\sqrt{\eta}R} \right) \Bigg] \Bigg/
$$

$$
\left[(\alpha_{0,n}^2\eta + k_C R^2)\,(e^{\frac{2\sqrt{\alpha_{0,n}^2\eta+k_C R^2}\,(b+a)}{\sqrt{\eta}R}} - 1) J_1(\alpha_{0,n})\alpha_{0,n} \right] \Bigg\}\; J_0\left(\frac{\alpha_{0,n} r}{R} \right) \tag{A.3.2}
$$

3.5.2 Maple Code: twoangisteady.mws

```
> restart:with(VectorCalculus):with(inttrans);
[addtable, fourier, fouriercos, fouriersin, hankel, hilbert,
invfourier, invhilbert, invlaplace,invmellin, laplace, mellin,
savetable]
> eq1:=0=eta*Laplacian(c[1](r,z),cylindrical[r,theta,z])+k[2]*c[10]-
k[C]*c[1](r,z);
>
```

$$
eq1 := 0 = \frac{\eta\left(\frac{\partial}{\partial r} c_1(r,z) + r\left(\frac{\partial^2}{\partial r^2} c_1(r,z)\right) + r\left(\frac{\partial^2}{\partial z^2} c_1(r,z)\right)\right)}{r} + k_2 c_{10} - k_C c_1(r,z)
$$

```
> eq2:=0=eta*Laplacian(c[2](r,z),cylindrical[r,theta,z])+k[2]*c[20]-
k[C]*c[2](r,z);
>
```

$$
eq2 := 0 = \frac{\eta\left(\frac{\partial}{\partial r} c_2(r,z) + r\left(\frac{\partial^2}{\partial r^2} c_2(r,z)\right) + r\left(\frac{\partial^2}{\partial z^2} c_2(r,z)\right)\right)}{r} + k_2 c_{20} - k_C c_2(r,z)
$$

```
> with(PDETools);
```

```
[CanonicalCoordinates, ChangeSymmetry, CharacteristicQ,
   CharacteristicQInvariants, ConservedCurrentTest, ConservedCurrents,
   ConsistencyTest, D_Dx, DeterminingPDE, Eta_k, Euler, FromJet,
   FunctionFieldSolutions, InfinitesimalGenerator, Infinitesimals,
   IntegratingFactorTest, IntegratingFactors, InvariantEquation,
   InvariantSolutions, InvariantTransformation, Invariants, Laplace,
   Library, PDEplot, PolynomialSolutions, ReducedForm,
   SimilaritySolutions, SimilarityTransformation, Solve,
   SymmetryCommutator, SymmetryGauge, SymmetrySolutions, SymmetryTest,
   SymmetryTransformation, TWSolutions, ToJet, build, casesplit,
   charstrip, dchange, dcoeffs, declare, diff_table, difforder,
   dplolyform, dsubs,mapde, separability, splitstrip, splitsys,
   undeclare]
>
> eq11:=c[1](r,z)=H[1](r):
> eq11A:=eval(subs(eq11,eq1)):
> eq11B:=dsolve(eq11A):
>
> eq11B1:=subs(_C1=0,eq11B):
> eq11B2:=subs(r=R,rhs(eq11B1))=0:
> eq11B3:=isolate(eq11B2,_C2):
> eq11B4:=subs(eq11B3,eq11B1):
>
> eq11C:=c[1](r,z)=rhs(eq11B4)+F[1](r,z):
> eq11D:=eval(subs(eq11C,eq1)):
> eq11E:=expand(simplify(lhs(eq11D)-rhs(eq11D)))=0:
> eq12:=c[2](r,z)=H[2](r):
> eq12A:=eval(subs(eq12,eq2)):
>
> eq12B:=dsolve(eq12A):
> eq12B1:=subs(_C1=0,eq12B):
> eq12B2:=subs(r=R,rhs(eq12B1))=0:
> eq12B3:=isolate(eq12B2,_C2):
> eq12B4:=subs(eq12B3,eq12B1):
> eq12C:=c[2](r,z)=rhs(eq12B4)+F[2](r,z):
> eq12D:=eval(subs(eq12C,eq2)):
> eq12E:=expand(simplify(lhs(eq12D)-rhs(eq12D)))=0:
> eq13:=pdsolve(eq11E,HINT=f[1](r)*g[1](z)):
> eq13A:=factor(build(eq13)):
> eq13B:=subs(_C2=0,_C1=1,eq13A):
> eq14:=pdsolve(eq12E,HINT=f[2](r)*g[2](z)):
>
eq14A:=subs(_C1=E1,_C2=E2,_C3=E3,_C4=E4,factor(build(eq14
))):
> eq14B:=subs(E2=0,E1=1,eq14A):
> eq14C:=eval(subs(z=0,diff(rhs(eq14B),z)=diff(rhs(eq13B),z))):
> eq14D:=isolate(eq14C,E3):
> eq14E:=subs(eq14D,eq14B):
> eq15:=subs(r=R,rhs(eq13B))=0:
> eq15A:=(-_c[1])^(1/2)*R=alpha[0,n]:
```

```
> eq15B:=isolate(eq15A,_c[1]):
> eq16:=simplify(subs(eq15B,eq13B),power,symbolic):
>
> eq16A:=simplify(subs(eq15B,eq14E),power,symbolic):
> eq17:=c[1](r,z)=rhs(eq11B4)+Sum(subs(_C3=A[n],_C4=E[n],rhs(eq16)),
n=1..infinity):
> eq17A:=c[2](r,z)=rhs(eq12B4)+Sum(subs(_C3=A[n],E4=B[n],_C4=E[n],
rhs(eq16A)),n=1..infinity):
> eq18:=eval(subs(z=0,rhs(eq17))=subs(z=0,rhs(eq17A))):
> eq18A:=factor(subs(BesselJ(0,alpha[0,m])=0,
expand(int(rhs(eq11B4)*r*BesselJ(0,alpha[0,m]/R*r),r=0..R))))+
Sum(expand(Int(r*BesselJ(0,alpha[0,m]/R*r)*eval(subs(z=0,_C3=A[n],
_C4=E[n],rhs(eq16))),r=0..R)),n=1..infinity)=factor(subs(BesselJ(0,
alpha[0,m])=0,expand(int(rhs(eq12B4)*r*BesselJ(0,alpha[0,m]/R*r),
r=0..R))))+Sum(expand(Int(r*BesselJ(0,alpha[0,m]/R*r)*eval(subs(z=0,
_C3=A[n],E4=B[n],_C4=E[n],rhs(eq16A))),r=0..R)),n=1..infinity):
> eq18B:=Int(r*BesselJ(0,alpha[0,m]/R*r)*BesselJ(0,alpha[0,n]/R*r),
r =0 .. R)=int(r*BesselJ(0,alpha[0,m]/R*r)*BesselJ(0,alpha[0,n]/R*r),
r =0 .. R):
> eq18C:=subs(BesselJ(0,alpha[0,m])=0,BesselJ(0,alpha[0,n])=0,eq18B):
> eq18D:=Int(r*BesselJ(0,alpha[0,n]/R*r)*BesselJ(0,alpha[0,n]/R*r),
r =0 .. R)=int(r*BesselJ(0,alpha[0,n]/R*r)^2,r =0 .. R):
> eq18E:=subs(BesselJ(0,alpha[0,n])=0,eq18D):
> eq18F:=lhs(eq18C)=rhs(eq18E)*delta[n,m]:
> eq18G:=subs(eq18F,eq18A):
> eq18H:=subs(m=n,k[2]*c[10]*R^4/(k[C]*R^2+alpha[0,m]^2*eta)/
alpha[0,m]*BesselJ(1,alpha[0,m])+(1/2*E[n]*R^2*BesselJ(1,alpha[0,n])^2) =
k[2]*c[20]*R^4/(k[C]*R^2+alpha[0,m]^2*eta)/alpha[0,m]*BesselJ(1,
alpha[0,m])+(1/2*B[n]*R^2*BesselJ(1,alpha[0,n])^2)):
>
> eq19:=subs(z=-a,rhs(eq17A)):
> eq19A:=factor(subs(BesselJ(0,alpha[0,m])=0,
expand(int(rhs(eq12B4)*r*BesselJ(0,alpha[0,m]/R*r),r=0..R))))+
Sum(expand(Int(r*BesselJ(0,alpha[0,m]/R*r)*eval(subs(z=-a,_C3=A[n],
E4=B[n],_C4=E[n],rhs(eq16A))),r=0..R)),n=1..infinity)=0:
> eq19B:=subs(eq18F,eq19A):
> eq19C:=subs(m=n,k[2]*c[20]*R^4/(k[C]*R^2+alpha[0,m]^2*eta)/
alpha[0,m]*BesselJ(1,alpha[0,m])+(1/2*(-I*A[n]*sinh(1/eta^(1/2)*
(eta*alpha[0,n]^2+k[C]*R^2)^(1/2)/R*a)+B[n]*cosh(1/eta^(1/2)*
(eta*alpha[0,n]^2+k[C]*R^2)^(1/2)/R*a))*R^2*BesselJ(1,alpha[0,n])^2)) =0:
> ñas:=subs(z=b,rhs(eq17)):
> ñasA:=factor(subs(BesselJ(0,alpha[0,m])=0,
expand(int(rhs(eq11B4)*r*BesselJ(0,alpha[0,m]/R*r),r=0..R))))+
Sum(expand(Int(r*BesselJ(0,alpha[0,m]/R*r)*eval(subs(z=b,_C3=A[n],
E4=B[n],_C4=E[n],rhs(eq16))),r=0..R)),n=1..infinity)=0:
> ñasB:=subs(eq18F,ñasA):
> ñasC:=subs(m=n,k[2]*c[10]*R^4/(k[C]*R^2+alpha[0,m]^2*eta)/
alpha[0,m]*BesselJ(1,alpha[0,m])+(1/2*(A[n]*sinh(1/eta^(1/2)*
```

```
(alpha[0,n]^2*eta+k[C]*R^2)^(1/2)/R*b)*I+E[n]*cosh(1/eta^(1/2)*
(alpha[0,n]^2*eta+k[C]*R^2)^(1/2)/R*b))*R^2*BesselJ(1,alpha[0,n])^2) =0):
>
eq20 := simplify(solve({eq18H,eq19C,ñasC},{A[n],B[n],E[n]})):
> eq21:=subs(eq20,eq17):
> eq21A:=subs(eq20,eq17A):
>
>
>
```

PROBLEMS

3.1. Find the total flux J using the following parameter values:
$\eta = 0.06048\ \text{cm}^2/\text{day}$,
$R = 0.235\ \text{cm}$,
$a = b = 0.15\ \text{cm}$,
$k_2 = 0.040608\ (1/\text{day})$,
$k_c = 19.9584\ (1/\text{day})$,
$c_{10} = 57638.7\ \text{ng/ml}$,
$c_{20} = 115277\ \text{ng/ml}$

3.2. Find the total flux J using the following parameter values:
$\eta = 0.06048\ \text{cm}^2/\text{day}$,
$R = 0.315\ \text{cm}$,
$a = b = 0.18\ \text{cm}$,
$k_2 = 0.040608\ (1/\text{day})$,
$k_c = 19.9584\ (1/\text{day})$,
$c_{10} = 57638.7\ \text{ng/ml}$,
$c_{20} = 115277\ \text{ng/ml}$

3.3. Plot the concentrations $c_1(0.2, z)$ and $c_2(0.2, z)$ using the following parameter values:
$\eta = 0.06048\ \text{cm}^2/\text{day}$,
$R = 0.235\ \text{cm}$,
$a = b = 0.15\ \text{cm}$,
$k_2 = 0.040608\ (1/\text{day})$,
$k_c = 19.9584\ (1/\text{day})$,
$c_{10} = 60000\ \text{ng/ml}$,
$c_{20} = 130000\ \text{ng/ml}$

3.4. Plot the concentrations $c_1(r, 0.10)$ and $c_2(r, -0.10)$ using the following parameter values:
$\eta = 0.070\ \text{cm}^2/\text{day}$,
$R = 0.235\ \text{cm}$,
$a = b = 0.15\ \text{cm}$,
$k_2 = 0.045\ (1/\text{day})$,

$k_c = 19.86$ (1/day),
$c_{10} = 60000$ ng/ml,
$c_{20} = 130000$ ng/ml

3.5. To assess the convergence of the procedure outlined in Section 3.2, calculate the total flux J for $N = 5$ and $N = 8$ using the following parameter values:
$\eta = 0.0.06048$ cm^2/day,
$R = 0.315$ cm,
$a = b = 0.14$ cm,
$k_2 = 0.040608$ (1/day),
$k_c = 19.9584$ (1/day),
$c_{10} = 58000$ ng/ml,
$c_{20} = 115277$ ng/ml
Note: N is the number of terms in the series solution.

REFERENCES

Chen, R. R., Silva, E. A., Yuen, W. W., Brock, A. A., Fischbach, C., Lin, A. S., Mooney, D. J. (2007a). Integrated approach to designing growth factor delivery systems. FASEB J, 21(14), 3896–3903.

Chen, R. R., Silva, E. A., Yuen, W. W., & Mooney, D. J. (2007b). Spatio-temporal VEGF and PDGF delivery patterns blood vessel formation and maturation. Pharm Res, 24(2), 258–264.

Grazul-Bilska, A. T., Johnson, M. L., Borowicz, P. P., Minten, M., Bilski, J. J., Wroblewski, R., Reynolds, L. P. (2011). Placental development during early pregnancy in sheep: cell proliferation, global methylation, and angiogenesis in the fetal placenta. Reproduction, 141(4), 529–540.

Reynolds, L. P., & Redmer, D. A. (2001). Angiogenesis in the placenta. Biol Reprod, 64(4), 1033–1040.

4

STEADY-STATE TWO-DIMENSIONAL DIFFUSION IN A HOLLOW SPHERE

4.1 SEPARATION OF VARIABLES AND LEGENDRE POLYNOMIALS IN 2-D SPHERICAL COORDINATES

The Laplace equation in spherical coordinates adapted to a problem with axial symmetry takes the form

$$\frac{2\left(\frac{\partial}{\partial r}u(r,\varphi)\right)}{r}+\frac{\partial^2}{\partial r^2}u(r,\varphi)+\frac{\cos(\varphi)\left(\frac{\partial}{\partial\varphi}u(r,\varphi)\right)}{r^2\sin(\varphi)}+\frac{\frac{\partial^2}{\partial\varphi^2}u(r,\varphi)}{r^2}=0 \tag{4.1}$$

where r measures the radius; φ indicates the zenith angle or latitude. The angle θ, which is not included here, represents the azimuthal angle or longitude. It is customary to change the convention of the notation for the zenith and azimuthal angles (See Chapters 7 and 9). The reader should become familiar with both notations.

A particular solution of Eq. (4.1) has the form $u(r,\varphi)=f(r)+g(\varphi)$ where the functions $f(r)$ and $g(\varphi)$ are defined by

$$\frac{d^2f(r)}{dr^2}=\frac{c_1}{r^2}-\frac{2}{r}\frac{df(r)}{dr} \tag{4.2}$$

and

$$\frac{d^2g(\varphi)}{d\varphi^2}=-c_1-\frac{\cos(\varphi)}{\sin(\varphi)}\frac{dg(\varphi)}{d\varphi} \tag{4.3}$$

Closed-form Solutions for Drug Transport through Controlled-Release Devices in Two and Three Dimensions, First Edition. Laurent Simon and Juan Ospina.

where c_1 is an arbitrary constant. The solution $u(r, \varphi)$ is obtained after solving for Eqs. (4.2) and (4.3):

$$u(r,\varphi) = c_1 \ln(r) - \frac{C_1}{r} + C_2 + c_1 \ln(\sin(\varphi)) + C_3 \ln\left(\frac{1-\cos(\varphi)}{\sin(\varphi)}\right) \tag{4.4}$$

The solution (4.4) is written as an **additive separation of variables** (Cherniavsky, 2010). Similar to problems in Chapters 1 and 2, Eq. (4.1) can be solved using the **multiplicative separation of variables** method such that $u(r, \varphi) = f(r)g(\varphi)$. In this case, $f(r)$ and $g(\varphi)$ satisfy the following ordinary differential equations:

$$\frac{d^2f(r)}{dr^2} = \frac{c_1 f(r)}{r^2} - \frac{2}{r}\frac{df(r)}{dr} \tag{4.5}$$

and

$$\frac{d^2g(\varphi)}{d\varphi^2} = -c_1 g(\varphi) - \frac{\cos(\varphi)}{\sin(\varphi)}\frac{dg(\varphi)}{d\varphi} \tag{4.6}$$

where c_1 represents an arbitrary constant. The solution is

$$u(r,\varphi) = \frac{1}{\sqrt{r}}\left(C_1 r^{\frac{1}{2}\sqrt{1+4c_1}} + C_2 r^{-\frac{1}{2}\sqrt{1+4c_1}}\right) \times \left(C_3 \mathrm{P}_{-\frac{1}{2}+\frac{1}{2}\sqrt{1+4c_1}}(\cos(\varphi)) + C_4 \mathrm{Q}_{-\frac{1}{2}+\frac{1}{2}\sqrt{1+4c_1}}(\cos(\varphi))\right) \tag{4.7}$$

Equation (4.7) contains the **Legendre polynomial** denoted by P_n and the **Lengendre function of the second kind**, Q_n. These functions are defined only when

$$-\frac{1}{2} + \frac{1}{2}\sqrt{1+4c_1} = n \tag{4.8}$$

where $n = 0, 1, 2, 3, \ldots$ Eq. (4.8) is solved for c_1 to give

$$c_1 = \frac{(2n+1)^2 - 1}{4} \tag{4.9}$$

which leads to

$$u(r,\varphi) = (C_1 r^n + C_2 r^{-1-n})(C_3 \mathrm{P}_n(\cos(\varphi)) + C_4 \mathrm{Q}_n(\cos(\varphi))) \tag{4.10}$$

Applying the **principle of superposition** to Eq. (4.10), we get

$$u(r,\varphi) = \sum_{n=0}^{\infty} (A_n r^n + B_n r^{-1-n})(E_n \mathrm{P}_n(\cos(\varphi)) + F_n \mathrm{Q}_n(\cos(\varphi))) \tag{4.11}$$

in a discretized form. In cases where Eqs. (4.4) and (4.11) are both solutions, their sum is also a solution:

$$u(r,\varphi) = c_1 \ln(r) - \frac{C_1}{r} + C_2 + c_1 \ln(\sin(\varphi)) + C_3 \ln\left(\frac{1-\cos(\varphi)}{\sin(\varphi)}\right) + \sum_{n=0}^{\infty} (A_n r^n + B_n r^{-1-n})(E_n \mathrm{P}_n(\cos(\varphi)) + F_n \mathrm{Q}_n(\cos(\varphi))) \tag{4.12}$$

To fully grasp the structure of Eq. (4.12), we take a look at the specific forms of the Legendre functions for lower values of n. The first ten P_n polynomials are

$$\begin{aligned}
P_0(x) &= 1\\
P_2(x) &= -\frac{1}{2} + \frac{3x^2}{2}\\
P_3(x) &= \frac{5}{2}x^3 - \frac{3}{2}x\\
P_4(x) &= \frac{3}{8} + \frac{35}{8}x^4 - \frac{15}{4}x^2\\
P_5(x) &= \frac{63}{8}x^5 - \frac{35}{4}x^3 + \frac{15}{8}x\\
P_6(x) &= -\frac{5}{16} + \frac{231}{16}x^6 - \frac{315}{16}x^4 + \frac{105}{16}x^2\\
P_7(x) &= \frac{429}{16}x^7 - \frac{693}{16}x^5 + \frac{315}{16}x^3 - \frac{35}{16}x\\
P_8(x) &= \frac{35}{128} + \frac{6435}{128}x^8 - \frac{3003}{32}x^6 + \frac{3465}{64}x^4 - \frac{315}{32}x^2\\
P_9(x) &= \frac{12155}{128}x^9 - \frac{6435}{32}x^7 + \frac{9009}{64}x^5 - \frac{1155}{32}x^3 + \frac{315}{128}x\\
P_{10}(x) &= -\frac{63}{256} + \frac{46189}{256}x^{10} - \frac{109395}{256}x^8 + \frac{45045}{128}x^6 - \frac{15015}{128}x^4 + \frac{3465}{256}x^2
\end{aligned} \tag{4.13}$$

The first ten Q_n functions are given in Eqs. (4.14) and (4.15):

$$\begin{aligned}
Q_0(x) &= \frac{1}{2}\ln(x+1) - \frac{1}{2}\ln(x-1)\\
Q_1(x) &= \frac{1}{2}x\ln(x+1) - \frac{1}{2}x\ln(x-1) - 1\\
Q_3(x) &= \frac{5}{4}x^3\ln(x+1) - \frac{5}{4}x^3\ln(x-1) - \frac{3}{4}x\ln(x+1) + \frac{3}{4}x\ln(x-1) + \frac{2}{3} - \frac{5x^2}{2}\\
Q_4(x) &= \frac{3}{16}\ln(x+1) - \frac{3}{16}\ln(x-1) + \frac{35}{16}x^4\ln(x+1) - \frac{35}{16}x^4\ln(x-1)\\
&\quad - \frac{15}{8}x^2\ln(x+1) + \frac{15}{8}x^2\ln(x-1) - \frac{35x^3}{8} + \frac{55x}{24}\\
Q_5(x) &= \frac{63}{16}x^5\ln(x+1) - \frac{63}{16}x^5\ln(x-1) - \frac{35}{8}x^3\ln(x+1) + \frac{35}{8}x^3\ln(x-1)\\
&\quad + \frac{15}{16}x\ln(x+1) - \frac{15}{16}x\ln(x-1) - \frac{8}{15} - \frac{63x^4}{8} + \frac{49x^2}{8}\\
Q_6(x) &= -\frac{5}{32}\ln(x+1) + \frac{5}{32}\ln(x-1) + \frac{231}{32}x^6\ln(x+1) - \frac{231}{32}x^6\ln(x-1)\\
&\quad - \frac{315}{32}x^4\ln(x+1) + \frac{315}{32}x^4\ln(x-1) + \frac{105}{32}x^2\ln(x+1) - \frac{105}{32}x^2\ln(x-1)\\
&\quad - \frac{231x^5}{16} + \frac{119x^3}{8} - \frac{231x}{80}
\end{aligned}$$

$$Q_7(x) = \frac{429}{32}x^7\ \ln(x+1) - \frac{429}{32}x^7\ \ln(x-1) - \frac{693}{32}x^5\ \ln(x+1) + \frac{693}{32}x^5\ \ln(x-1)$$
$$+ \frac{315}{32}x^3\ \ln(x+1) - \frac{315}{32}x^3\ \ln(x-1) - \frac{35}{32}x\ \ln(x+1) + \frac{35}{32}x\ \ln(x-1) + \frac{16}{35} - \frac{429x^6}{16}$$
$$+ \frac{275x^4}{8} - \frac{849x^2}{80}$$
$$Q_8(x) = \frac{35}{256}\ln(x+1) - \frac{35}{256}\ln(x-1) + \frac{6435}{256}x^8\ \ln(x+1) - \frac{6435}{256}x^8\ \ln(x-1)$$
$$- \frac{3003}{64}x^6\ \ln(x+1) + \frac{3003}{64}x^6\ \ln(x-1) + \frac{3465}{128}x^4\ \ln(x+1) - \frac{3465}{128}x^4\ \ln(x-1)$$
$$- \frac{315}{64}x^2\ \ln(x+1) + \frac{315}{64}x^2\ \ln(x-1) - \frac{6435x^7}{128} + \frac{9867x^5}{128} - \frac{4213x^3}{128} + \frac{15159x}{4480} \tag{4.14}$$

$$Q_9(x) = \frac{12155}{256}x^9\ \ln(x+1) - \frac{12155}{256}x^9\ \ln(x-1) - \frac{6435}{64}x^7\ \ln(x+1) + \frac{6435}{64}x^7\ \ln(x-1)$$
$$+ \frac{9009}{128}x^5\ \ln(x+1) - \frac{9009}{128}x^5\ \ln(x-1) - \frac{1155}{64}x^3\ \ln(x+1) + \frac{1155}{64}x^3\ \ln(x-1)$$
$$+ \frac{315}{256}x\ \ln(x+1) - \frac{315}{256}x\ \ln(x-1) - \frac{128}{315} - \frac{12155x^8}{128}$$
$$+ \frac{65065x^6}{384} - \frac{11869x^4}{128} + \frac{14179x^2}{896}$$
$$Q_{10}(x) = -\frac{63}{512}\ln(x+1) + \frac{63}{512}\ln(x-1) + \frac{46189}{512}x^{10}\ \ln(x+1) - \frac{46189}{512}x^{10}\ \ln(x-1)$$
$$- \frac{109395}{512}x^8\ \ln(x+1) + \frac{109395}{512}x^8\ \ln(x-1) + \frac{45045}{256}x^6\ \ln(x+1)$$
$$- \frac{45045}{256}x^6\ \ln(x-1) - \frac{15015}{256}x^4\ \ln(x+1) + \frac{15015}{256}x^4\ \ln(x-1) + \frac{3465}{512}x^2\ \ln(x+1)$$
$$- \frac{3465}{512}x^2\ \ln(x-1) - \frac{46189x^9}{256} + \frac{70499x^7}{192} - \frac{157157x^5}{640} + \frac{26741x^3}{448} - \frac{61567x}{16128} \tag{4.15}$$

The function Q_n exhibits logarithmic singularities at the points $x = 1$ and $x = -1$. According to Eq. (4.15), Q_n also has logarithmic singularities at $\varphi = 0$ and $\varphi = \pi$. Consequently, when solving a problem in a region that includes $\varphi = 0$ and $\varphi = \pi$, this function must be excluded from the solution. When the domain is the interior of a sphere, the Eq. (4.12) reduces to

$$u(r,\varphi) = \sum_{n=0}^{\infty} A_n r^n \mathrm{P}_n(\cos(\varphi)) \tag{4.16}$$

If the problem is defined at the exterior of the sphere, the Eq. (4.12) becomes

$$u(r,\varphi) = \sum_{n=0}^{\infty} B_n r^{-1-n} \mathrm{P}_n(\cos(\varphi)) \tag{4.17}$$

In the region between two concentric spheres, the solution (4.12) reduces to

$$u(r,\varphi) = \sum_{n=0}^{\infty}(A_n r^n + B_n r^{-1-n})\mathrm{P}_n(\cos(\varphi)) \tag{4.18}$$

To illustrate the behavior of $\mathrm{P}_n(\cos(\varphi))$, it is useful to draw the curves corresponding to the first ten Legendre polynomials (Figs. 4.1–4.3). As demonstrated in Chapters 1 through 3, the integration constants in Eqs. (4.16), (4.17), and (4.18) are determined by boundary conditions. The following orthogonality relation holds for Legendre polynomials:

$$\int_{-1}^{1} P_m(x)P_n(x)dx = \frac{2}{2n+1}\delta_{mn} \tag{4.19}$$

Figures 4.4 and 4.5 represent the first two **spherical harmonics** $Y_0^0(\varphi,\theta)$ and $Y_1^0(\varphi,\theta)$, where $Y_n^0(\varphi,\theta)$ is defined by

$$Y_n^0(\varphi,\theta) = P_n(\cos(\varphi)) \tag{4.20}$$

4.2 MODEL FOR 2-D DIFFUSION IN A SPHERE

Interests in the modeling of drug transport in a spherical geometry are spurred, in part, from the need to fabricate drug-loaded microspheres (Kim & Simon, 2012) and efforts devoted toward better control of drug delivered to the brain (Siepmann et al., 2006) and tumors (Lankelma et al., 2000). Many applications involve only the radial dimension. The concentration distribution in a hollow sphere is provided when the surfaces are maintained at constant concentrations

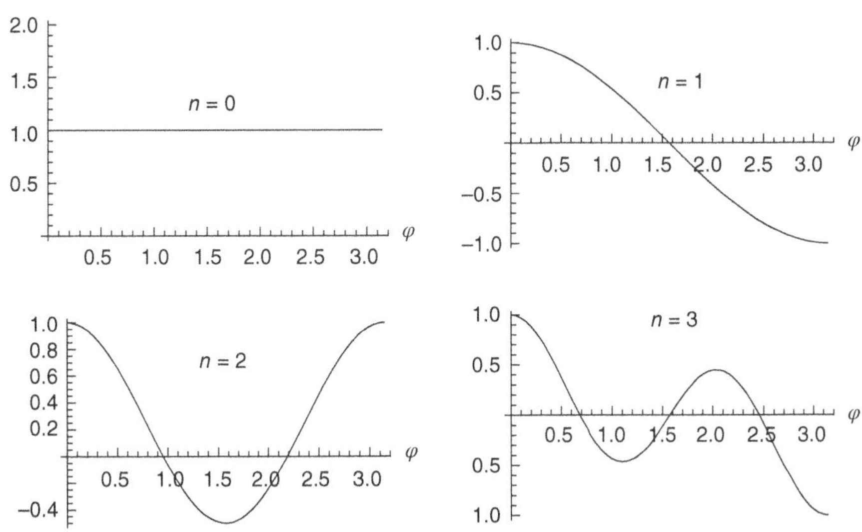

Figure 4.1 Lengendre polynomials $P_n(x)$ for $n = 0$, 1, 2, and 3.

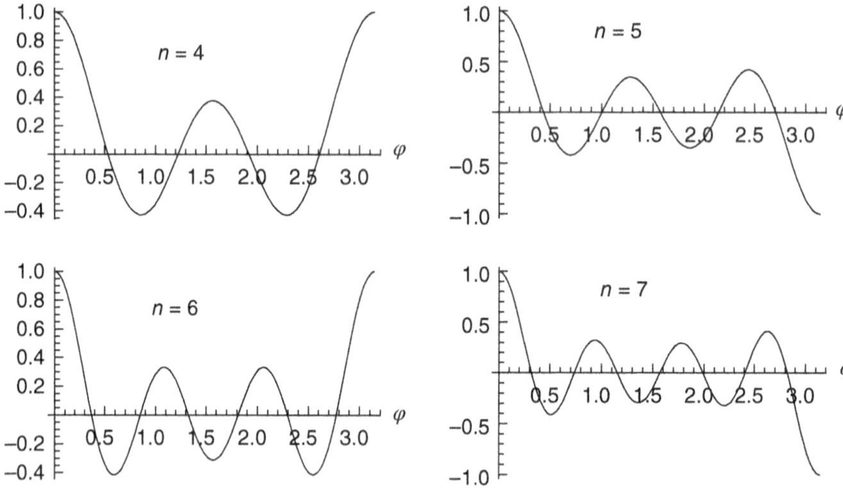

Figure 4.2 Lengendre polynomials $P_n(x)$ for $n = 4, 5, 6$, and 7.

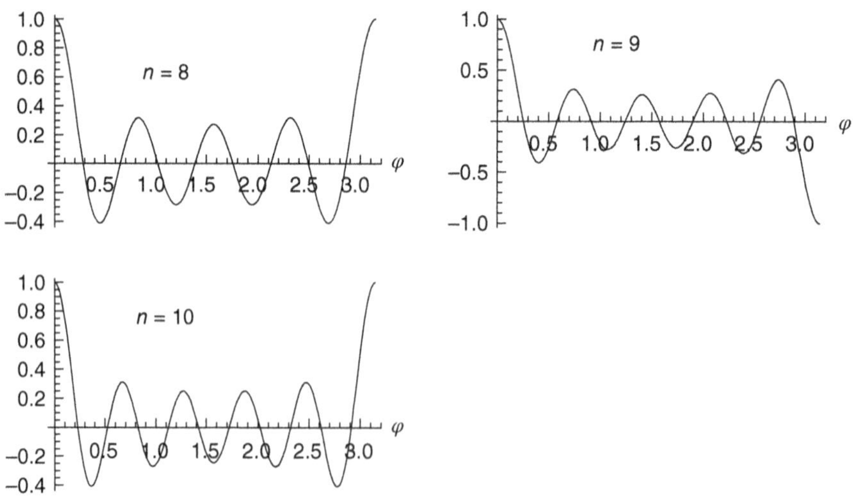

Figure 4.3 Lengendre polynomials $P_n(x)$ for $n = 8, 9$, and 10.

(Carslaw & Jaeger, 1959). The total amount of substances entering or leaving the device is provided in the literature (Crank, 1975). As researchers continue to develop new drug-release systems and focus on target sites of complex geometries, it becomes necessary to develop mathematical tools that can assist with the analysis. For 2-D systems in general, numerical techniques are proposed routinely to monitor concentration profiles (George, 2005; George et al., 2004). In spite of their efficiency, such approaches make it difficult to relate properties, such as diffusion coefficient, to performance criteria, for example, the delivery rate and the

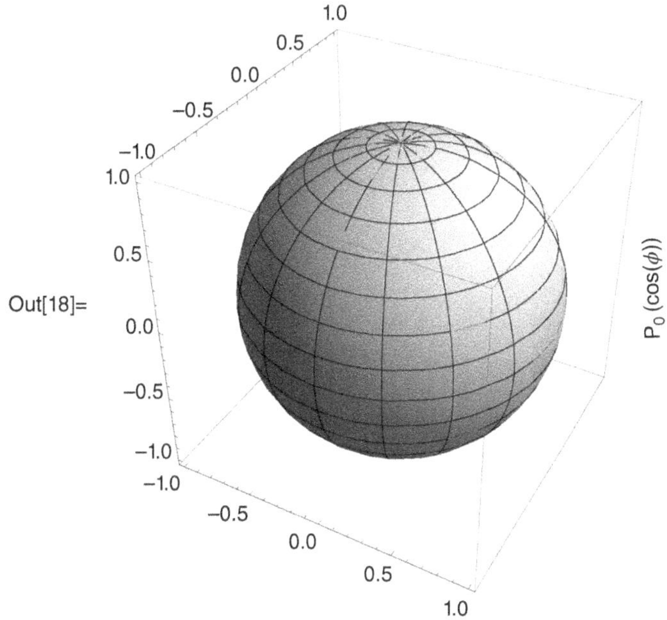

Figure 4.4 Spherical harmonic $Y_0^0(\varphi, \theta)$

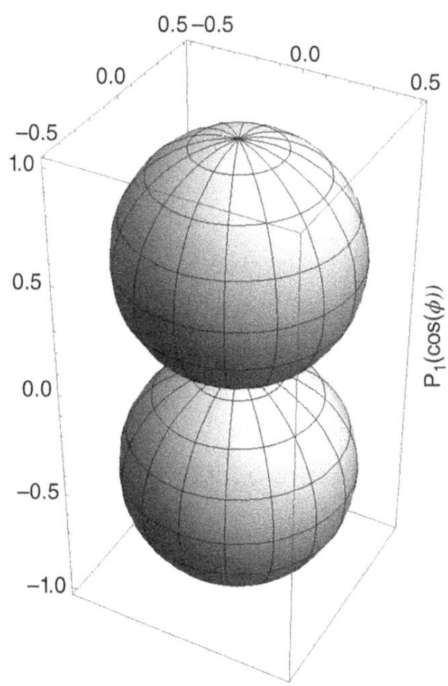

Figure 4.5 Spherical harmonic $Y_1^0(\varphi, \theta)$

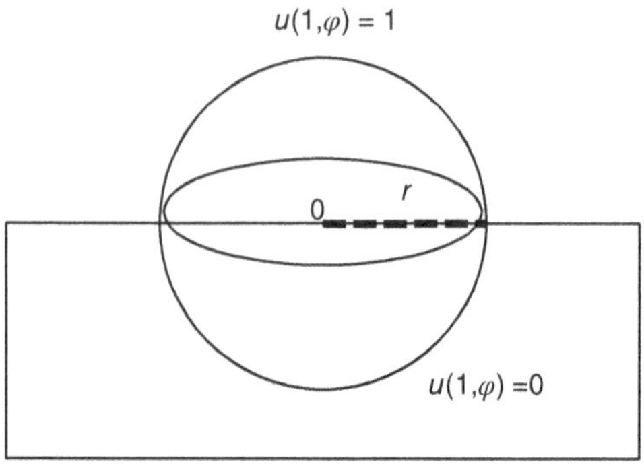

Figure 4.6 Schematic of the hollow sphere system. The upper and lower hemispheres are maintained at different drug concentrations.

time required to reach a steady-state flux. In addition, the formation of an analytical framework could help with the design of experiments and the extraction of meaningful parameters (Simon & Ospina, 2013).

Consider a hollow sphere with the upper and lower hemispheres maintained at different constant drug concentrations (Fig. 4.6). The upper hemisphere is set equal to 1 while the lower hemisphere is kept at 0.

$$\frac{\partial^2 u}{\partial r^2} + \frac{2}{r}\frac{\partial u}{\partial r} + \frac{1}{r^2}\frac{\partial^2 u}{\partial \varphi^2} + \frac{\cos\varphi}{r^2 \sin\varphi}\frac{\partial u}{\partial \varphi} = 0 \tag{4.21}$$

$$u(1,\varphi) = \begin{cases} 1, 0 \le \varphi < \dfrac{\pi}{2} \\ 0, \dfrac{\pi}{2} < \varphi \le \pi \end{cases} \tag{4.22}$$

$$\left.\frac{\partial u(r,\varphi)}{\partial r}\right|_{r=0} = 0 \tag{4.23}$$

The problem is analyzed in two dimensions because the boundary conditions are independent of θ. As a result, the solution will be independent of θ as well. Therefore, only the θ-independent spherical harmonic polynomials will be considered. Solutions to these problems are usually described by series whose coefficients are determined by the boundary values. The goal is to determine the drug concentration inside the hollow sphere.

4.3 ANALYTICAL SOLUTION OF 2-D DIFFUSION IN SPHERICAL COORDINATES

To solve the problem defined by Eqs. (4.21)–(4.23), we use the solution (4.16), which satisfies the boundary condition (4.23) automatically. After applying the boundary condition (4.22), we have

$$\sum_{n=0}^{\infty} A_n P_n(\cos(\varphi)) = \begin{cases} 1, 0 \le \varphi < \dfrac{\pi}{2} \\ 0, \dfrac{\pi}{2} < \varphi \le \pi \end{cases} \tag{4.24}$$

We multiply both sides of Eq. (4.24) by $\mathrm{P}_m(\cos(\varphi))\sin(\varphi)$ and integrate from $\varphi = 0$ to $\varphi = \pi$ to yield

$$\sum_{n=0}^{\infty} A_n \int_0^{\pi} [\mathrm{P}_n(\cos(\varphi))\mathrm{P}_m(\cos(\varphi))\sin(\varphi)]d\varphi = \begin{cases} \int_0^{\pi} \mathrm{P}_m(\cos(\varphi))\sin(\varphi)d\varphi & 0 \le \varphi < \frac{\pi}{2} \\ 0 & \frac{\pi}{2} < \varphi \le \pi \end{cases} \tag{4.25}$$

After making the change of variable $x = \cos(\varphi)$, Eq. (4.25) becomes

$$\sum_{n=0}^{\infty} A_n \int_{-1}^{1} [\mathrm{P}_n(x)\mathrm{P}_m(x)]dx = \begin{cases} \int_{-1}^{1} \mathrm{P}_m(x)dx & 0 \le x < 1 \\ 0 & -1 < x < 0 \end{cases} \tag{4.26}$$

or

$$\sum_{n=0}^{\infty} \frac{2A_n\delta_{mn}}{2n+1} = \begin{cases} \int_{-1}^{1} \mathrm{P}_m(x)dx & 0 \le x < 1 \\ 0 & -1 < x < 0 \end{cases} \tag{4.27}$$

after using the relation (4.19). We apply a basic property of the Kronecker delta to get

$$\frac{2A_m}{2m+1} = \begin{cases} \int_{-1}^{1} \mathrm{P}_m(x)dx & 0 \le x < 1 \\ 0 & -1 < x < 0 \end{cases} \tag{4.28}$$

The constant A_m is isolated from Eq. (4.28) and after replacing m by n, we obtain

$$A_n = \begin{cases} \frac{2n+1}{2} \int_{-1}^{1} \mathrm{P}_n(x)dx & 0 \le x < 1 \\ 0 & -1 < x < 0 \end{cases} \tag{4.29}$$

which yields

$$A_n = \frac{2n+1}{2} \int_0^1 \mathrm{P}_n(x)dx \tag{4.30}$$

Consequently, the following expression is developed after replacing Eq. (4.30) in (4.16):

$$u(r,\varphi) = \sum_{n=0}^{\infty} \frac{(2n+1)r^n}{2} \mathrm{P}_n(\cos(\varphi)) \int_0^1 \mathrm{P}_n(x)dx \tag{4.31}$$

The first twenty terms of the solution (4.31) are given in the Appendix. To verify that Eq. (A.4.1) satisfies the boundary condition (4.22), we replace r by 1 and draw the resulting

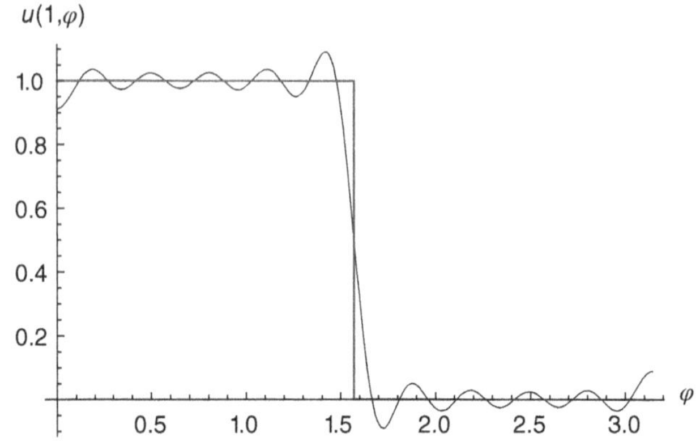

Figure 4.7 Concentration $u(1, \varphi)$ as a function of φ.

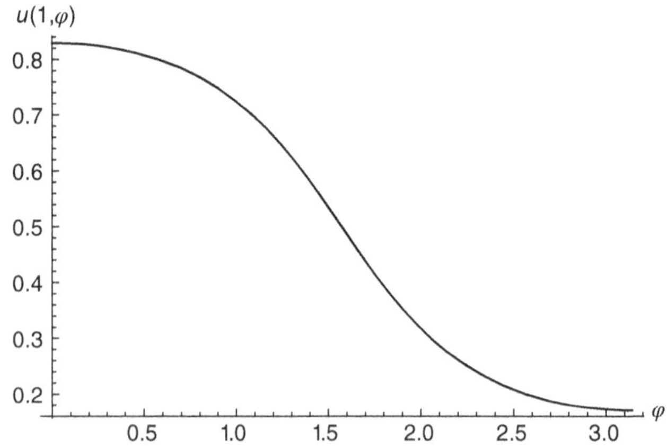

Figure 4.8 Concentration $u(1/2, \varphi)$ as a function of φ.

expression as a function of φ (Fig. 4.7). As observed, Eq. (4.31) yields a satisfactory approximation when the first twenty terms of the series are used. The concentration is $1/2$ at the center of the sphere and can be determined at any position along the radius. Figure 4.8 shows the profile when $r = 1/2$ (See Eq. (A.4.2)).

Assuming that the surface of the sphere is maintained at a constant concentration $u(R, \varphi) = 1$, a three-dimensional surface plot should produce a perfect sphere because the function is 1 for any value of φ. If $u(R, \varphi)$ is selected such that $u(R, \varphi) = 1 + 2\sin(\varphi)$, the concentration at the boundary depends on φ, which would result in a deformed sphere. These surfaces represent the nonhomogeneous distribution of the concentration at a fixed value of r. Figure 4.9 shows the concentration corresponding to the boundary condition (4.21), that is, the north hemisphere $(0 \leq \varphi < \pi/2)$ is maintained at 1 while the south hemisphere $(\pi/2 < \varphi \leq \pi)$ is kept at 0.

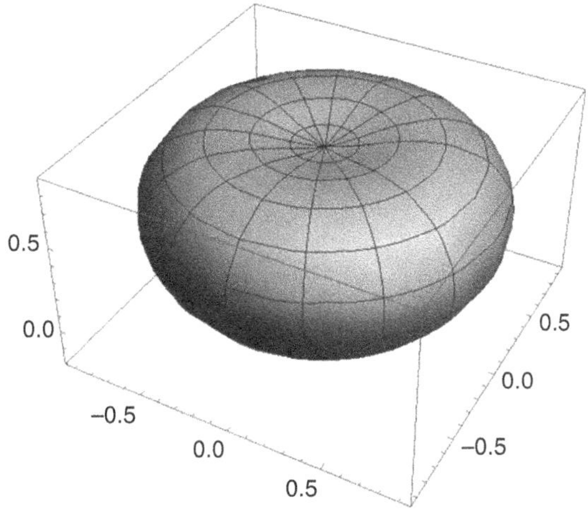

Figure 4.9 Three-dimensional concentration plot at $r = 1$.

4.4 SUMMARY

A drug-released problem in 2-D spherical coordinates was solved applying separation of variables and Legendre polynomials (P_n). The solution was developed in terms of P_n due to the singularity of Legendre functions of the second kind Q_n. Simulations show that the boundary conditions are satisfied with the first twenty terms of the series. These studies may help elucidate molecular transport in spherical geometry, guide the preparation of drug-loaded microsphere, and assist in the delivery of medicine to the brain.

4.5 APPENDIX: MAPLE, MATHEMATICA, AND MAXIMA CODE LISTINGS

4.5.1 Solution $u(r, \varphi)$ using the first twenty terms

$$u(r,\ \phi) = \frac{1}{2} - \frac{94809}{34359738368}\ \cos(\phi)\ (-10039179150\ \cos(\phi)^8 + 4418157975\ \cos(\phi)^{18}$$
$$- 20419054425\ \cos(\phi)^{16} + 39671305740\ \cos(\phi)^{14} - 267711444\ \cos(\phi)^4$$
$$+ 2230928700\ \cos(\phi)^6 + 14549535\ \cos(\phi)^2 - 42075627300\ \cos(\phi)^{12}$$
$$+ 26466926850\ \cos(\phi)^{10} - 230945)\ r^{19}$$
$$+ \frac{25025}{4294967296}\ \cos(\phi)\ (-535422888\ \cos(\phi)^6$$
$$+ 1859107250\ \cos(\phi)^8 + 109395 - 3650610600\ \cos(\phi)^{10}$$
$$- 2404321560\ \cos(\phi)^{14} + 4071834900\ \cos(\phi)^{12}$$
$$- 5542680\ \cos(\phi)^2 + 81477396\ \cos(\phi)^4$$

$$
\begin{aligned}
&+ \ 583401555 \ \cos(\phi)^{16}) \ r^{17} - \frac{13299}{134217728} \ \cos(\phi) \ (255255 \ \cos(\phi)^2 \\
&+ 9694845 \ \cos(\phi)^1 + \ 50702925 \ \cos(\phi)^{10} - 35102025 \ \cos(\phi)^{12} \\
&+ 14549535 \ \cos(\phi)^6 - 6435 - 2909907 \ \cos(\phi)^4 \\
&- 37182145 \ \cos(\phi)^8) \ r^{15} + \frac{891}{4194304} \ \cos(\phi) \ (-90090 \ \cos(\phi)^2 \\
&+ \ 4849845 \ \cos(\phi)^8 + 1300075 \ \cos(\phi)^{12} - 4056234 \ \cos(\phi)^{10} + 3003 \\
&- 2771340 \ \cos(\phi)^6 + 765765 \ \cos(\phi)^4) \ r^{13} - \frac{483}{524288} \ \cos(\phi) \ (15015 \ \cos(\phi)^2 \\
&- 90090 \ \cos(\phi)^4 - 693 + 218790 \ \cos(\phi)^6 - 230945 \ \cos(\phi)^8 \\
&+ 88179 \ \cos(\phi)^{10}) \ r^{11} + \frac{133}{65536} \ \cos(\phi) \ (-4620 \ \cos(\phi)^2 \\
&+ 12155 \ \cos(\phi)^8 + 315 + 18018 \ \cos(\phi)^4 \\
&- \ 25740 \ \cos(\phi)^6) \ r^9 - \frac{75}{4096} \ \cos(\phi) \ (315 \ \cos(\phi)^2 - 693 \ \cos(\phi)^4 - 35 \\
&+ 429 \ \cos(\phi)^6) \ r^7 + \frac{11}{256} \ \cos(\phi) \ (-70 \ \cos(\phi)^2 + 15 + 63 \ \cos(\phi)^4) \ r^5 \\
&- \frac{7}{32} \ \cos(\phi) \ (5 \ \cos(\phi)^2 - 3) \ r^3 + \frac{3}{4} \ r \ \cos(\phi)
\end{aligned}
\tag{A.4.1}
$$

4.5.2 Function $u(1/2, \varphi)$ using the first twenty terms

---

$$
\begin{aligned}
u\left(\frac{1}{2}, \ \phi\right) = \frac{1}{2} - \frac{94809}{18014398509481984} \ \cos(\phi) \ (4418157975 \ \cos(\phi)^{18} \\
- \ 20419054425 \ \cos(\phi)^{16} + 39671305740 \ \cos(\phi)^{14} - 42075627300 \ \cos(\phi)^{12} \\
+ \ 26466926850 \ \cos(\phi)^{10} - 10039179150 \ \cos(\phi)^8 + 2230928700 \ \cos(\phi)^6 \\
- \ 267711444 \ \cos(\phi)^4 + 14549535 \ \cos(\phi)^2 - 230945) \\
+ \ \frac{25025}{562949953421312} \ \cos(\phi) \ (583401555 \ \cos(\phi)^{16} - 2404321560 \ \cos(\phi)^{14} \\
+ \ 4071834900 \ \cos(\phi)^{12} - 3650610600 \ \cos(\phi)^{10} + 1859107250 \ \cos(\phi)^8 \\
- 535422888 \ \cos(\phi)^6 + 81477396 \ \cos(\phi)^4 - 5542680 \ \cos(\phi)^2 + 109395) \\
- \ \frac{13299}{4398046511104} \ \cos(\phi) \ (9694845 \ \cos(\phi)^{14} - 35102025 \ \cos(\phi)^{12} \\
+ \ 50702925 \ \cos(\phi)^{10} - 37182145 \ \cos(\phi)^8 \\
+ 14549535 \ \cos(\phi)^6 - 2909907 \ \cos(\phi)^4
\end{aligned}
$$

$$+ 255255 \cos(\phi)^2 - 6435) + \frac{891}{34359738368} \cos(\phi) (1300075 \cos(\phi)^{12}$$
$$- 4056234 \cos(\phi)^{10} + 4849845 \cos(\phi)^8$$
$$- 2771340 \cos(\phi)^6 + 765765 \cos(\phi)^4 - 90090 \cos(\phi)^2 + 3003)$$
$$- \frac{483}{1073741824} \cos(\phi) (88179 \cos(\phi)^{10} - 230945 \cos(\phi)^8$$
$$+ 218790 \cos(\phi)^6 - 90090 \cos(\phi)^4 + 15015 \cos(\phi)^2 - 693)$$
$$+ \frac{133}{33554432} \cos(\phi) (12155 \cos(\phi)^8 - 25740 \cos(\phi)^6 + 18018 \cos(\phi)^4$$
$$- 4620 \cos(\phi)^2 + 315) - \frac{75}{524288} \cos(\phi) (429 \cos(\phi)^6 - 693 \cos(\phi)^4$$
$$+ 315 \cos(\phi)^2 - 35) + \frac{11}{8192} \cos(\phi) (63 \cos(\phi)^4 - 70 \cos(\phi)^2 + 15)$$
$$- \frac{7}{256} \cos(\phi) (5 \cos(\phi)^2 - 3) + \frac{3}{8} \cos(\phi) \tag{A.4.2}$$

4.5.3 Maple Code: twospheresteady.mws

```
> restart:with(VectorCalculus):with(PDETools);
```

[CanonicalCoordinates, ChangeSymmetry, CharacteristicQ, CharacteristicQInvariants, ConservedCurrentTest, ConservedCurrents, ConsistencyTest, D_Dx, DeterminingPDE, Eta_k, Euler, FromJet, FunctionFieldSolutions, InfinitesimalGenerator, Infinitesimals, IntegratingFactorTest, IntegratingFactors, InvariantEquation, InvariantSolutions, InvariantTransformation, Invariants, Laplace, Library, PDEplot, PolynomialSolutions, ReducedForm, SimilaritySolutions, SimilarityTransformation, Solve, SymmetryCommutator, SymmetryGauge, SymmetrySolutions, SymmetryTest, SymmetryTransformation, TWSolutions, ToJet, build, casesplit, charstrip, dchange, dcoeffs, declare, diff_table, difforder, dplolyform, dsubs,mapde, separability, splitstrip, splitsys, undeclare]

```
> eq:=expand(Laplacian(C(r,phi),spherical[r,phi,theta]))=0;
```

$$eq := \frac{2\left(\frac{\partial}{\partial r} C(r, \phi)\right)}{r} + \frac{\partial^2}{\partial r^2} C(r, \phi) + \frac{\cos(\phi)\left(\frac{\partial}{\partial \phi} C(r, \phi)\right)}{r^2 \sin(\phi)} + \frac{\frac{\partial^2}{\partial \phi^2} C(r, \phi)}{r^2} = 0$$

```
> C(r,phi)=f(r)+g(phi):
> eq0:=pdsolve(eq,HINT=f(r)+g(phi)):
> eq01:=build(eq0):
> C(r,phi)=f(r)*g(phi):
> eq02:=pdsolve(eq,HINT=f(r)*g(phi)):
> eq03:=factor(build(eq02)):
> eq04:=-1/2+1/2*(1+4*_c[1])^(1/2)=n:
```

```
> eq05:=isolate(eq04,_c[1]):
> eq06:=factor(simplify(expand(simplify(subs(eq05,eq03),power,
   symoblic)),power,symbolic)) assuming n>0:
> eq07:=C(r,phi)=Sum(rhs(subs(_C1=A[n],_C2=B[n],_C3=E[n],_C4=F[n],
   eq06)),n=0..infinity):
> eq08:=C(r,phi)=rhs(eq01)+rhs(eq07):
> eq09A:=subs(F[n]=0,B[n]=0,E[n]=1,eq07):
> eq09B:=subs(F[n]=0,A[n]=0,E[n]=1,eq07):
> eq09C:=subs(F[n]=0,E[n]=1,eq07):
> eq4:=eq09A:
> eq9:=u(r,phi)=rhs(eq09A):
> eq10:=u(1,phi)=piecewise(phi<Pi/2 and phi>0,1,0):
> eq11:=u(1,phi)=rhs(subs(r=1,eq09A)):
> eq12:=subs(eq10,eq11):
> eq10A:=u(1,x)=piecewise(x<1 and x>0,1,0):
> LegendreP(m,cos(phi))*sin(phi):
> eq13:=Int(lhs(eq12)*LegendreP(m,cos(phi))*sin(phi),phi=0..Pi)
   =Sum((A[n])*Int(LegendreP(n,cos(phi))*LegendreP(m,cos(phi))
   *sin(phi),phi=0..Pi),n =0 .. infinity):
> eq13:=Int(rhs(eq10A)*LegendreP(m,x),x=-1..1)=Sum((A[n])*Int
   (LegendreP(n,x)*LegendreP(m,x),x=-1..1),n =0 .. infinity):
> eq14:=Int(LegendreP(n,x)*LegendreP(m,x),x = -1 .. 1)=(2/(2*n+1))
   *delta[n,m]:
> eq15:=subs(eq14,eq13):
> eq16:=lhs(eq15)=2*A[m]/(2*m+1):
> eq17:=isolate(eq16,A[m]):
> eq18:=subs(m=n,eq17):
> eq18A:=A[n]=((2*n+1)/2)*Int(LegendreP(n,x),x=0..1):
> eq19:=subs(x=y,eq18):
> eq20:=subs(eq18A,eq9);
```

$$\text{eq20} := u(r,\ \phi) = \sum_{n=0}^{\infty} \text{LegendreP}(n,\ \cos(\phi))\ r^n \left(n+\frac{1}{2}\right)\left(\int_0^1 \text{LegendreP}(n,\ x)\ dx\right)$$

4.5.4 Mathematica Code: twospheresteady.nb

eq = Laplacian[C[r, ϕ], {r, ϕ, θ}, "Spherical"]//Expand

$$\frac{\text{Cot}[\phi]C^{(0,1)}[r,\phi]}{r^2} + \frac{C^{(0,2)}[r,\phi]}{r^2} + \frac{2C^{(1,0)}[r,\phi]}{r} + C^{(2,0)}[r,\phi]$$

eq0 = eq/.C → Function[{r, ϕ}, f[r] + g[ϕ]];

$$\text{eq0A} = \frac{\text{Cot}[\phi]g'[\phi]}{r^2} + \frac{g''[\phi]}{r^2} == -ar^2;$$

$$\text{eq0B} = \frac{2f'[r]}{r} + f''[r] - \frac{a}{r^2} == 0;$$

eq0C = DSolve[eq0A, g, ϕ];

eq0D = (DSolve[eq0B, f, r]/.C[1]- > A[1])/.C[2] → B[1];

$$\text{eq0E} = C[2] - C[1]\text{Log}\left[\text{Cos}\left[\frac{\phi}{2}\right]\right] + C[1]\text{Log}\left[\text{Sin}\left[\frac{\phi}{2}\right]\right] + a\text{Log}[\text{Sin}[\phi]] - \frac{A[1]}{r} + B[1] + a\text{Log}[r];$$

eq0E1 = C[r, ϕ] == eq0E;

eq2 = eq/.C → Function[{r, ϕ}, f[r] * g[ϕ]];

```
eq3 = Expand[eq2/(f[r] * g[ϕ])] == 0;
eq4 = Cot[ϕ]g'[ϕ]/(r^2 g[ϕ]) + g''[ϕ]/(r^2 g[ϕ]) == -α/r^2;
eq4A = 2f'[r]/(rf[r]) - α/r^2 + f''[r]/f[r] == 0;
eq5 = DSolve[eq4, g, ϕ];
eq5A = (DSolve[eq4A, f, r]/.C[1]- > A[1])/.C[2] → B[1];

eq6 = (r^(1/2 i(i - Sqrt[-1 - 4α])) A[1] + r^(1/2 i(i + Sqrt[-1 - 4α])) B[1]) *

(C[1] LegendreP[1/2 (-1 + Sqrt[1 + 4α]), Cos[ϕ]] + C[2] *

 LegendreQ[1/2 (-1 + Sqrt[1 + 4α]), Cos[ϕ]]);
eq6A = 1/2(-1 + Sqrt[1 + 4α]) == n;
eq6B = Solve[eq6A, α];
eq6C = eq6/.α → n + n^2;
eq6D = Simplify[eq6C, n > 0];
eq6E = (((eq6D/.A[1] → A_n)/.B[1] → B_n)/.C[1]- > C[n])/.C[2] → F[n];

eq6F = Sum[r^n(C[n]LegendreP[n, Cos[ϕ]] + F[n]LegendreQ[n, Cos[ϕ]]) *

(A_n + r^(-1-2n) B_n), {n, 0, inf}];
eq6G = C[r, ϕ] == eq6F;
eq6H = C[r, ϕ] == eq0E1 + eq6F;
eq7 = eq6G;
eq8 = ((eq6G/.B_n → 0)/.F[n] → 0)/.C[n] → 1;
eq9 = eq8/.r → 1;

eq10 = Integrate[C[1, ϕ] * LegendreP[m, Cos[ϕ]] * Sin[ϕ], {ϕ, 0, π}]

 == Sum_{n=0}^{inf} Integrate[LegendreP[n, Cos[ϕ]] * LegendreP[m, Cos[ϕ]] * Sin[ϕ], {ϕ, 0, π}]A_n;

eq11 = Integrate[C[1, x] * LegendreP[m, x], {x, -1, 1}]

 == Sum_{n=0}^{inf} Integrate[LegendreP[n, x] * LegendreP[m, x], {x, -1, 1}]A_n;
eq12 = ∫_{-1}^{1} LegendreP[m, x]LegendreP[n, x] dx == (2/(2 * n + 1)) * δ_{n,m};
eq13 = ∫_{-1}^{1} C[1, x]LegendreP[m, x] dx == Sum_{n=0}^{∞} (2δ_{n,m}/(1+2n)) A_n;
eq14 = ∫_{-1}^{1} C[1, x]LegendreP[m, x] dx == 2A_m/(1+2m);
eq15 = Solve[eq14, A_m];
eq16 = A_m == 1/2(1 + 2m) ∫_{-1}^{1} C[1, x]LegendreP[m, x] dx;
eq17 = eq16/.m → n;

eq18 = ∫_{-1}^{1} C[1, x]LegendreP[n, x] dx- > Factor[∫_0^1 C[1, x] * LegendreP[n, x]dx];
eq18A = eq17/.eq18;
eq19 = A_n- > 1/2(1 + 2n) ∫_0^1 C[1, x]LegendreP[n, x] dx;
```

eq20 = eq8/.eq19

$$C[r,\phi] == \sum_{n=0}^{inf} \frac{1}{2}(1+2n)r^n \left(\int_0^1 C[1,x]\,\mathrm{LegendreP}[n,x]\,dx\right) \mathrm{LegendreP}[n,\mathrm{Cos}[\phi]]$$

$$eq20A = \sum_{n=0}^{inf} \frac{1}{2}(1+2n)r^n \left(\int_0^1 C[1,x]\,\mathrm{LegendreP}[n,x]\,dx\right) \mathrm{LegendreP}[n,\mathrm{Cos}[\phi]];$$

eq21 = (eq20/.inf → 20)/.C[1, x] → 1;

4.5.5 wxMaxima Code: twospheresteady.wxm

```
(%i1) load(vect);
(%i2) scalefactors([[r*sin(phi)*cos(theta),r*sin(phi)*sin(theta),
    r*cos(phi)],r,phi,theta])$
(%i3) depends(C,[r,phi])$
(%i4) eq0: express(laplacian(C));
```

$$(\%o4)\quad \frac{\frac{d}{d\,\theta}\frac{\frac{d}{d\,\theta}C}{|\sin(\phi)|}+\frac{d}{d\,r}\left(|\sin(\phi)|\;r^2\left(\frac{d}{d\,r}C\right)\right)+\frac{d}{d\,\phi}\left(|\sin(\phi)|\;\left(\frac{d}{d\,\phi}C\right)\right)}{|\sin(\phi)|\;r^2}$$

```
(%i5) eq0A: subst(abs(sin(phi))=sin(phi),eq0),simp, diff, expand;
```

$$(\%o5)\quad \frac{d^2}{d\,r^2}C+\frac{2\left(\frac{d}{d\,r}C\right)}{r}+\frac{\frac{d^2}{d\,\phi^2}C}{r^2}+\frac{\cos(\phi)\,\left(\frac{d}{d\,\phi}C\right)}{\sin(\phi)\;r^2}$$

```
Additive separation of variables: wxMaxima was not able to solve
   the equations;
(%i6) eq1: C=f(r)+g(phi)$
(%i7) eq2: (subst(eq1,eq0A))$
(%i8) eq3: eq2,simp, diff$
(%i9) eq4: 'diff(g(phi),phi,2)/r^2+(cos(phi)*('diff(g(phi),phi,1)))
    /(sin(phi)*r^2)=-alpha/r^2$
(%i10) eq5: desolve(eq4,g(phi))$
(%i11) eq6: 'diff(f(r),r,2)+(2*('diff(f(r),r,1)))/r-alpha/r^2=0$
(%i13) eq7: desolve(eq6,f(r))$
Multiplicative separation of variables: wxMaxima was not able to
    solve the equations.The Maple solution was used in these cases.
(%i14) eq8: C=f(r)*g(phi)$
(%i15) eq9: (subst(eq8,eq0A))$
(%i16) eq10: eq9,simp, diff$
(%i17) eq11: eq10/(f(r)*g(phi)), simp, expand$
(%i18) eq12: 'diff(g(phi),phi,2)/(g(phi)*r^2)+(cos(phi)*
    ('diff(g(phi),phi,1)))/(g(phi)*sin(phi)*r^2)=-alpha/r^2$
(%i22) eq12A: desolve(eq12,g(phi))$

desolve: can't handle this case.
-- an error. To debug this try: debugmode(true);

(%i25) eq13: g(phi)=_C3*legendre_p(-1/2+1/2*(1+4*_c[1])^(1/2),
   cos(phi))+_C4*legendre_q(-1/2+1/2*(1+4*_c[1])^(1/2),cos(phi))$
(%i30) eq14: sqrt(4*_c[1]+1)/2-1/2=n$
(%i31) assume_pos: n$
(%i32) eq14A: solve(eq14,_c[1])$
(%i33) eq14B: subst(eq14A,eq13)$
```

```
(%i34) eq14C: sqrt(4*(n^2+n)+1)/2-1/2=n$
(%i35) eq14C1: sqrt(4*(n^2+n)+1)/2=n+1/2$
(%i36) eq14D: subst(eq14C,eq14B)$
(%i37) eq15: subst(_C4=B[n],subst(_C3=A[n],eq14D))$
(%i38) eq16: 'diff(f(r),r,2)/f(r)+(2*('diff(f(r),r,1)))
   /(r*f(r))-alpha/r^2$
(%i39) eq17: desolve(eq16, f(r))$
```

desolve: can't handle this case.
-- an error. To debug this try: debugmode(true);

```
(%i40) eq18: f(r)=(_C1*r^(1/2*(1+4*_c[1])^(1/2)-1/2)
    +_C2*r^(-1/2-1/2*(1+4*_c[1])^(1/2)))$
(%i41) eq19: subst(eq14A,eq18)$
(%i42) eq20: subst(eq14C1,eq19)$
(%i43) eq21: subst(_C2=E[n],subst(_C1=C[n],eq20))$
(%i46) eq22: C(r,phi)=rhs(eq21)*rhs(eq15)$
(%i47) eq23: C(r,phi)=sum (rhs(eq22), n, 0, inf)$
The boundary equations are used
(%i48) eq24: u(r,phi)=subst(C[n]=1,subst(B[n]=0,
    subst(E[n]=0,rhs(eq23))))$
(%i49) eq25: subst(r=1,eq24)$
(%i50) eq26: integrate(u(1,phi)*legendre_p(m,cos(phi))*sin(phi),
   phi,0,%pi)=sum(A[n]*Integrate(p[n](cos(phi))*p[m](cos(phi))*
   sin(phi),phi,0,pi),n,0,inf)$
(%i51) eq26A: integrate(u(1,x)*legendre_p(m,x),x,-1,1)=sum(A[n]*
   integrate(p[m](x)*p[n](x),x,-1,1),n,0,inf)$
(%i52) eq27: integrate(p[m](x)*p[n](x),x,-1,1)
    =(2/(2*n+1))*delta[n,m]$
(%i53) eq28: subst(eq27,eq26A)$
(%i54) eq29: lhs(eq28)=2*(A[m])/(2*n+1)$
(%i55) eq30: solve(eq29,A[m]),factor$
(%i56) eq31: integrate(u(1,x)*legendre_p(m,x),x,-1,1)
   =integrate(u(1,x)*legendre_p(m,x),x,0,1)$
(%i57) eq32: subst(m=n,subst(eq31,eq30))$
(%i58) eq33: subst(eq32,eq24)$
(%i59) eq33A: subst(u(1,x)=1,eq33);
```

(%o59) $u(r,\ \phi) = \dfrac{\sum_{n=0}^{\infty}(2\ n+1)\ p_n(\cos(\phi))\ \int_0^1 p_n(x)dx\ r^n}{2}$

```
(%i71) eq34: subst(inf=5,subst(u(1,x)=1,eq33A)),simpsum$
(%i72) eq35: u(r,phi)=(3*cos(phi)*r*integrate(x,x,0,1)+5*
    (-3*(1-cos(phi))+(3*(1-cos(phi))^2)/2+1)*r^2*integrate
    (-3*(1-x)+(3*(1-x)^2)/2+1,x,0,1)+7*(-6*(1-cos(phi))-
    (5*(1-cos(phi))^3)/2+(15*(1-cos(phi))^2)/2+1)*r^3*integrate
    (-6*(1-x)-(5*(1-x)^3)/2+(15*(1-x)^2)/2+1,x,0,1)+9*(-10*(1-cos
    (phi))+(35*(1-cos(phi))^4)/8-(35*(1-cos(phi))^3)/2+(45*(1-cos
    (phi))^2)/2+1)*r^4*integrate(-10*(1-x)+(35*(1-x)^4)/8
    -(35*(1-x)^3)/2+(45*(1-x)^2)/2+1,x,0,1)+11*(-15*(1-cos(phi))
    -(63*(1-cos(phi))^5)/8+(315*(1-cos(phi))^4)/8-70*(1-cos(phi))^3
```

```
    + (105*(1-cos(phi))^2)/2+1)*r^5*integrate(-15*(1-x)
    -(63*(1-x)^5)/8+(315*(1-x)^4)/8-70*(1-x)^3+(105*(1-x)^2)/2
    +1,x,0,1)+1)/2,expand$
(%i62) eq36: subst(r=1,eq35);
```

(%o62) $u(1,\ \phi) = \frac{693\ \cos(\phi)^5}{256} - \frac{525\ \cos(\phi)^3}{128} + \frac{525\ \cos(\phi)}{256} + \frac{1}{2}$

```
(%i66) plot2d (rhs(eq36), [phi, 0, %pi]);
(%o66)
Three-dimensional plots of the first three Legendre polynomials
(%i68) plot3d (legendre_p(0,cos(theta)), [theta, 0, %pi],
[phi, 0, 2*%pi],
[transform_xy, spherical_to_xyz], [grid,30,60]);
(%o68)
(%i69) plot3d (legendre_p(1,cos(theta)), [theta, 0, %pi],
[phi, 0, 2*%pi],
[transform_xy, spherical_to_xyz], [grid,30,60]);
(%o69)
(%i70) plot3d (legendre_p(2,cos(theta)), [theta, 0, %pi],
[phi, 0, 2*%pi],
[transform_xy, spherical_to_xyz], [grid,30,60]);
(%o70)
```

PROBLEMS

4.1. Using the concentration

$$u(r,\varphi) = \sum_{n=0}^{\infty} \frac{(2n+1)r^n}{2} \mathrm{P}_n(\cos(\varphi)) \int_0^1 \mathrm{P}_n(x)dx \tag{1}$$

derive an expression for the flux through the spherical surface:

$$J(\varphi) = -\eta \left. \frac{\partial u(r,\varphi)}{\partial r} \right|_{r=1} \tag{2}$$

and plot $J(\varphi)$ for $\eta = 1$.

4.2. Compare the plots of $u(1/2, \varphi)$ using 5 and 10 terms in the solution.

4.3. Derive the solution $u(r, \varphi)$ if the condition at $r = 1$ is changed to

$$u(1,\varphi) = \begin{cases} u_1, 0 \le \varphi < \dfrac{\pi}{2} \\ u_2, \dfrac{\pi}{2} < \varphi \le \pi \end{cases}$$

where u_1 and u_2 are constants.

4.4. Develop an expression for the average drug concentration inside the sphere. Use the solution in Problem 4.3.

4.5. Show that the concentration $u(r, \varphi)$ in Problem 4.3 satisfies the boundary condition at $r = 1$ by plotting $u(1, \varphi)$. Use $u_1 = 2$ and $u_2 = 1$.

REFERENCES

Carslaw, H. S., & Jaeger, J. C. (1959). *Conduction of heat in solids* (2nd ed.). Oxford: Clarendon Press.

Cherniavsky, Y. (2010). A note on separation of variables. *Int J Math Educ Sci Technol*, 42(1), 129–131.

Crank, J. (1975). *The mathematics of diffusion* (2nd ed.). Oxford: Clarendon Press.

George, K. (2005). A two-dimensional mathematical model of non-linear dual-sorption of percutaneous drug absorption. *Biomed Eng Online*, 4, 40.

George, K., Kubota, K., & Twizell, E. H. (2004). A two-dimensional mathematical model of percutaneous drug absorption. *Biomed Eng Online*, 3(1), 18.

Kim, K. S., & Simon, L. (2012). Application of a dissolution-diffusion model to the release of 5-fluorouracil from polymer microspheres. *Chem Eng Commun*, 199(5), 587–599.

Lankelma, J., Fernandez Luque, R., Dekker, H., Schinkel, W., & Pinedo, H. M. (2000). A mathematical model of drug transport in human breast cancer. *Microvasc Res*, 59(1), 149–161.

Siepmann, J., Siepmann, F., & Florence, A. T. (2006). Local controlled drug delivery to the brain: mathematical modeling of the underlying mass transport mechanisms. *Int J Pharm*, 314(2), 101–119.

Simon, L., & Ospina, J. (2013). Two-dimensional transport analysis of transdermal drug absorption with a non-perfect sink boundary condition at the skin-capillary interface. *Math Biosci*, 244(1), 58–67.

5

STEADY-STATE THREE-DIMENSIONAL DRUG DIFFUSION THROUGH MEMBRANES FROM DISTRIBUTED SOURCES

5.1 SEPARATION OF VARIABLES IN 3-D CARTESIAN COORDINATES

In previous chapters, the method of separation of variables was applied to two-dimensional systems. As devices become more complex, efficient mathematical tools are required to analyze transport in more realistic 3-D geometries. In cases when edge effects are significant, a three-dimensional diffusion equation is critical to depict transport (Ainaoui et al., 2001; Vergnaud & Rosca, 2005). The higher dimension may also be necessary to visualize drug distribution within the target site.

The Laplace equation in three-dimensional Cartesian coordinates takes the form

$$\frac{\partial^2}{\partial x^2}c(x, y, z) + \frac{\partial^2}{\partial y^2}c(x, y, z) + \frac{\partial^2}{\partial z^2}c(x, y, z) = 0 \tag{5.1}$$

One possible solution of Eq. (5.1) is obtained by using the **additive separation of variables** method where $c(x, y, z)$ is written as $c(x, y, z) = f(x) + g(y) + h(z)$ and the functions $f(x)$, $g(y)$ and $h(z)$ satisfy, respectively, the following ordinary differential equations:

$$\frac{\mathrm{d}^2 f(x)}{\mathrm{d}x^2} = -c_2 - c_3 \tag{5.2}$$

$$\frac{\mathrm{d}^2 g(y)}{\mathrm{d}y^2} = c_2 \tag{5.3}$$

and

$$\frac{\mathrm{d}^2 h(z)}{\mathrm{d}z^2} = c_3. \tag{5.4}$$

Closed-form Solutions for Drug Transport through Controlled-Release Devices in Two and Three Dimensions, First Edition. Laurent Simon and Juan Ospina.

where c_2 and c_3 are constants. Based on the solution of the system formed by Eqs. (5.2), (5.3) and (5.4), we have

$$c(x, y, z) = -c_2\frac{1}{2}x^2 - c_3\frac{1}{2}x^2 + C_1x + C_2 + c_2\frac{1}{2}y^2 + C_3y + c_3\frac{1}{2}z^2 + C_5z \tag{5.5}$$

As shown in previous chapters, Eq. (5.1) can be solved using the **multiplicative separation of variables**. We have $c(x, y, z) = f(x)g(y)h(z)$ with the functions satisfying the following ordinary differential equations:

$$\frac{\mathrm{d}^2 f(x)}{\mathrm{d}x^2} = c_4 f(x) \tag{5.6}$$

$$\frac{\mathrm{d}^2 g(y)}{\mathrm{d}y^2} = c_5 g(y) \tag{5.7}$$

and

$$\frac{\mathrm{d}^2 h(z)}{\mathrm{d}z^2} = -c_4 h(z) - c_5 h(z) \tag{5.8}$$

The solution $c(x, y, z)$ is developed after solving Eqs. (5.6), (5.7) and (5.8):

$$\begin{aligned} c(x, y, z) = {} & (C_6 e^{y\sqrt{c_5}} + C_7 e^{-y\sqrt{c_5}})(C_8 e^{x\sqrt{c_4}} + C_9 e^{-x\sqrt{c_4}}) \times \\ & (C_{10} \sin(z\sqrt{c_4 + c_5}) + C_{11} \cos(z\sqrt{c_4 + c_5})) \end{aligned} \tag{5.9}$$

The separation constants c_4 and c_5 are determined by the boundary conditions of the problem. The **principle of superposition** can be applied to Eq. (5.9) and the particular form depends on whether the separation constants are continuous or discrete. Four combinations are possible: both c_4 and c_5 are continuous; c_4 is continuous while c_5 is discrete; c_4 is discrete while c_5 is continuous; and both separation constants are discrete. These four combinations are represented by the following equations:

$$\begin{aligned} c(x, y, z) = {} & \int\int (C_6(\alpha, \beta) e^{y\sqrt{\beta}} + C_7(\alpha, \beta) e^{-y\sqrt{\beta}}) \\ & (C_8(\alpha, \beta) e^{x\sqrt{\alpha}} + C_9(\alpha, \beta) e^{-x\sqrt{\alpha}}) \times \\ & (C_{10}(\alpha, \beta) \sin(z\sqrt{\alpha + \beta}) + C_{11}(\alpha, \beta) \cos(z\sqrt{\alpha + \beta}))\ d\alpha\ d\beta \end{aligned} \tag{5.10}$$

$$\begin{aligned} c(x, y, z) = {} & \sum_{\beta} \int (C_6(\alpha, \beta) e^{y\sqrt{\beta}} + C_7(\alpha, \beta) e^{-y\sqrt{\beta}}) \times \\ & (C_8(\alpha, \beta) e^{x\sqrt{\alpha}} + C_9(\alpha, \beta) e^{-x\sqrt{\alpha}}) \times \\ & (C_{10}(\alpha, \beta) \sin(z\sqrt{\alpha + \beta}) + C_{11}(\alpha, \beta) \cos(z\sqrt{\alpha + \beta}))\ d\alpha \end{aligned} \tag{5.11}$$

$$\begin{aligned} c(x, y, z) = {} & \sum_{\alpha} \int (C_6(\alpha, \beta) e^{y\sqrt{\beta}} + C_7(\alpha, \beta) e^{-y\sqrt{\beta}}) \times \\ & (C_8(\alpha, \beta) e^{x\sqrt{\alpha}} + C_9(\alpha, \beta) e^{-x\sqrt{\alpha}}) \times \\ & (C_{10}(\alpha, \beta) \sin(z\sqrt{\alpha + \beta}) + C_{11}(\alpha, \beta) \cos(z\sqrt{\alpha + \beta}))\ d\beta \end{aligned} \tag{5.12}$$

and

$$c(x,y,z)=\sum_{\alpha}\left[\sum_{\beta}\left(C_6(\alpha,\beta)e^{y\sqrt{\beta}}+C_7(\alpha,\beta)e^{-y\sqrt{\beta}}\right)\times\right.$$
$$(C_8(\alpha,\beta)e^{x\sqrt{\alpha}}+C_9(\alpha,\beta)e^{-x\sqrt{\alpha}})\times$$
$$\left.(C_{10}(\alpha,\beta)\sin(z\sqrt{\alpha+\beta})+C_{11}(\alpha,\beta)\cos(z\sqrt{\alpha+\beta}))\right] \tag{5.13}$$

respectively. The types of boundary conditions determine which of the last four equations is best suited to solve the problem. Some solutions may be generated by adding Eq. (5.5) to Eq. (5.10)–(5.12) or (5.13). In the next application, the latter form will be used:

$$c(x,y,z)=\sum_{\alpha}\left[\sum_{\beta}\left(C_6(\alpha,\beta)e^{y\sqrt{\beta}}+C_7(\alpha,\beta)e^{-y\sqrt{\beta}}\right)\times\right.$$
$$(C_8(\alpha,\beta)e^{x\sqrt{\alpha}}+C_9(\alpha,\beta)e^{-x\sqrt{\alpha}})\times$$
$$\left.(C_{10}(\alpha,\beta)\sin(z\sqrt{\alpha+\beta})+C_{11}(\alpha,\beta)\cos(z\sqrt{\alpha+\beta}))\right]$$
$$-c_2\frac{1}{2}x^2-c_3\frac{1}{2}x^2+C_1x+C_2+c_2\frac{1}{2}y^2+C_3y+c_3\frac{1}{2}z^2+C_5z \tag{5.14}$$

5.2 TRANSPORT ACROSS THE MEMBRANE

Consider Fig. 5.1 where a spatially-distributed source is in contact with a membrane (Ley et al., 2006). The concentration in the membrane, in equilibrium with the source, is represented by C_{eq}. The steady-state case is analyzed:

$$D_x\frac{\partial^2 C}{\partial x^2}+D_y\frac{\partial^2 C}{\partial y^2}+D_z\frac{\partial^2 C}{\partial z^2}=0;\ 0\le x\le\frac{b}{2},\ 0\le y\le\frac{b}{2},\ 0\le z\le L \tag{5.15}$$

where C is the drug concentration in the membrane. The diffusion coefficients in the $x-$, $y-$ and $z-$ directions are labeled D_x, D_y and D_z, respectively. Another 3-D analysis of diffusion through a membrane can be found in Itoh et al. (1995). Additional examples include patch formulations and powered material (Ley et al., 2006).

For this problem, it is assumed that $D_x = D_y \neq D_z$ (Ley et al., 2006). The applied drug distribution is such that the flux through the membrane surface is zero, except at the source regions:

$$C(x,y,0)=C_{eq};\ 0\le x\le\frac{d_s}{2},\ 0\le y\le\frac{d_s}{2} \tag{5.16}$$

$$\left.\frac{\partial C(x,y,z)}{\partial z}\right|_{z=0}=0;\ \frac{d_s}{2}<x\le\frac{b}{2},\ \frac{d_s}{2}<y\le\frac{b}{2} \tag{5.17}$$

$$\left.\frac{\partial C(x,y,z)}{\partial x}\right|_{x=0}=0;\ 0\le y\le\frac{b}{2},\ 0\le z\le L \tag{5.18}$$

$$\left.\frac{\partial C(x,y,z)}{\partial x}\right|_{x=\frac{b}{2}}=0;\ 0\le y\le\frac{b}{2},\ 0\le z\le L \tag{5.19}$$

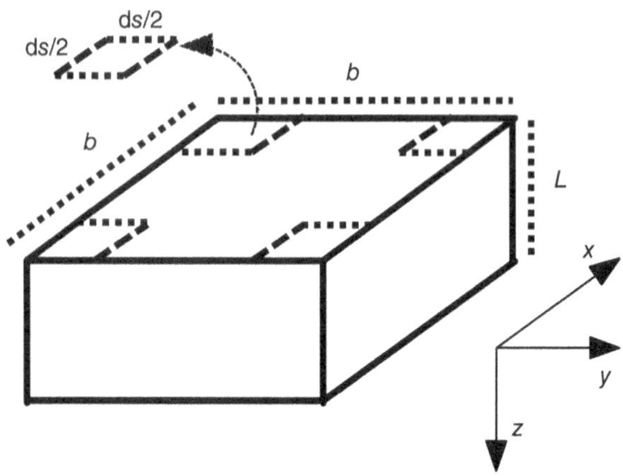

Figure 5.1 Diagram of the membrane surface and the spatially distributed source. The percentage of the area in contact with the medication is $d_s^{\,2}/b^2$.

$$\left.\frac{\partial C(x,y,z)}{\partial y}\right|_{y=0} = 0;\ 0 \le x \le \frac{b}{2},\ 0 \le z \le L \tag{5.20}$$

$$\left.\frac{\partial C(x,y,z)}{\partial y}\right|_{y=\frac{b}{2}} = 0;\ 0 \le x \le \frac{b}{2},\ 0 \le z \le L \tag{5.21}$$

$$\left.\frac{\partial C(x,y,z)}{\partial z}\right|_{z=L} + wC(x,y,L) = 0;\ 0 \le x \le \frac{b}{2},\ 0 \le y \le \frac{b}{2} \tag{5.22}$$

The no-flux statement at $z = L$ is equivalent to setting w equal to zero in Eq. (5.22), which was the equation adopted in Ley et al. (2006). The Robin boundary condition is used here to expand the applicability of the solution procedure. With the following transformation of variables,

$$\theta = \frac{C}{C_{eq}},\ \xi = \frac{x}{b/2},\ \eta = \frac{y}{b/2},\ \zeta = \frac{z}{L},\ \delta = \frac{d_s}{b},\ \beta = \frac{b}{2L}\sqrt{\frac{D_z}{D_x}} \tag{5.23}$$

we have

$$\frac{1}{\beta^2}\left(\frac{\partial^2\theta}{\partial\xi^2} + \frac{\partial^2\theta}{\partial\eta^2}\right) + \frac{\partial^2\theta}{\partial\zeta^2} = 0;\ 0 \le \xi \le 1,\ 0 \le \eta \le 1,\ 0 \le \zeta \le 1 \tag{5.24}$$

$$\theta(\xi,\eta,0) = 1;\ 0 \le \xi \le \delta,\ 0 \le \eta \le \delta \tag{5.25}$$

$$\left.\frac{\partial\theta(\xi,\eta,\zeta)}{\partial\zeta}\right|_{\zeta=0} = 0;\ \delta < \xi \le 1,\ \delta < \eta \le 1 \tag{5.26}$$

$$\left.\frac{\partial\theta(\xi,\eta,\zeta)}{\partial\xi}\right|_{\xi=0} = 0;\ 0 \le \eta \le 1,\ 0 \le \zeta \le 1 \tag{5.27}$$

$$\left.\frac{\partial\theta(\xi,\eta,\zeta)}{\partial\xi}\right|_{\xi=1} = 0;\ 0 \le \eta \le 1,\ 0 \le \zeta \le 1 \tag{5.28}$$

$$\left.\frac{\partial\theta\,(\xi,\eta,\zeta)}{\partial\eta}\right|_{\eta=0} = 0;\ 0 \le \xi \le 1,\ 0 \le \zeta \le 1 \tag{5.29}$$

$$\left.\frac{\partial\theta\,(\xi,\eta,\zeta)}{\partial\eta}\right|_{\eta=1} = 0;\ 0 \le \xi \le 1,\ 0 \le \zeta \le 1 \tag{5.30}$$

$$\left.\frac{\partial\theta\,(\xi,\eta,\zeta)}{\partial\zeta}\right|_{\zeta=1} + w\theta(\xi,\eta,1) = 0;\ 0 \le \xi \le 1,\ 0 \le \eta \le 1 \tag{5.31}$$

5.3 ANALYTICAL SOLUTION OF THE DIFFUSION MODEL IN 3-D CARTESIAN SYSTEMS

To solve the problem defined by Eqs. (5.24)–(5.31), we will look for a solution similar in structure to Eq. (5.14). Consider the following form: $\theta(\xi,\eta,\zeta) = f(\xi)g(\zeta)h(\eta) + k(\zeta)$ where $f(\xi)$, $g(\zeta)$, $h(\eta)$ and $k(\zeta)$ satisfy the following differential equations:

$$\frac{\mathrm{d}^2k(\zeta)}{\mathrm{d}\zeta^2} = 0 \tag{5.32}$$

$$\frac{\mathrm{d}^2f(\xi)}{\mathrm{d}\xi^2} = c_1 f(\xi) \tag{5.33}$$

$$\frac{\mathrm{d}^2g(\eta)}{\mathrm{d}\eta^2} = c_2 g(\eta) \tag{5.34}$$

and

$$\frac{\mathrm{d}^2h(\zeta)}{\mathrm{d}\zeta^2} = -\frac{c_1 h(\zeta)}{\beta^2} - \frac{c_2 h(\zeta)}{\beta^2} \tag{5.35}$$

The solution of Eqs. (5.32)–(5.35) leads to

$$\begin{aligned}\theta(\xi,\eta,\zeta) = {} & C_7\zeta + C_8 + (C_3\mathrm{e}^{\eta\sqrt{c_2}} + C_4\mathrm{e}^{-\eta\sqrt{c_2}})(C_1\mathrm{e}^{\xi\sqrt{c_1}} + C_2\mathrm{e}^{-\xi\sqrt{c_1}})\times \\ & \left(C_5\sin\left(\frac{\zeta\sqrt{c_1+c_2}}{\beta}\right) + C_6\cos\left(\frac{\zeta\sqrt{c_1+c_2}}{\beta}\right)\right)\end{aligned} \tag{5.36}$$

An application of the boundary condition (5.27) to Eq. (5.36) gives

$$\begin{aligned}& (C_3\mathrm{e}^{\eta\sqrt{c_2}} + C_4\mathrm{e}^{-\eta\sqrt{c_2}})(C_1\sqrt{c_1} - C_2\sqrt{c_1})\times \\ & \left(C_5\sin\left(\frac{\zeta\sqrt{c_1+c_2}}{\beta}\right) + C_6\cos\left(\frac{\zeta\sqrt{c_1+c_2}}{\beta}\right)\right) = 0\end{aligned} \tag{5.37}$$

From Eq. (5.37), we deduce $C_2 = C_1$. As a result, Eq. (5.36) becomes

$$\begin{aligned}\theta(\xi,\eta,\zeta) = {} & C_7\zeta + C_8 + 2(C_3\mathrm{e}^{\eta\sqrt{c_2}} + C_4\mathrm{e}^{-\eta\sqrt{c_2}})C_1\cosh(\xi\sqrt{c_1})\times \\ & \left(C_5\sin\left(\frac{\zeta\sqrt{c_1+c_2}}{\beta}\right) + C_6\cos\left(\frac{\zeta\sqrt{c_1+c_2}}{\beta}\right)\right)\end{aligned} \tag{5.38}$$

Without any loss of generality, we set $C_1 = 1$ yielding

$$\theta(\xi,\eta,\zeta) = C_7\zeta + C_8 + 2(C_3 e^{\eta\sqrt{c_2}} + C_4 e^{-\eta\sqrt{c_2}})\cosh(\xi\sqrt{c_1})\times \left(C_5 \sin\left(\frac{\zeta\sqrt{c_1+c_2}}{\beta}\right) + C_6\cos\left(\frac{\zeta\sqrt{c_1+c_2}}{\beta}\right)\right) \tag{5.39}$$

The use of the boundary condition (5.28) leads to

$$2(C_3 e^{\eta\sqrt{c_2}} + C_4 e^{-\eta\sqrt{c_2}})\sqrt{c_1}\sinh(\sqrt{c_1})\left(C_5 \sin\left(\frac{\zeta\sqrt{c_1+c_2}}{\beta}\right) + C_6\cos\left(\frac{\zeta\sqrt{c_1+c_2}}{\beta}\right)\right) = 0 \tag{5.40}$$

which means

$$\sinh(\sqrt{c_1}) = 0 \tag{5.41}$$

from which we deduce

$$\sqrt{c_1} = n\pi i \tag{5.42}$$

or

$$c_1 = -n^2\pi^2 \tag{5.43}$$

Equation (5.39) becomes

$$\theta(\xi,\eta,\zeta) = C_7\zeta + C_8 + 2(C_3 e^{\eta\sqrt{c_2}} + C_4 e^{-\eta\sqrt{c_2}})\cos(n\pi\xi)\times \left(C_5 \sin\left(\frac{\zeta\sqrt{-n^2\pi^2+c_2}}{\beta}\right) + C_6\cos\left(\frac{\zeta\sqrt{-n^2\pi^2+c_2}}{\beta}\right)\right) \tag{5.44}$$

As a result, the boundary condition (5.29) yields

$$2(C_3\sqrt{c_2} - C_4\sqrt{c_2})\cos(n\pi\xi)\times \left(C_5 \sin\left(\frac{\zeta\sqrt{-n^2\pi^2+c_2}}{\beta}\right) + C_6\cos\left(\frac{\zeta\sqrt{-n^2\pi^2+c_2}}{\beta}\right)\right) = 0 \tag{5.45}$$

Isolating C_4, we have $C_3 = C_4$ and, therefore,

$$\theta(\xi,\eta,\zeta) = C_7\zeta + C_8 + 4C_3\cosh(\eta\sqrt{c_2})\cos(n\pi\xi)\times \left(C_5 \sin\left(\frac{\zeta\sqrt{-n^2\pi^2+c_2}}{\beta}\right) + C_6\cos\left(\frac{\zeta\sqrt{-n^2\pi^2+c_2}}{\beta}\right)\right) \tag{5.46}$$

or

$$\theta(\xi,\eta,\zeta) = C_7\zeta + C_8 + 4\cosh(\eta\sqrt{c_2})\cos(n\pi\xi)\times \left(C_5 \sin\left(\frac{\zeta\sqrt{-n^2\pi^2+c_2}}{\beta}\right) + C_6\cos\left(\frac{\zeta\sqrt{-n^2\pi^2+c_2}}{\beta}\right)\right) \tag{5.47}$$

after setting $C_3 = 1$. Application of Eq. (5.30) to Eq. (5.47) results in

$$4\sqrt{c_2}\sinh(\sqrt{c_2})\cos(n\pi\xi)\left(C_5\sin\left(\frac{\zeta\sqrt{-n^2\pi^2+c_2}}{\beta}\right)+C_6\cos\left(\frac{\zeta\sqrt{-n^2\pi^2+c_2}}{\beta}\right)\right)=0 \tag{5.48}$$

Eq. (5.48) implies that

$$\sinh(\sqrt{c_2}) = 0 \tag{5.49}$$

from which we have

$$\sqrt{c_2} = m\pi i \tag{5.50}$$

or

$$c_2 = -m^2\pi^2 \tag{5.51}$$

Substitution of c_2 in Eq. (5.47) gives

$$\theta(\xi,\eta,\zeta) = C_7\zeta + C_8 + 4\cos(m\pi\eta)\cos(n\pi\xi)\times$$
$$\left(C_5\sin\left(\frac{\pi\zeta\sqrt{-n^2-m^2}}{\beta}\right)+C_6\cos\left(\frac{\pi\zeta\sqrt{-n^2-m^2}}{\beta}\right)\right) \tag{5.52}$$

Applying the principle of superposition to Eq. (5.52), we write

$$\theta(\xi,\eta,\zeta) = C_7\zeta + C_8 + \sum_{m=0}^{\infty}\left[\sum_{n=0}^{\infty}(4\cos(m\pi\eta)\cos(n\pi\xi)\times\right.$$
$$\left.\left(A_{n,m}\sin\left(\frac{\pi\zeta\sqrt{-n^2-m^2}}{\beta}\right)+B_{n,m}\cos\left(\frac{\pi\zeta\sqrt{-n^2-m^2}}{\beta}\right)\right))\right] \tag{5.53}$$

which can take the following form:

$$\theta(\xi,\eta,\zeta) = 4B_{0,0}$$
$$+\sum_{n=1}^{\infty}\left(4i\cos(n\pi\zeta)A_{n,0}\sinh\left(\frac{n\pi\zeta}{\beta}\right)+4\cos(n\pi\xi)B_{n,0}\cosh\left(\frac{n\pi\zeta}{\beta}\right)\right)$$
$$+\sum_{m=1}^{\infty}\left(4i\cos(m\pi\eta)A_{0,m}\sinh\left(\frac{m\pi\zeta}{\beta}\right)+4\cos(m\pi\eta)B_{0,m}\cosh\left(\frac{m\pi\zeta}{\beta}\right)\right)$$
$$+\sum_{m=1}^{\infty}\left(\sum_{n=1}^{\infty}\left(\begin{matrix}4\cos(m\pi\eta)\cos(n\pi\xi)A_{n,m}\sin\left(\frac{\pi\zeta\sqrt{-n^2-m^2}}{\beta}\right)+\\4\cos(m\pi\eta)\cos(n\pi\xi)B_{n,m}\cos\left(\frac{\pi\zeta\sqrt{-n^2-m^2}}{\beta}\right)\end{matrix}\right)\right)+C_7\zeta+C_8 \tag{5.54}$$

The application of the boundary condition (5.31) to Eq. (5.54) gives

$$w\left(\begin{matrix}4B_{0,0}+4\cos(n\pi\xi)\left(A_{n,0}\sinh\left(\frac{n\pi}{\beta}\right)i+B_{n,0}\cosh\left(\frac{n\pi}{\beta}\right)\right)\\+4\cos(m\pi\eta)\left(A_{0,m}\sinh\left(\frac{m\pi}{\beta}\right)i+B_{0,m}\cosh\left(\frac{m\pi}{\beta}\right)\right)\\+4\cos(m\pi\eta)\cos(n\pi\xi)\left(A_{n,m}\sin\left(\frac{\pi\sqrt{-n^2-m^2}}{\beta}\right)+B_{n,m}\cos\left(\frac{\pi\sqrt{-n^2-m^2}}{\beta}\right)\right)\\+C_7+C_8\end{matrix}\right)$$

$$+\,4\cos(n\pi\xi)\left(\frac{A_{n,0}\cosh\left(\frac{n\pi}{\beta}\right)n\pi i}{\beta}+\frac{B_{n,0}\sinh\left(\frac{n\pi}{\beta}\right)n\pi}{\beta}\right)$$

$$+\,4\cos(m\pi\eta)\left(\frac{A_{0,m}\cosh\left(\frac{m\pi}{\beta}\right)m\pi i}{\beta}+\frac{B_{0,m}\sinh\left(\frac{m\pi}{\beta}\right)m\pi}{\beta}\right)$$

$$+\,4\cos(m\pi\eta)\cos(n\pi\xi)\left(\frac{A_{n,m}\cos\left(\frac{\pi\sqrt{-n^2-m^2}}{\beta}\right)\pi\sqrt{-n^2-m^2}}{\beta}-\frac{B_{n,m}\sin\left(\frac{\pi\sqrt{-n^2-m^2}}{\beta}\right)\pi\sqrt{-n^2-m^2}}{\beta}\right)+C_7=0 \tag{5.55}$$

where the symbols of sums were deleted to facilitate reading and manipulation. From Eq. (5.55), we deduce

$$B_{n,0}=-\frac{iA_{n,0}\left(\cosh\left(\frac{\pi n}{\beta}\right)\pi n+w\beta\sinh\left(\frac{\pi n}{\beta}\right)\right)}{\sinh\left(\frac{\pi n}{\beta}\right)\pi n+w\beta\cosh\left(\frac{\pi n}{\beta}\right)} \tag{5.56}$$

$$B_{0,m}=-\frac{IA_{0,m}\left(\cosh\left(\frac{\pi m}{\beta}\right)\pi m+w\beta\sinh\left(\frac{\pi m}{\beta}\right)\right)}{w\beta\cosh\left(\frac{\pi m}{\beta}\right)+\sinh\left(\frac{\pi m}{\beta}\right)\pi m} \tag{5.57}$$

$$B_{n,m}=\frac{A_{n,m}\left(\cos\left(\frac{\pi\sqrt{-n^2-m^2}}{\beta}\right)\pi\sqrt{-n^2-m^2}+w\beta\sin\left(\frac{\pi\sqrt{-n^2-m^2}}{\beta}\right)\right)}{-w\beta\cos\left(\frac{\pi\sqrt{-n^2-m^2}}{\beta}\right)+\sin\left(\frac{\pi\sqrt{-n^2-m^2}}{\beta}\right)\pi\sqrt{-n^2-m^2}} \tag{5.58}$$

$$B_{0,0}=-\frac{C_8}{4}-\frac{C_7}{4} \tag{5.59}$$

Substitution of Eqs. (5.56)–(5.59) into Eq. (5.54) gives

$$\theta(\xi,\eta,\zeta)=-C_7+\sum_{n=1}^{\infty}\left[4i\cos(\pi n\xi)A_{n,0}\sinh\left(\frac{\pi n(\zeta)}{\beta}\right)-\frac{4i\cos(\pi n\xi)A_{n,0}\left(\cosh\left(\frac{\pi n}{\beta}\right)\pi n+w\beta\sinh\left(\frac{\pi n}{\beta}\right)\right)\cosh\left(\frac{\pi n\zeta}{\beta}\right)}{\sinh\left(\frac{\pi n}{\beta}\right)\pi n+w\beta\cosh\left(\frac{\pi n}{\beta}\right)}\right]$$

$$+\sum_{m=1}^{\infty}\left[4i\cos(\pi m\eta)A_{0,m}\sinh\left(\frac{\pi m\zeta}{\beta}\right)-\frac{4i\cos(\pi m\eta)A_{0,m}\left(\cosh\left(\frac{\pi m}{\beta}\right)\pi m+w\beta\sinh\left(\frac{\pi m}{\beta}\right)\right)\cosh\left(\frac{\pi m\zeta}{\beta}\right)}{w\beta\cosh\left(\frac{\pi m}{\beta}\right)+\sinh\left(\frac{\pi m}{\beta}\right)\pi m}\right]$$

$$+\sum_{m=1}^{\infty}\left(\begin{array}{c}\sum_{n=1}^{\infty}[4\cos(\pi m\eta)\cos(\pi n\xi)A_{n,m}\sin\left(\frac{\pi\zeta\sqrt{-m^2-n^2}}{\beta}\right)\\ +4\cos(\pi m\eta)\cos(\pi n\xi)A_{n,m}\times\\ \left(\cos\left(\frac{\pi\sqrt{-m^2-n^2}}{\beta}\right)\pi\sqrt{-m^2-n^2}+w\beta\sin\left(\frac{\pi\sqrt{-m^2-n^2}}{\beta}\right)\right)\\ \times\frac{\cos\left(\frac{\pi\zeta\sqrt{-m^2-n^2}}{\beta}\right)}{\left(-w\beta\cos\left(\frac{\pi\sqrt{-m^2-n^2}}{\beta}\right)+\sin\left(\frac{\pi\sqrt{-m^2-n^2}}{\beta}\right)\pi\sqrt{-m^2-n^2}\right)}]\end{array}\right)+C_7\zeta \tag{5.60}$$

The boundary conditions (5.25) and (5.25) are combined as

$$\alpha(\xi,\eta)\theta(\xi,\eta,0)+k(\xi,\eta)\frac{\partial}{\partial\zeta}\theta(\xi,\eta,\zeta)\bigg|_{\zeta=0}=f(\xi,\eta) \tag{5.61}$$

where

$$\alpha(\xi,\eta)=(\mathrm{H}(\xi)-\mathrm{H}(\xi-\delta))(\mathrm{H}(\eta)-\mathrm{H}(\eta-\delta)) \tag{5.62}$$

$$k(\xi,\eta)=1-\alpha(\xi,\eta) \tag{5.63}$$

and

$$f(\xi,\eta)=\alpha(\xi,\eta) \tag{5.64}$$

where H(.) is the **unit step function**, defined as

$$H(x)=\begin{cases}0 & x<0\\ 1 & x\geq 0\end{cases} \tag{5.65}$$

When we apply the boundary condition (5.62) to Eq.(5.60), we obtain

$$\begin{array}{l}\alpha(\xi,\eta)\left(\begin{array}{c}-C_7-\sum_{n=1}^{\infty}\left[\frac{4\,\mathrm{i}\cos(\pi n\xi)A_{n,0}\left(\cosh\left(\frac{\pi n}{\beta}\right)\pi n+w\beta\sinh\left(\frac{\pi n}{\beta}\right)\right)}{\sinh\left(\frac{\pi n}{\beta}\right)\pi n+w\beta\cosh\left(\frac{\pi n}{\beta}\right)}\right]\\ -\sum_{m=1}^{\infty}\left[\frac{4\,\mathrm{i}\cos(\pi m\eta)A_{0,m}\left(\cosh\left(\frac{\pi m}{\beta}\right)\pi m+w\beta\sinh\left(\frac{\pi m}{\beta}\right)\right)}{w\beta\cosh\left(\frac{\pi m}{\beta}\right)+\sinh\left(\frac{\pi m}{\beta}\right)\pi m}\right]\\ +\sum_{m=1}^{\infty}\left(\sum_{n=1}^{\infty}\left[\frac{\begin{array}{c}4\cos(\pi m\eta)\cos(\pi n\xi)A_{n,m}\times\\ \left(\cos\left(\frac{\pi\sqrt{-m^2-n^2}}{\beta}\right)\pi\sqrt{-m^2-n^2}+w\beta\sin\left(\frac{\pi\sqrt{-m^2-n^2}}{\beta}\right)\right)\end{array}}{-w\beta\cos\left(\frac{\pi\sqrt{-m^2-n^2}}{\beta}\right)+\sin\left(\frac{\pi\sqrt{-m^2-n^2}}{\beta}\right)\pi\sqrt{-m^2-n^2}}\right]\right)\end{array}\right)\\ +k(\xi,\eta)\left(\begin{array}{c}\sum_{n=1}^{\infty}\left[\frac{4\,\mathrm{i}\cos(\pi n\xi)A_{n,0}\pi n}{\beta}\right]+\sum_{m=1}^{\infty}\left[\frac{4\,\mathrm{i}\cos(\pi m\eta)A_{0,m}\pi m}{\beta}\right]\\ +\sum_{m=1}^{\infty}\left(\sum_{n=1}^{\infty}\left[4\frac{\cos(\pi m\eta)\cos(\pi n\xi)A_{n,m}\pi\sqrt{-m^2-n^2}}{\beta}\right]\right)+C_7\end{array}\right)\\ =f(\xi,\eta)\end{array} \tag{5.66}$$

Using the properties of the double Fourier series, Eq. (5.66) provides a system of infinite equations in an infinite number of unknowns: $A_{n,0}$, $A_{0,m}$, $A_{n.m}$ and C_7. The solution $\theta(\xi,\eta,\zeta)$ is fully specified from knowledge of these coefficients. Two quantities of interest are the average dimensionless concentration in the membrane:

$$\theta_{ave} = \int_0^1 \int_0^1 \int_0^1 \theta(\xi, \eta, \zeta) d\xi d\eta d\zeta \tag{5.67}$$

and the average dimensionless flux across a plane at ζ:

$$\Psi_{ave} = -\int_0^1 \int_0^1 \left. \frac{\partial \theta}{\partial \zeta} \right|_{\zeta} d\xi d\eta \tag{5.68}$$

Figures 5.2 and 5.3 show θ_{ave} and Ψ_{ave} at $\zeta = 1$ as a function of β. Expressions for θ_{ave} and Ψ_{ave} when $\zeta = 1$ are given in sections 5.2 and 5.3 of the appendix. A first-order approximation (i.e., upper limit of summation in Eq. (5.60) is 1) is considered. In this case, w is set equal to 0 and the percentage of the membrane surface covered by the source ($\varepsilon = d_s^2/b^2 = \delta^2$) is employed instead of δ. It can be shown that when the source is applied to the entire membrane surface, or $\varepsilon = 1$, the following values are obtained by using the first-order solution: $\theta_{ave} = 1/2$ and $\Psi_{ave} = 1$.

5.4 SUMMARY

The model of a membrane, where the surface is in contact with a distributed source of drug, was analyzed. Laplace's equation in three dimensions was written to describe steady-state diffusion in the medium. The closed-form solution, based on the method of separation of variables, helped predict the average flux and concentration of a drug in the membrane. These results may elucidate drug-release mechanisms in complex systems and clarify the transport of chemicals applied to membranes in the form of a powder.

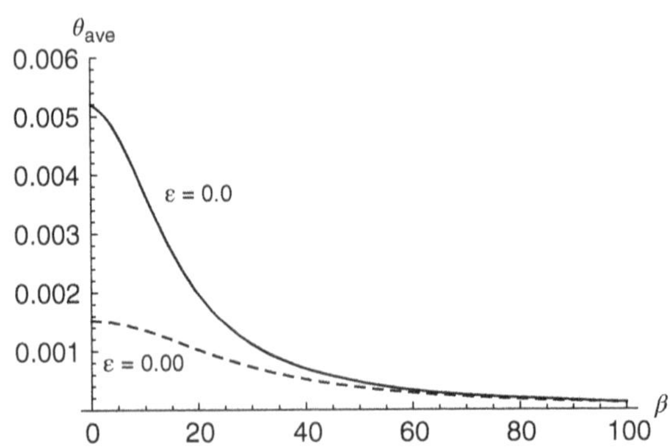

Figure 5.2 Average dimensionless concentration in the membrane θ_{ave} as a function of β when $w = 0$.

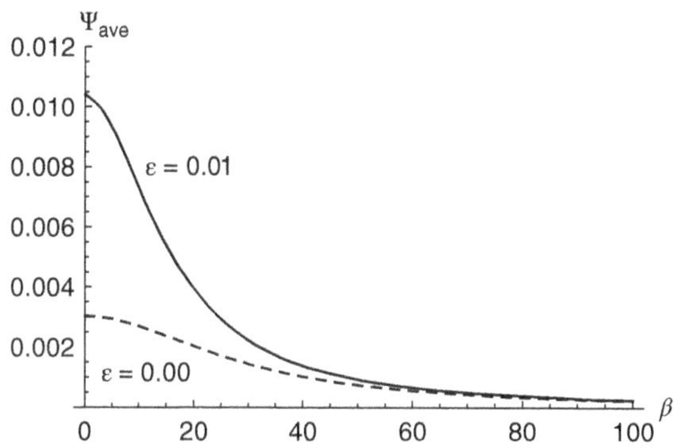

Figure 5.3 Average dimensionless flux concentration Ψ_{ave} at $\zeta = 1$ as a function of β when $w = 0$.

5.5 APPENDIX: MAPLE CODE LISTINGS

5.5.1 Maple Code: threeDcartsteady.mws

```
> restart:with(inttrans):with(PDETools);
[CanonicalCoordinates, ChangeSymmetry, CharacteristicQ,
   CharacteristicQInvariants, ConservedCurrentTest, ConservedCurrents,
   ConsistencyTest, D_Dx, DeterminingPDE, Eta_k, Euler, FromJet,
   FunctionFieldSolutions, InfinitesimalGenerator, Infinitesimals,
   IntegratingFactorTest, IntegratingFactors, InvariantEquation,
   InvariantSolutions, InvariantTransformation, Invariants, Laplace,
   Library, PDEplot, PolynomialSolutions, ReducedForm,
   SimilaritySolutions, SimilarityTransformation, Solve,
   SymmetryCommutator, SymmetryGauge, SymmetrySolutions, SymmetryTest,
   SymmetryTransformation, TWSolutions, ToJet, build, casesplit,
   charstrip, dchange, dcoeffs, declare, diff_table, difforder,
   dplolyform, dsubs,mapde, separability, splitstrip, splitsys,
   undeclare]
> aux:=diff(C(x,y,z),x,x)+diff(C(x,y,z),y,y)+diff(C(x,y,z),z,z)=0;
```

$$aux := \frac{\partial^2}{\partial x^2} C(x, y, z) + \frac{\partial^2}{\partial y^2} C(x, y, z) + \frac{\partial^2}{\partial z^2} C(x, y, z) = 0$$

```
Additive Separation of Variables:
> aux1:=pdsolve(aux,HINT=f(x)+g(y)+h(z)):
> aux2:=build(aux1):
Multiplicative Separation of Variables:
> aux3:=pdsolve(aux,HINT=f(x)*g(y)*h(z)):
> aux4:=factor(build(aux3)):
> aux5:=exp(_c[2]^(1/2)*y)^2*_C3+_C4=exp(_c[2]^(1/2)*y)*
   (_C3*exp(_c[2]^(1/2)*y)+_C4*exp(-_c[2]^(1/2)*y)):
> aux6:=_C1*exp(_c[1]^(1/2)*x)^2+_C2=exp(_c[1]^(1/2)*x)*
   (_C1*exp(_c[1]^(1/2)*x)+_C2*exp(-_c[1]^(1/2)*x)):
```

```
> aux7:=subs(aux5,aux6,aux4):
> aux8:=subs(_C1=C[1](alpha,beta),_C2=C[2](alpha,beta),_C3=C[3]
  (alpha,beta),_C3=C[3](alpha,beta),_C4=C[4](alpha,beta),_C5=C[5]
  (alpha,beta),_C6=C[6](alpha,beta),subs(_c[1]=alpha,_c[2]=beta,aux7)):
Superposition Principle
> aux9:=C(x,y,z)=Int(Int(rhs(aux8),alpha),beta):
> aux10:=C(x,y,z)=Sum(Int(rhs(aux8),alpha),beta):
> aux11:=C(x,y,z)=Sum(Int(rhs(aux8),beta),alpha):
> aux12:=C(x,y,z)=Sum(Sum(rhs(aux8),beta),alpha):
General Solutions
> aux13:=C(x,y,z)=rhs(aux9)+rhs(aux2):
> aux14:=C(x,y,z)=rhs(aux10)+rhs(aux2):
> aux15:=C(x,y,z)=rhs(aux11)+rhs(aux2):
> aux16:=C(x,y,z)=rhs(aux12)+rhs(aux2):
Solving the Boundary-Value Problem:
> eq:=(1/beta^2)*(diff(theta(xi,eta,zeta),xi,xi)+diff(theta(xi,eta,
   zeta),eta,eta))+diff(theta(xi,eta,zeta),zeta,zeta)=0:
> theta(xi,zeta,eta)=f(xi)*g(zeta)*h(eta)+ k(zeta):
> eq0:=eval(subs(theta(xi,eta,zeta)=k(zeta),eq)):
> eq0A:=dsolve(eq0):
>
> eq0B:=subs(_C1=C7,_C2=C8,eq0A):
>
> eq1:=pdsolve(eq,HINT=f(xi)*g(eta)*h(zeta)):
> eq2:=factor(build(eq1)):
> eq2A:=_C1*exp(_c[1]^(1/2)*xi)^2+_C2=exp(_c[1]^(1/2)*xi)*
    (_C1*exp(_c[1]^(1/2)*xi)+_C2*exp(-_c[1]^(1/2)*xi)):
> eq2B:=exp(_c[2]^(1/2)*eta)^2*_C3+_C4=exp(_c[2]^(1/2)*eta)*
    (exp(_c[2]^(1/2)*eta)*_C3+_C4*exp(-_c[2]^(1/2)*eta)):
> eq3:=subs(eq2A,eq2B,eq2):
> eq3A:=lhs(eq3)=rhs(eq3)+rhs(eq0B):
Boundary Condition
> eq4:=eval(subs(xi=0,diff(rhs(eq3A),xi)))=0:
> eq5:=isolate(eq4,_C2):
> eq6:=subs(eq5,eq3A):
> eq7:=_C1*exp(_c[1]^(1/2)*xi)+_C1*exp(-_c[1]^(1/2)*xi)=convert
    (factor(_C1*exp(_c[1]^(1/2)*xi)+_C1*exp(-_c[1]^(1/2)*xi)),cos):
> eq8:=subs(eq7,eq6):
> eq9:=subs(_C1=1,eq8):
Boundary Condition
> eq10:=subs(xi=1,diff(rhs(eq9),xi))=0:
> eq11:=sinh(_c[1]^(1/2))=0:
> eq12:=_c[1]^(1/2)=I*n*Pi:
> eq13:=isolate(eq12,_c[1]):
> eq14:=simplify(subs(eq13,eq9),power,symbolic):
Boundary Condition
> eq15:=eval(subs(eta=0,diff(rhs(eq14),eta)))=0:
> eq16:=isolate(eq15,_C4):
> eq17:=subs(eq16,eq14):
```

```
> eq18:=_C3*exp(_c[2]^(1/2)*eta)+_C3*exp(-_c[2]^(1/2)*eta)=convert
    (factor(_C3*exp(_c[2]^(1/2)*eta)+_C3*exp(-_c[2]^(1/2)*eta)),cos):
> eq19:=subs(eq18,eq17):
> eq20:=subs(_C3=1,eq19):
Boundary Condition
> eq21:=subs(eta=1,diff(rhs(eq20),eta))=0:
> eq22:=sinh(_c[2]^(1/2))=0:
> eq23:=_c[2]^(1/2)=I*m*Pi:
> eq24:=isolate(eq23,_c[2]):
> eq25:=simplify(subs(eq24,eq20),power,symbolic):
> theta(xi,eta,zeta) = C7*zeta+C8+factor(4*cos(m*Pi*eta)*cos(n*Pi*xi)*
    _C5*sin(1/beta*Pi*(-n^2-m^2)^(1/2)*zeta)+4*cos(m*Pi*eta)*
    cos(n*Pi*xi)*_C6*cos(1/beta*Pi*(-n^2-m^2)^(1/2)*zeta)):
> eq26:=subs(theta=theta[n,m],_C5=A[n,m],_C6=B[n,m],eq25):
> eq26A:=theta[n,m](xi,eta,zeta) =4*cos(m*Pi*eta)*cos(n*Pi*xi)*A[n,m]*
    sin(1/beta*Pi*(-n^2-m^2)^(1/2)*zeta)+4*cos(m*Pi*eta)*cos(n*Pi*xi)
    *B[n,m]*cos(1/beta*Pi*(-n^2-m^2)^(1/2)*zeta):
> theta[n,m](xi,eta,zeta) = C7*zeta+C8+factor(4*cos(m*Pi*eta)*
    cos(n*Pi*xi)*A[n,m]*sin(1/beta*Pi*(-n^2-m^2)^(1/2)*zeta)
    +4*cos(m*Pi*eta)*cos(n*Pi*xi)*B[n,m]*cos(1/beta*Pi*(-n^2-m^2)
    ^(1/2)*zeta)):
> eq27:=theta(xi,eta,zeta)=Sum(Sum(rhs(eq26),n=0..infinity),
    m=0..infinity):
> theta(xi,eta,zeta) = C7*zeta+C8+Sum(Sum(factor(4*cos(m*Pi*eta)*
    cos(n*Pi*xi)*A[n,m]*sin(1/beta*Pi*(-n^2-m^2)^(1/2)*zeta)
    +4*cos(m*Pi*eta)*cos(n*Pi*xi)*B[n,m]*cos(1/beta*Pi*(-n^2-m^2)
    ^(1/2)*zeta)),n =0 .. infinity),m =0 .. infinity):
> eq27A:=lhs(eq27)=eval(subs(n=0,m=0,rhs(eq26A)))+Sum
    (simplify(eval(subs(m=0,rhs(eq26A))),power,symbolic),n=1..infinity)
    +Sum(simplify(eval(subs(n=0,rhs(eq26A))),power,symbolic),
    m=1..infinity)
    +Sum(Sum(rhs(eq26A),n=1..infinity),m=1..infinity)+C7*zeta+C8:
> eq27B:=theta(xi,eta,zeta) =4*B[0,0]+4*cos(n*Pi*xi)*(A[n,0]*sinh
    (1/beta*Pi*n*zeta)*I+B[n,0]*cosh(1/beta*Pi*n*zeta))+(4*cos(m*Pi
    *eta)*(A[0,m]*sinh(1/beta*Pi*m*zeta)*I+B[0,m]*cosh(1/beta*Pi*
    m*zeta)))+(4*cos(m*Pi*eta)*cos(n*Pi*xi)*(A[n,m]*sin(1/beta*Pi*
    (-n^2-m^2)^(1/2)*zeta)+B[n,m]*cos(1/beta*Pi*(-n^2-m^2)
    ^(1/2)*zeta)))+C7*zeta+C8:
Boundary Condition
> eq28:=w*theta(xi,eta,1)+Eval(diff(theta(xi,eta,zeta),zeta),zeta=1)=0:
> eq29:=w*subs(zeta=1,rhs(eq27B))+subs(zeta=1,diff(rhs(eq27B),zeta))=0:
> eq29A:=subs(cos(n*Pi*xi)=X,cos(m*Pi*eta)=Y,eq29):
> eq29B:=w*(4*B[0,0]+4*X*(A[n,0]*sinh(1/beta*Pi*n)*I+B[n,0]*
    cosh(1/beta*Pi*n))+4*Y*(A[0,m]*sinh(1/beta*Pi*m)*I+B[0,m]*
    cosh(1/beta*Pi*m))+4*Z*(A[n,m]*sin(1/beta*Pi*(-n^2-m^2)^(1/2))
    +B[n,m]*cos(1/beta*Pi*(-n^2-m^2)^(1/2)))+_C7+_C8)
    +4*X*(A[n,0]*cosh(1/beta*Pi*n)/beta*Pi*n*I+B[n,0]*sinh(1/beta*Pi*n)
    /beta*Pi*n)+4*Y*(A[0,m]*cosh(1/beta*Pi*m)/beta*Pi*m*I+B[0,m]*sinh
    (1/beta*Pi*m)/beta*Pi*m)+4*Z*(A[n,m]*cos(1/beta*Pi*(-n^2-m^2)
```

```
   ^(1/2))/beta*Pi*(-n^2-m^2)^(1/2)-B[n,m]*sin(1/beta*Pi*
   (-n^2-m^2)^(1/2))/beta*Pi*(-n^2-m^2)^(1/2))+_C7 =0:
> eq29C:=collect(eq29B,[X,Y,Z]):
> eq30:=factor(isolate(coeff(lhs(eq29C),X,1)=0,B[n,0])):
> eq30A:=factor(isolate(coeff(lhs(eq29C),Y,1)=0,B[0,m])):
> eq30B:=factor(isolate(coeff(lhs(eq29C),Z,1)=0,B[n,m])):
> eq30C:=isolate(4*B[0,0]+C8+C7=0,B[0,0]):
> eq31:=subs(eq30,eq30A,eq30B,eq30C,eq27A):
>
>
Boundary Conditions
> eq33B:=alpha(xi,eta)*theta(xi,eta,0)+k(xi,eta)*Eval(diff(theta(xi,
   eta,zeta),zeta),zeta=0)=f(xi,eta):
> eq33C:=alpha(xi,eta)=(Heaviside(xi)-Heaviside(xi-delta))*
   (Heaviside(eta)-Heaviside(eta-delta)):
> eq33D:=k(xi,eta)=1-rhs(eq33C):
> eq33E:= f(xi,eta)=rhs(eq33C):
> eq33F:=eval(subs(zeta=0,eq31)):
> eq33G:=subs(eq31,subs(eq33F,eq33B)):
> eq33H:=eval(subs(Eval=eval,eq33G)):
>
```

5.5.2 Expression for θ_{ave} when $w = 0$

$$\theta = \frac{1}{2}\Big(\Big(-\pi^4 e^{\frac{2\pi}{\beta}} + 2\pi\beta e^{\frac{2\pi}{\beta}} + 2\pi^2 - 2\pi^2 e^{\frac{2\pi}{\beta}} + \pi^4 + 2\pi\beta$$
$$+ \pi^3\sin(\pi\sqrt{\varepsilon})\sqrt{\varepsilon}\cos(\pi\sqrt{\varepsilon})e^{\frac{2\pi}{\beta}} + 2\pi^2 e^{\frac{2\pi}{\beta}}\cos(\pi\sqrt{\varepsilon})^2 + \pi^4\varepsilon e^{\frac{2\pi}{\beta}}$$
$$- 2\pi\beta\cos(\pi\sqrt{\varepsilon})^2 - \pi^3\beta\varepsilon - \pi^2\sin(\pi\sqrt{\varepsilon})\beta\sqrt{\varepsilon}\cos(\pi\sqrt{\varepsilon}) - \pi^3\beta\varepsilon e^{\frac{2\pi}{\beta}}$$
$$-\pi^3\sin(\pi\sqrt{\varepsilon})\sqrt{\varepsilon}\cos(\pi\sqrt{\varepsilon}) - \pi^2\sin(\pi\sqrt{\varepsilon})\beta\sqrt{\varepsilon}\cos(\pi\sqrt{\varepsilon})\, e^{\frac{2\pi}{\beta}}$$
$$- 2\pi^2\cos(\pi\sqrt{\varepsilon})^2 - \pi^4\varepsilon - 2\pi\beta e^{\frac{2\pi}{\beta}}\cos(\pi\sqrt{\varepsilon})^2)\varepsilon)/\Big(-\pi^4 e^{\frac{2\pi}{\beta}} - 2\pi\beta e^{\frac{2\pi}{\beta}}$$
$$- 2\pi^2 + 2\pi^2 e^{\frac{2\pi}{\beta}} + \pi^4 - 2\pi\beta + \pi^3\sin(\pi\sqrt{\varepsilon})\sqrt{\varepsilon}\cos(\pi\sqrt{\varepsilon})e^{\frac{2\pi}{\beta}}$$
$$-2\pi^2 e^{\frac{2\pi}{\beta}}\cos(\pi\sqrt{\varepsilon})^2 + \pi^4\varepsilon e^{\frac{2\pi}{\beta}} + 2\pi\beta\cos(\pi\sqrt{\varepsilon})^2 - \pi^3\beta\varepsilon$$
$$- \pi^2\sin(\pi\sqrt{\varepsilon})\beta\sqrt{\varepsilon}\cos(\pi\sqrt{\varepsilon}) - \pi^3\beta\varepsilon e^{\frac{2\pi}{\beta}} - \pi^3\sin(\pi\sqrt{\varepsilon})\sqrt{\varepsilon}\cos(\pi\sqrt{\varepsilon})$$
$$- \pi^2\sin(\pi\sqrt{\varepsilon})\beta\sqrt{\varepsilon}\cos(\pi\sqrt{\varepsilon})\, e^{\frac{2\pi}{\beta}} + 2\pi^2\cos(\pi\sqrt{\varepsilon})^2 - \pi^4\varepsilon$$
$$+ 2\pi\beta e^{\frac{2\pi}{\beta}}\cos(\pi\sqrt{\varepsilon})^2)$$

5.5.3 Expression for Ψ_{ave} at $\zeta = 1$ when $w = 0$

$$\Phi = \Big(\Big(-\pi^4 e^{\frac{2\pi}{\beta}} + 2\pi\beta e^{\frac{2\pi}{\beta}} + 2\pi^2 - 2\pi^2 e^{\frac{2\pi}{\beta}} + \pi^4 + 2\pi\beta$$
$$+ \pi^3\sin(\pi\sqrt{\varepsilon})\sqrt{\varepsilon}\cos(\pi\sqrt{\varepsilon})\, e^{\frac{2\pi}{\beta}} + 2\pi^2 e^{\frac{2\pi}{\beta}}\cos(\pi\sqrt{\varepsilon})^2 + \pi^4\varepsilon e^{\frac{2\pi}{\beta}}$$
$$- 2\pi\beta\cos(\pi\sqrt{\varepsilon})^2 - \pi^3\beta\varepsilon - \pi^2\sin(\pi\sqrt{\varepsilon})\beta\sqrt{\varepsilon}\cos(\pi\sqrt{\varepsilon}) - \pi^3\beta\varepsilon e^{\frac{2\pi}{\beta}}$$
$$- \pi^3\sin(\pi\sqrt{\varepsilon})\sqrt{\varepsilon}\cos(\pi\sqrt{\varepsilon}) - \pi^2\sin(\pi\sqrt{\varepsilon})\beta\sqrt{\varepsilon}\cos(\pi\sqrt{\varepsilon})\, e^{\frac{2\pi}{\beta}}$$

$$- 2\,\pi^2 \cos(\pi\sqrt{\varepsilon})^2 - \pi^4\varepsilon - 2\,\pi\beta\, e^{\frac{2\pi}{\beta}} \cos(\pi\sqrt{\varepsilon})^2)\varepsilon)/\left(-\pi^4 e^{\frac{2\pi}{\beta}} - 2\pi\beta e^{\frac{2\pi}{\beta}}\right.$$
$$- 2\,\pi^2 + 2\,\pi^2 e^{\frac{2\pi}{\beta}} + \pi^4 - 2\pi\beta + \pi^3 \sin(\pi\sqrt{\varepsilon})\sqrt{\varepsilon}\cos(\pi\sqrt{\varepsilon}) e^{\frac{2\pi}{\beta}}$$
$$- \pi^2 \sin(\pi\sqrt{\varepsilon})\,\beta\sqrt{\varepsilon}\cos(\pi\sqrt{\varepsilon}) - \pi^3 \beta\varepsilon e^{\frac{2\pi}{\beta}} - \pi^3 \sin(\pi\sqrt{\varepsilon})\sqrt{\varepsilon}\cos(\pi\sqrt{\varepsilon})$$
$$- \pi^2 \sin(\pi\sqrt{\varepsilon})\beta\sqrt{\varepsilon}\cos(\pi\sqrt{\varepsilon}) - \pi^3 \beta\varepsilon e^{\frac{2\pi}{\beta}} - \pi^3 \sin(\pi\sqrt{\varepsilon})\sqrt{\varepsilon}\cos(\pi\sqrt{\varepsilon})$$
$$- \pi^2 \sin(\pi\sqrt{\varepsilon})\,\beta\sqrt{\varepsilon}\cos(\pi\sqrt{\varepsilon})\, e^{\frac{2\pi}{\beta}} + 2\,\pi^2 \cos(\pi\sqrt{\varepsilon})^2 - \pi^4\varepsilon$$
$$\left. + 2\pi\beta\, e^{\frac{2\pi}{\beta}} \cos(\pi\sqrt{\varepsilon})^2\right)$$

PROBLEMS

5.1. Develop an expression for the average dimensionless concentration in the membrane volume.

5.2. Develop an expression for the average dimensionless flux of drug across a plane located at ζ.

5.3. Calculate the average dimensionless concentration in the membrane volume, using a first-order approximation (i.e., upper limit of summation is 1). The following parameters should are given: $w = 0$, $\delta = 0.1$ and $\beta = 1$.

5.4. Calculate the average dimensionless flux at $\zeta = 1$, using a first-order approximation (i.e., upper limit of summation is 1). The following parameters are given: $w = 0$, $\delta = 0.1$ and $\beta = 1$.

5.5. Plot the average dimensionless concentration in the membrane volume as a function of β (domain = [0.001, 1000]), using a first-order approximation (i.e., upper limit of summation is 1). The following parameters are given: $w = 0$ and $\delta = 0.1$.

REFERENCES

Ainaoui, A., Siepmann, J., Bodmeier, R., & Vergnaud, J. M. (2001). Calculation of the dimensions of dosage forms with release controlled by diffusion for in vivo use. *Eur J Pharm Biopharm*, 51(1), 17–24.

Itoh, N., Tian-Hua, W., & Haraya, K. (1995). Two- and three-dimensional analysis of diffusion through a dense membrane supported on a porous material. *J Membrane Sci*, 99(2), 175–183.

Ley, E. E., Goodyer, C. E., & Bunge, A. L. (2006). Mathematical models of diffusion through membranes from spatially distributed sources. *J Membrane Sci*, 283(1–2), 399–410.

Vergnaud, J. M., & Rosca, I.-D. (2005). *Assessing bioavailability of drug delivery systems: mathematical and numerical treatment*. Boca Raton, FL: CRC: Taylor & Francis.

6

CONSTANT DRUG RELEASE FROM A HOLLOW CYLINDER OF FINITE LENGTH IN THREE DIMENSIONS

6.1 SEPARATION OF VARIABLES IN 3-D CYLINDRICAL COORDINATES

The Laplace equation in 3-D cylindrical coordinates takes the form

$$\frac{\frac{\partial u(r,\theta,z)}{\partial r}+r\left(\frac{\partial^2 u(r,\theta,z)}{\partial r^2}\right)+\frac{\frac{\partial^2 u(r,\theta,z)}{\partial \theta^2}}{r}+r\left(\frac{\partial^2 u(r,\theta,z)}{\partial z^2}\right)}{r}=0 \tag{6.1}$$

A possible solution for Eq. (6.1) is obtained after using the **additive separation of variables** method where $u(r,\theta,z)=f(r)+g(\theta)+h(z)$. The functions $f(r)$, $g(\theta)$ and $h(z)$ satisfy the following ordinary differential equations:

$$\frac{\mathrm{d}^2 f(r)}{\mathrm{d}r^2}=-\frac{c_2}{r^2}-c_3-\frac{\frac{\mathrm{d}f(r)}{\mathrm{d}r}}{r} \tag{6.2}$$

$$\frac{\mathrm{d}^2 g(\theta)}{\mathrm{d}\theta^2}=c_2 \tag{6.3}$$

and

$$\frac{\mathrm{d}^2 h(z)}{\mathrm{d}z^2}=c_3 \tag{6.4}$$

Closed-form Solutions for Drug Transport through Controlled-Release Devices in Two and Three Dimensions, First Edition. Laurent Simon and Juan Ospina.

where c_2 and c_3 are separation constants. Solving Eqs. (6.2), (6.3) and (6.4), we have

$$u(r,\theta,z) = -\frac{1}{4}c_3r^2 - \frac{1}{2}c_2[\ln(r)]^2 + C_1\ln(r) + C_2 + \frac{1}{2}c_2\theta^2 + C_3\theta + \frac{1}{2}c_3z^2 + C_5z \tag{6.5}$$

Another possible solution for Eq. (6.1) is derived by using the **multiplicative separation of variables** method where $u(r,\theta,z) = f(r)g(\theta)h(z)$ and the functions $f(r)$, $g(\theta)$ and $h(z)$ satisfy the following ordinary differential equations:

$$\frac{d^2f(r)}{dr^2} = -\frac{c_2f(r)}{r^2} - c_3f(r) - \frac{\frac{df(r)}{dr}}{r} \tag{6.6}$$

$$\frac{d^2g(\theta)}{d\theta^2} = c_2g(\theta) \tag{6.7}$$

and

$$\frac{d^2h(z)}{dz^2} = c_3h(z) \tag{6.8}$$

Therefore, $u(r,\theta,z)$ is obtained after solving the above equations:

$$u(r,\theta,z) = (F_1e^{z\sqrt{\alpha}} + F_2e^{-z\sqrt{\alpha}})(B_1e^{\theta\sqrt{\beta}} + B_2e^{-\theta\sqrt{\beta}})(A_1J_{\sqrt{-\beta}}(r\sqrt{\alpha}) + A_2Y_{\sqrt{-\beta}}(r\sqrt{\alpha})) \tag{6.9}$$

In general, Eq. (6.9) needs to satisfy the periodicity condition where $u(r,\theta,z) = u(r,\theta + 2\pi, z)$ which implies that

$$\sqrt{\beta} = mi \tag{6.10}$$

where m is an integer, i.e., $m = -\infty, \ldots, -1, 0, \ldots, \infty$. From Eq. (6.10), we get

$$\beta = -m^2 \tag{6.11}$$

As a result,

$$u(r,\theta,z) = (F_1e^{z\sqrt{\alpha}} + F_2e^{-z\sqrt{\alpha}})(B_1e^{m\theta i} + B_2e^{-m\theta i})(A_1J_m(r\sqrt{\alpha}) + A_2Y_m(r\sqrt{\alpha})) \tag{6.12}$$

Application of the **principle of superposition** leads to

$$\begin{aligned} u(r,\theta,z) = \sum_{m=0}^{\infty}\int & (F_{1,m}(\alpha)e^{z\sqrt{\alpha}} + F_{2,m}(\alpha)e^{-z\sqrt{\alpha}}) \\ & \times (B_{1,m}(\alpha)e^{m\theta i} + B_{2,m}(\alpha)e^{-m\theta i}) \\ & \times (A_{1,m}(\alpha)J_m(r\sqrt{\alpha}) + A_{2,m}(\alpha)Y_m(r\sqrt{\alpha}))d\alpha \end{aligned} \tag{6.13}$$

when the separation constant α is continuous. We obtain

$$\begin{aligned} u(r,\theta,z) = \sum_{m=0}^{\infty}\Bigg(\sum_{\alpha} & \left(F_{1,m,\alpha}(\alpha)e^{z\sqrt{\alpha}} + F_{2,m,\alpha}(\alpha)e^{-z\sqrt{\alpha}}\right) \\ & \times (B_{1,m,\alpha}(\alpha)e^{m\theta i} + B_{2,m,\alpha}(\alpha)e^{-m\theta i}) \\ & \times (A_{1,m,\alpha}(\alpha)J_m(r\sqrt{\alpha}) + A_{2,m,\alpha}(\alpha)Y_m(r\sqrt{\alpha}))\Bigg) \end{aligned} \tag{6.14}$$

for discrete values of α. The choice of Eq. (6.13) or (6.14) is determined by the boundary conditions of the problem. The additive combination of (6.5) to (6.13) is also a solution:

$$\begin{aligned} u(r,\theta,z) = & -\frac{1}{4}c_3r^2 - \frac{1}{2}c_2[\ln(r)]^2 + C_1\ln(r) + C_2 + \frac{1}{2}c_2\theta^2 + C_3\theta + \frac{1}{2}c_3z^2 + C_5z \\ & + \sum_{m=0}^{\infty}\int (F_{1,m}(\alpha)e^{z\sqrt{\alpha}} + F_{2,m}(\alpha)e^{-z\sqrt{\alpha}}) \\ & \times (B_{1,m}(\alpha)e^{m\theta i} + B_{2,m}(\alpha)e^{-m\theta i}) \\ & \times (A_{1,m}(\alpha)J_m(r\sqrt{\alpha}) + A_{2,m}(\alpha)Y_m(r\sqrt{\alpha}))d\alpha \end{aligned} \tag{6.15}$$

Similarly, if Eq. (6.14) is used, we have

$$\begin{aligned} u(r,\theta,z) = & -\frac{1}{4}c_3r^2 - \frac{1}{2}c_2[\ln(r)]^2 + C_1\ln(r) + C_2 + \frac{1}{2}c_2\theta^2 + C_3\theta + \frac{1}{2}c_3z^2 + C_5z \\ & + \sum_{m=0}^{\infty}\left(\sum_{\alpha}\left(F_{1,m,\alpha}(\alpha)\,e^{z\sqrt{\alpha}} + F_{2,m,\alpha}(\alpha)e^{-z\sqrt{\alpha}}\right)\right. \\ & \times (B_{1,m,\alpha}(\alpha)e^{m\theta i} + B_{2,m,\alpha}(\alpha)e^{-m\theta i}) \\ & \times (A_{1,m,\alpha}(\alpha)J_m(r\sqrt{\alpha}) + A_{2,m,\alpha}(\alpha)Y_m(r\sqrt{\alpha}))) \end{aligned} \tag{6.16}$$

6.2 MODEL FOR 3-D DRUG RELEASE FROM A HOLLOW CYLINDER

Analysis of 3-D models usually involves numerical methods using advanced computational software. This approach gives invaluable insight into the transport phenomena and helps estimate the performance of new devices. In the case of nonlinear models, techniques, such as finite difference methods, are required because analytical solutions are not readily available. As mentioned in Chapter 5, the three-dimensional framework is necessary to describe certain processes (Ainaoui et al., 2001; Vergnaud & Rosca, 2005). Consider the delivery system developed in (Rastogi et al., 2010, 2012) where researchers studied the release kinetics of fluorescein, crystal violet and ethinyl estradiol from polyimide microtubes. The drug was only allowed to escape through a hole because the rest of the tube was coated with an impermeable membrane. All three dimensions were needed to fully describe diffusion through the apparatus. Analytical methods can be developed to estimate the size of cylindrical dosage forms (Siepmann et al., 1998).

This chapter considers drug transport through the walls of a hollow cylinder with a finite length (Fig 6.1):

$$\frac{1}{r}\frac{\partial}{\partial r}\left(r\frac{\partial u}{\partial r}\right) + \frac{1}{r^2}\frac{\partial^2 u}{\partial \theta^2} + \frac{\partial^2 u}{\partial z^2} = 0 \qquad a \le r \le b,\;\; 0 \le z \le h,\; 0 \le \theta \le 2\pi \tag{6.17}$$

The boundary conditions are given below:

$$u(a,\theta,z) = 0 \tag{6.18}$$

$$u(b,\theta,z) = u_b\theta e^z \tag{6.19}$$

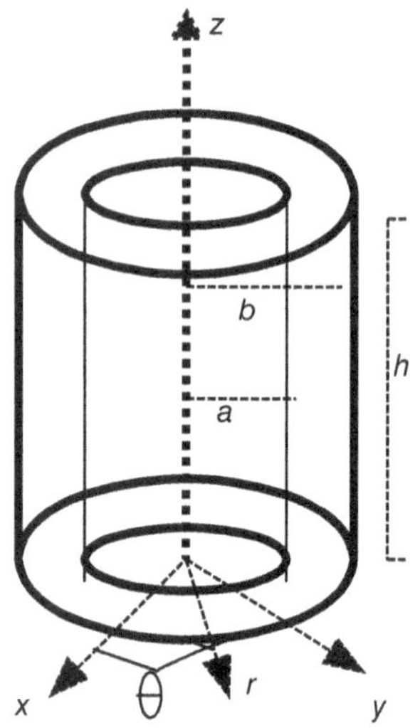

Figure 6.1 Diagram of the drug delivery device in cylindrical coordinates.

$$u(r,\theta,0)=0 \tag{6.20}$$

$$u(r,\theta,h)=0 \tag{6.21}$$

$$u(r,\theta,z)=u(r,\theta+2\pi,z) \tag{6.22}$$

where u_b is a constant concentration. Eq. (6.19) is used to promote three-dimensional diffusion though the device. This condition was used when investigating heat conduction through a hollow cylinder (Shafei & Nekoo, 2013). A periodic boundary condition Eq. (6.22) was applied.

6.3 ANALYTICAL SOLUTION OF THE TRANSPORT MODEL IN 3-D CYLINDRICAL COORDINATES

To solve the system formed by Eqs. (6.17)–(6.22), we look for a solution of the form (6.14) that automatically satisfies Eq. (6.22). Beginning with Eq. (6.12), the boundary condition (6.20) is used:

$$(F_1+F_2)(B_1\mathrm{e}^{m\theta i}+B_2\mathrm{e}^{-m\theta i})(A_1J_m(r\sqrt{\alpha})+A_2Y_m(r\sqrt{\alpha}))=0 \tag{6.23}$$

From Eq. (6.23), we deduce that

$$F_2=-F_1 \tag{6.24}$$

leading to

$$u(r,\theta,z)=2F_1\sinh(z\sqrt{\alpha})(B_1\mathrm{e}^{m\theta i}+B_2\mathrm{e}^{-m\theta i})(A_1J_m(r\sqrt{\alpha})+A_2Y_m(r\sqrt{\alpha})) \tag{6.25}$$

Without loss of generality, we write $F_1 = 1$ and obtain

$$u(r,\theta,z) = 2\sinh(z\sqrt{\alpha})(B_1 e^{m\theta i} + B_2 e^{-m\theta i})(A_1 J_m(r\sqrt{\alpha}) + A_2 Y_m(r\sqrt{\alpha})) \tag{6.26}$$

Furthermore, application of the boundary condition (6.21) yields

$$2\sinh(h\sqrt{\alpha})(B_1 e^{m\theta i} + B_2 e^{-m\theta i})(A_1 J_m(r\sqrt{\alpha}) + A_2 Y_m(r\sqrt{\alpha})) = 0 \tag{6.27}$$

which means that

$$\sinh(h\sqrt{\alpha}) = 0 \tag{6.28}$$

or, equivalently,

$$h\sqrt{\alpha} = n\pi i \tag{6.29}$$

where $n = 0, 1, 2, \ldots$ The constant α is

$$\alpha = -\frac{n^2\pi^2}{h^2} \tag{6.30}$$

Therefore,

$$u(r,\theta,z) = 2i\sin\left(\frac{zn\pi}{h}\right)(B_1 e^{m\theta i} + B_2 e^{-m\theta i})\left(A_1 J_m\left(\frac{rn\pi i}{h}\right) + A_2 Y_m\left(\frac{rn\pi i}{h}\right)\right) \tag{6.31}$$

Application of the boundary condition (6.18) gives

$$2i\sin\left(\frac{zn\pi}{h}\right)(B_1 e^{m\theta i} + B_2 e^{-m\theta i})\left(A_1 J_m\left(\frac{an\pi i}{h}\right) + A_2 Y_m\left(\frac{an\pi i}{h}\right)\right) = 0 \tag{6.32}$$

Eq. (6.32) allows the computation of A_2 as

$$A_2 = -A_1 \frac{J_m\left(\frac{an\pi i}{h}\right)}{Y_m\left(\frac{an\pi i}{h}\right)} \tag{6.33}$$

which gives

$$u(r,\theta,z) = 2iA_1 \sin\left(\frac{zn\pi}{h}\right)\frac{(B_1 e^{m\theta i} + B_2 e^{-m\theta i})}{Y_m\left(\frac{an\pi i}{h}\right)} \times \left(J_m\left(\frac{rn\pi i}{h}\right)Y_m\left(\frac{an\pi i}{h}\right) - J_m\left(\frac{an\pi i}{h}\right)Y_m\left(\frac{rn\pi i}{h}\right)\right) \tag{6.34}$$

after substitution into Eq. (6.31), or

$$u(r,\theta,z) = 2i\sin\left(\frac{zn\pi}{h}\right)\frac{(B_1 e^{m\theta i} + B_2 e^{-m\theta i})}{Y_m\left(\frac{an\pi i}{h}\right)} \times \left(J_m\left(\frac{rn\pi i}{h}\right)Y_m\left(\frac{an\pi i}{h}\right) - J_m\left(\frac{an\pi i}{h}\right)Y_m\left(\frac{rn\pi i}{h}\right)\right) \tag{6.35}$$

after setting $A_1 = 1$. Before applying the principle of superposition, we use the discrete form:

$$u_{n,m}(r,\theta,z) = 2i\sin\left(\frac{zn\pi}{h}\right)\frac{(A_{n,m}\cos(m\theta)+B_{n,m}\sin(m\theta))}{Y_m\left(\frac{an\pi i}{h}\right)} \times\left(J_m\left(\frac{rn\pi i}{h}\right)Y_m\left(\frac{an\pi i}{h}\right)-J_m\left(\frac{an\pi i}{h}\right)Y_m\left(\frac{rn\pi i}{h}\right)\right) \quad (6.36)$$

As a result,

$$u_{n,m}(r,\theta,z) = 2i\sin\left(\frac{zn\pi}{h}\right)\frac{(A_{n,m}\cos(m\theta)+B_{n,m}\sin(m\theta))}{Y_m\left(\frac{an\pi i}{h}\right)} \times\left(J_m\left(\frac{rn\pi i}{h}\right)Y_m\left(\frac{an\pi i}{h}\right)-J_m\left(\frac{an\pi i}{h}\right)Y_m\left(\frac{rn\pi i}{h}\right)\right) \quad (6.37)$$

$$u(r,\theta,z) = \sum_{m=0}^{\infty}\left[\sum_{n=1}^{\infty}\left(2i\sin\left(\frac{zn\pi}{h}\right)(A_{n,m}\cos(m\theta)+B_{n,m}\sin(m\theta))\right)\frac{1}{Y_m\left(\frac{an\pi i}{h}\right)} \times\left(J_m\left(\frac{rn\pi i}{h}\right)Y_m\left(\frac{an\pi i}{h}\right)-J_m\left(\frac{an\pi i}{h}\right)Y_m\left(\frac{rn\pi i}{h}\right)\right)\right] \quad (6.38)$$

which can be rewritten as

$$u(r,\theta,z) = \sum_{n=1}^{\infty}2iA_{n,0}\sin\left(\frac{n\pi z}{h}\right)\left(\frac{I_0\left(\frac{n\pi r}{h}\right)Y_0\left(\frac{in\pi a}{h}\right)-I_0\left(\frac{n\pi a}{h}\right)Y_0\left(\frac{in\pi r}{h}\right)}{Y_m\left(\frac{an\pi i}{h}\right)}\right) + \sum_{m=1}^{\infty}\left(\sum_{n=1}^{\infty}2i\left(A_{n,m}\cos(m\theta)+B_{n,m}\sin(m\theta)\right)\sin\left(\frac{n\pi z}{h}\right) \times\left(\frac{J_m\left(\frac{in\pi r}{h}\right)Y_m\left(\frac{in\pi a}{h}\right)-J_m\left(\frac{in\pi a}{h}\right)Y_m\left(\frac{in\pi r}{h}\right)}{Y_m\left(\frac{an\pi i}{h}\right)}\right)\right) \quad (6.39)$$

The boundary condition (6.19) becomes

$$\sum_{n=1}^{\infty}2iA_{n,0}\sin\left(\frac{n\pi z}{h}\right)\left(\frac{I_0\left(\frac{n\pi b}{h}\right)Y_0\left(\frac{in\pi a}{h}\right)-I_0\left(\frac{n\pi a}{h}\right)Y_0\left(\frac{in\pi b}{h}\right)}{Y_m\left(\frac{an\pi i}{h}\right)}\right) + \sum_{m=1}^{\infty}\left(\sum_{n=1}^{\infty}2i\left(A_{n,m}\cos(m\theta)+B_{n,m}\sin(m\theta)\right)\sin\left(\frac{n\pi z}{h}\right) \times\left(\frac{J_m\left(\frac{in\pi b}{h}\right)Y_m\left(\frac{in\pi a}{h}\right)-J_m\left(\frac{in\pi a}{h}\right)Y_m\left(\frac{in\pi b}{h}\right)}{Y_m\left(\frac{an\pi i}{h}\right)}\right)\right) = u_b\theta e^z \quad (6.40)$$

After multiplying both sides of Eq. (6.40) by $\sin\left(\frac{n\pi z}{h}\right)$, integrating from $z = 0$ to $z = h$ and from $\theta = 0$ to $\theta = 2\pi$, we have

$$2\,ihA_{n,0}\pi\left(\frac{I_0\left(\frac{n\pi b}{h}\right)Y_0\left(\frac{in\pi a}{h}\right)-I_0\left(\frac{n\pi a}{h}\right)Y_0\left(\frac{in\pi b}{h}\right)}{Y_0\left(\frac{in\pi a}{h}\right)}\right) = -\frac{2n\pi^3u_bh(-1+(-1)^n e^h)}{h^2+n^2\pi^2} \tag{6.41}$$

This equation is used to calculate $A_{n,0}$

$$A_{n,0} = \frac{n\pi^2u_b(-1+(-1)^n e^h)Y_0\left(\frac{in\pi a}{h}\right)i}{(n^2\pi^2+h^2)\left(I_0\left(\frac{n\pi b}{h}\right)Y_0\left(\frac{in\pi a}{h}\right)-I_0\left(\frac{n\pi a}{h}\right)Y_0\left(\frac{in\pi b}{h}\right)\right)} \tag{6.42}$$

To compute $B_{n,m}$, we first multiply both sides of Eq. (6.40) by $\sin\left(\frac{n\pi z}{h}\right)\sin(m\theta)$, integrate from $z = 0$ and $z = h$ and from $\theta = 0$ to $\theta = 2\pi$, to get

$$-\frac{ih}{m}\left(\frac{-B_{n,m}J_m\left(\frac{in\pi b}{h}\right)Y_m\left(\frac{in\pi a}{h}\right)m\pi+B_{n,m}J_m\left(\frac{in\pi a}{h}\right)Y_m\left(\frac{in\pi b}{h}\right)m\pi}{Y_m\left(\frac{in\pi a}{h}\right)}\right) = \frac{2n\pi u_bh(-\pi m+\pi m(-1)^n e^h)}{(h^2+n^2\pi^2)m^2} \tag{6.43}$$

and

$$B_{n,m} = \frac{2in\pi u_bh(-\pi m+\pi m(-1)^n e^h)mY_m\left(\frac{in\pi a}{h}\right)}{h\left(-J_m\left(\frac{in\pi b}{h}\right)Y_m\left(\frac{in\pi a}{h}\right)m\pi+J_m\left(\frac{in\pi a}{h}\right)Y_m\left(\frac{in\pi b}{h}\right)m\pi\right)(h^2+n^2\pi^2)m^2} \tag{6.44}$$

The coefficient $A_{n,m}$ is computed in a similar fashion. We first multiply both sides of Eq. (6.40) by $\sin\left(\frac{n\pi z}{h}\right)\cos(m\theta)$, integrate from $z = 0$ to $z = h$ and from $\theta = 0$ to $\theta = 2\pi$ to have

$$\frac{-ih}{m}\left(\frac{-A_{n,m}J_m\left(\frac{in\pi b}{h}\right)Y_m\left(\frac{in\pi a}{h}\right)m\pi+A_{n,m}J_m\left(\frac{in\pi a}{h}\right)Y_m\left(\frac{in\pi b}{h}\right)m\pi}{Y_m\left(\frac{in\pi a}{h}\right)}\right) = 0 \tag{6.45}$$

Which allows the calculation of $A_{n,m}$:

$$A_{n,m} = 0 \tag{6.46}$$

The following expression is derived for $u(r, \theta, z)$:

$$u(r,\theta,z) = \sum_{n=1}^{\infty}\left[\frac{-2\sin\left(\frac{n\pi z}{h}\right) n\pi^2 u_b(-1+(-1)^n e^h)}{n^2\pi^2+h^2} \times \left(\frac{-I_0\left(\frac{n\pi r}{h}\right) Y_0\left(\frac{in\pi a}{h}\right) + I_0\left(\frac{n\pi a}{h}\right) Y_0\left(\frac{in\pi r}{h}\right)}{-I_0\left(\frac{n\pi b}{h}\right) Y_0\left(\frac{in\pi a}{h}\right) + I_0\left(\frac{n\pi a}{h}\right) Y_0\left(\frac{in\pi b}{h}\right)}\right)\right]$$

$$+\sum_{m=1}^{\infty}\left(\sum_{n=1}^{\infty}\left[\frac{4\sin\left(\frac{n\pi z}{h}\right) u_b n\pi(-1+(-1)^n e^h)\sin(m\theta)}{m(n^2\pi^2+h^2)} \times \left(\frac{-J_m\left(\frac{in\pi r}{h}\right) Y_m\left(\frac{in\pi a}{h}\right) + J_m\left(\frac{in\pi a}{h}\right) Y_m\left(\frac{in\pi r}{h}\right)}{-J_m\left(\frac{in\pi b}{h}\right) Y_m\left(\frac{in\pi a}{h}\right) + J_m\left(\frac{in\pi a}{h}\right) Y_m\left(\frac{in\pi b}{h}\right)}\right)\right]\right) \quad (6.47)$$

after inserting $A_{n,0}$, $B_{n,m}$ and $A_{n,m}$ into Eq. (6.39). Figures 6.2, 6.3 and 6.4 show $u(r, \theta, 0.5)$, $u(r, \pi, z)$ and $u(0.5, \theta, z)$, respectively. The upper limit of summation was set at 10 for the plots. The following parameter values were selected: $a = 0.1$,

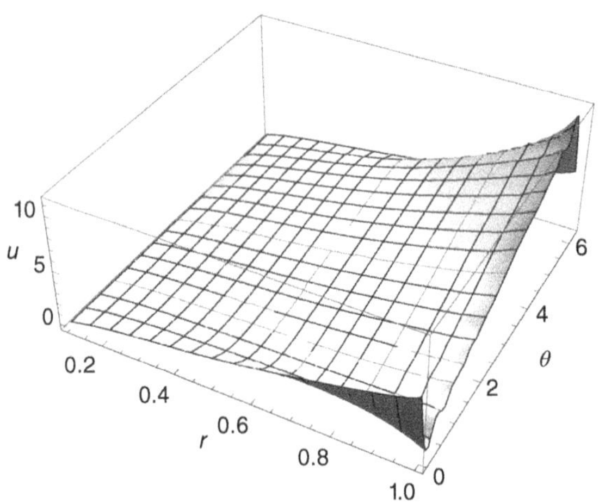

Figure 6.2 Three-dimensional plot of the concentration $u(r, \theta, 0.5)$. The following parameter values were selected: $a = 0.1$, $b = 1$, $u_b = 1$ and $h = 1$.

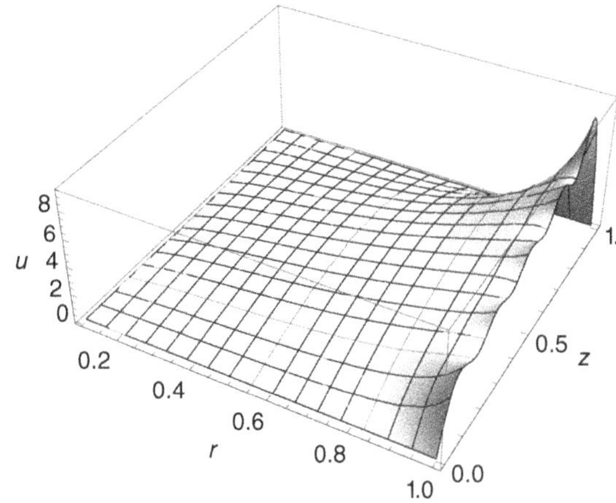

Figure 6.3 Three-dimensional plot of the concentration $u(r, \pi, z)$. The following parameter values were selected: $a = 0.1$, $b = 1$, $u_b = 1$ and $h = 1$.

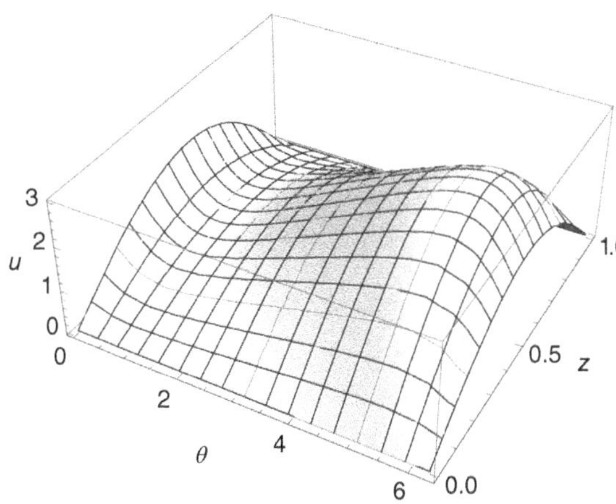

Figure 6.4 Three-dimensional plot of the concentration $u(0.5, \theta, z)$. The following parameter values were selected: $a = 0.1$, $b = 1$, $u_b = 1$ and $h = 1$

$b = 1$, $u_b = 1$ and $h = 1$. We can also verify that the boundary conditions, especially, Eq. (6.22), are satisfied. The following results were obtained: $u(a, \theta, z) = 0$; Eq. (6.19) was maintained, as shown by Figs. 6.5 and 6.6.; $u(r, \theta, 0) = 0$; $u(r, \theta, h) = 0$ and $u(r, \theta, z) - u(r, \theta + 2\pi, z) = 0$.

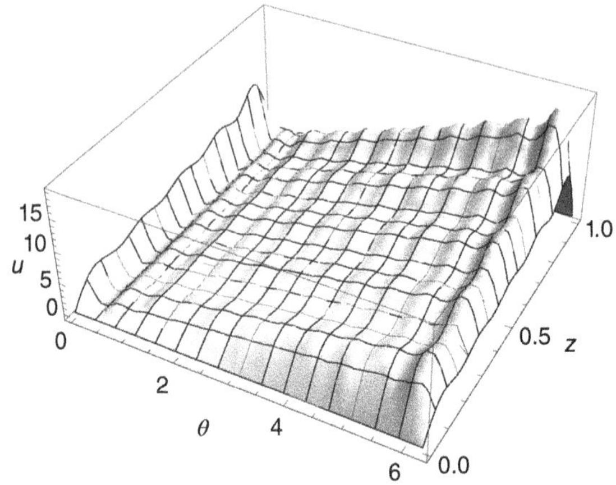

Figure 6.5 Three-dimensional plot representing the concentration $u(b, \theta, z)$. The following parameter values were selected: $a = 0.1$, $b = 1$, $u_b = 1$ and $h = 1$.

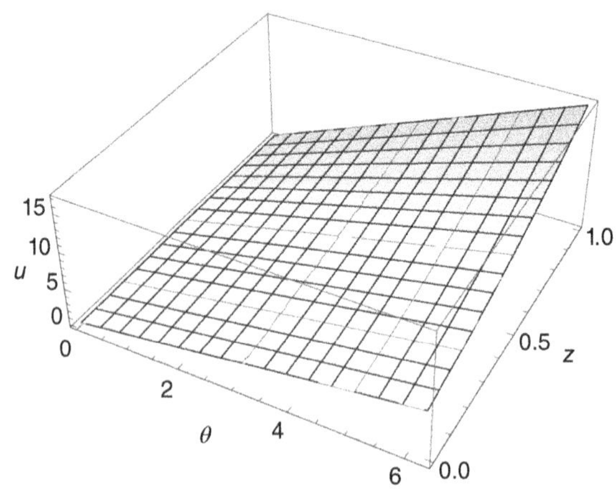

Figure 6.6 Three-dimensional plot of the boundary condition $u(b, \theta, z) = u_b \theta e^z$. The following parameter values were selected: $a = 0.1$, $b = 1$, $u_b = 1$ and $h = 1$.

6.4 SUMMARY

This chapter deals with the solution of the Laplace equation in 3-D cylindrical coordinates by the method of multiplicative separation of variables. The technique is applied to investigate drug transport through the walls of a hollow cylinder with a finite length. A closed-form solution, which included the Bessel function, was derived. Through simulations, we verified that the boundary conditions were satisfied. The method outlined may be used to elucidate drug transport in cylindrical release devices.

6.5 APPENDIX: MAPLE CODE LISTINGS

6.5.1 Maple Code: cyl3Dsteady.mws

```
> restart:with(VectorCalculus):with(PDETools);
```

[CanonicalCorrdinates, ChangeSymmetry, CharacteristicQ, CharacteristicQInvariants,ConservedCurrentTest, ConservedCurrents, ConsistencyTest, D_Dx, DeterminingPDE,ETa_k, Euler, FromJet, FunctionFieldSolutions, InfinitesimalGenerator, Infinitesimals, IntegratingFactorTest, IntegratingFactors, InvariantEquation, InvariantSolutions, InvariantTransformation, Invariants, Laplace, Library, PDEplot, PolynomialSolutions,ReducedForm, SimilaritySolutions, SimilartyTransformation, Solve, SymmetryCommutator, SymmetryGauge, SymmetrySollutions, SymmetryTest, SymmetryTransformation, TWSolutions, ToJet, build, casesplit, charstrip, dchange, dcoeffs, declare, diff_table, difforder, dployform, dsubs, mapde, separability, splitstrip, splitsys, undeclare]

```
> eq:=Laplacian(u(r,theta,z),cylindrical[r,theta,z])=0;
```

$$eq := \frac{\frac{\partial}{\partial r}u(r,\theta,z) + r\left(\frac{\partial^2}{\partial r^2}u(r,\theta,z)\right) + \frac{\frac{\partial^2}{\partial \theta^2}u(r,\theta,z)}{r} + r\left(\frac{\partial^2}{\partial z^2}u(r,\theta,z)\right)}{r} = 0$$

```
> u(r,theta,z)=f(r)*g(theta)*h(z):
> eq0:=pdsolve(eq,HINT=f(r)+g(theta)+h(z)):
> eq0A:=build(eq0):
> eq0B:=subs(_C4=0,_C6=0,eq0A):
> eq1:=pdsolve(eq,HINT=f(r)*g(theta)*h(z)):
> eq2:=factor(build(eq1)):
>
eq3:=exp(_c[3]^(1/2)*z)^2*_C5+_C6=exp(_c[3]^(1/2)*z)*(exp(_c[3]^
(1/2)*z)*_C5+_C6*exp(-_c[3]^(1/2)*z)):
>
eq3A:=_C3*exp(_c[2]^(1/2)*theta)^2+_C4=exp(_c[2]^(1/2)*theta)*(_C3
x*exp(_c[2]^(1/2)*theta)+_C4*exp(-_c[2]^(1/2)*theta)):
> eq4:=subs(eq3,eq3A,eq2):
>
eq4A:=subs(_c[3]=alpha,_c[2]=beta,_C1=A[1],_C2=A[2],_C3=B[1],_C4=B[2],
_C5=F[1],_C6=F[2],eq4):
> eq5:=sqrt(beta)=I*m:
> eq6:=isolate(eq5,beta):
>
> eq7:=simplify(subs(eq6,eq4A),power,symbolic):
>
eq7A:=subs(F[1]=F[1,m](alpha),F[2]=F[2,m](alpha),A[1]=A[1,m](alpha),
A[2]=A[2,m](alpha),B[1]=B[1,m](alpha),B[2]=B[2,m](alpha),eq7):
> eq7B:=lhs(eq7A)=Sum(Int(rhs(eq7A),alpha),m=0..infinity):
> eq7C:=lhs(eq7B)=rhs(eq7B)+rhs(eq0B):
```

```
> eq8:=eval(subs(z=0,rhs(eq7)))=0:
> eq9:=isolate(eq8,F[2]):
> eq10:=subs(eq9,eq7):
>
eq11:=F[1]*exp(alpha^(1/2)*z)-F[1]*exp(-alpha^(1/2)*z)=convert(factor
(F[1]*exp(alpha^(1/2)*z)-F[1]*exp(-alpha^(1/2)*z)),sin):
> eq12:=subs(eq11,eq10):
> eq13:=subs(F[1]=1,eq12):
> eq14:=subs(z=h,rhs(eq13))=0:
> eq15:=sinh(alpha^(1/2)*h)=0:
> eq16:=alpha^(1/2)*h=I*n*Pi:
> eq17:=isolate(eq16,alpha):
> eq18:=simplify(subs(eq17,eq13),power,symbolic):
> eq19:=subs(r=a,rhs(eq18))=0:
> eq20:=isolate(eq19,A[2]):
> eq21:=factor(subs(eq20,eq18)):
> eq22:=subs(A[1]=1,eq21):
> eq23:=B[1]*exp(m*theta*I)+B[2]*exp(-I*m*theta)=A[n,m]*cos(m*theta)
+B[n,m]*sin(m*theta):
> eq24:=subs(eq23,u=u[n,m],eq22):
> eq25:=u(r,theta,z)=Sum(Sum(rhs(eq24),n=1..infinity),m=0..infinity):
> eq26:=u(r,theta,z)=Sum(eval(subs(m=0,rhs(eq24))),n=1..infinity)
+Sum(Sum(rhs(eq24),n=1..infinity),m=1..infinity):
> eq27:=subs(r=b,rhs(eq26))=u[b]*theta*exp(z):
> sin(n*Pi/h*z):
> eq28:=int(int(subs(r=b,eval(subs(m=0,rhs(eq24)))))*sin(n*Pi/h*z),
z=0..h),theta=0..2*Pi)=int(int(rhs(eq27)*sin(n*Pi/h*z),z=0..h),
theta=0..2*Pi) assuming n::integer:
> eq28A:=isolate(eq28,A[n,0]):
> sin(n*Pi/h*z)*sin(m*theta):
>
eq29:=int(int(subs(r=b,rhs(eq24))*sin(n*Pi/h*z)*sin(m*theta),z=0..h),
theta=0..2*Pi)=int(int(rhs(eq27)*sin(n*Pi/h*z)*sin(m*theta),z=0..h),
theta=0..2*Pi) assuming n::integer:
>
> eq29A:=simplify(eq29,power,symbolic) assuming m::integer:
> eq29B:=isolate(eq29A,B[n,m]):
> sin(n*Pi/h*z)*cos(m*theta):
>
eq30:=int(int(subs(r=b,rhs(eq24))*sin(n*Pi/h*z)*cos(m*theta),z=0..h),
theta=0..2*Pi)=int(int(rhs(eq27)*sin(n*Pi/h*z)*cos(m*theta),z=0..h),
theta=0..2*Pi) assuming n::integer:
> eq30A:=simplify(eq30,power,symbolic) assuming m::integer:
> eq30B:=isolate(eq30A,A[n,m]):
>
>
> eq31:=factor(subs(eq28A,eq29B,eq30B,eq26));
```

$$
\begin{aligned}
\mathrm{eq31} :=\ & u(r,\ \theta,\ z) = \sum_{n=1}^{\infty}\Bigg(-\Big(2\ \sin\Big(\frac{n\pi z}{h}\Big)\ n\pi^2 u_b\,(e^h(-1)^n - 1)\ (\mathrm{BesselI}(0,\\
& \frac{n\pi r}{h}\Big)\ \mathrm{BesselY}\Big(0,\ \frac{In\pi a}{h}\Big) - \mathrm{BesselI}\Big(0,\ \frac{n\pi a}{h}\Big)\ \mathrm{BesselY}\Big(0,\ \frac{In\pi r}{h}\Big)\Big)\Big)\Big/((\pi^2 n^2\\
& +h^2)\ \Big(\mathrm{BesselI}\Big(0,\ \frac{n\pi b}{h}\Big)\ \mathrm{BesselY}\Big(0,\ \frac{In\pi a}{h}\Big) - \mathrm{BesselI}\Big(0,\ \frac{n\pi a}{h}\Big)\ \mathrm{BesselY}(0,\\
& \frac{In\pi b}{h}\Big)\Big)\Big)\Bigg) + \sum_{m=1}^{\infty}\Bigg(\sum_{m=1}^{\infty}\Big(-\Big(4\ \sin\Big(\frac{n\pi z}{h}\Big)\ u_b\, n\pi\ (2\ e^h(-1)^{1+n} + e^h(-1)^n\\
& + 1)\ \sin(m\theta)\ \Big(\mathrm{BesselJ}\Big(m,\ \frac{In\pi r}{h}\Big)\ \mathrm{BesselY}\Big(m,\ \frac{In\pi a}{h}\Big) - \mathrm{BesselJ}\Big(m,\\
& \frac{In\pi a}{h}\Big)\ \mathrm{BesselY}\Big(m,\ \frac{In\pi r}{h}\Big)\Big)\Big)\Big/\Big(m\ (\pi^2 n^2 + h^2)\ \Big(\mathrm{BesselJ}\Big(m,\ \frac{In\pi b}{h}\Big)\ \mathrm{BesselY}(m,\\
& \frac{In\pi a}{h}\Big) - \mathrm{BesselJ}\ \Big(m,\ \frac{In\pi a}{h}\Big)\ \mathrm{BesselY}\Big(m,\ \frac{In\pi b}{h}\Big)\Big)\Big)\Big)\Bigg)
\end{aligned}
$$

>

PROBLEMS

6.1. Solve the hollow cylinder problem using the no-flux boundary condition at the lower radial boundary $r = 0$, i.e., $\left.\frac{\partial u(r,\theta,z)}{\partial r}\right|_{r=0} = 0$. The domains of interest are $0 \le r \le b,\ \ 0 \le z \le h,\ 0 \le \theta \le 2\pi$. The other boundary conditions remain the same. Note: The solution developed in Section 6.3 leads to an indeterminate form as a approaches zero.

6.2. Solve the hollow cylinder problem using no flux boundary conditions at $z = 0$ and $z = h$, i.e., $\left.\frac{\partial u(r,\theta,z)}{\partial z}\right|_{z=0} = \left.\frac{\partial u(r,\theta,z)}{\partial z}\right|_{z=h} = 0$. The other conditions remain the same. Note: The domains of interest are $a \le r \le b,\ \ 0 \le z \le h,\ \ 0 \le \theta \le 2\pi$.

6.3. Solve the hollow cylinder problem using $u(b, \theta, z) = u_b$. The other conditions remain the same. Note: The domains of interest are $a \le r \le b,\ \ 0 \le z \le h,\ \ 0 \le \theta \le 2\pi$.

6.4. Plot $u(r,\ \theta,\ 0.5)$ using the solution derived in Problem 6.1 and the following parameter values: $a = 0$, $b = 1$, $u_b = 1$ and $h = 1$.

6.5. Plot $u(r, \theta, 0.7)$ using the solution developed in Problem 6.3 and the following parameter values: $a = 0.5$, $b = 1$, $u_b = 1$ and $h = 2$.

REFERENCES

Ainaoui, A., Siepmann, J., Bodmeier, R., & Vergnaud, J. M. (2001). Calculation of the dimensions of dosage forms with release controlled by diffusion for in vivo use. *Eur J Pharm Biopharm*, 51(1), 17–24.

Rastogi, A., Luo, Z., Wu, Z., Ho, P. S., Bowman, P. D., & Stavchansky, S. (2010). Development and characterization of a scalable microperforated device capable of long-term zero order drug release. *Biomed Microdevices*, 12(5), 915–921.

Rastogi, A., Bowman, P. D., & Stavchansky, S.. (2012). Evaluation of a perforated drug delivery system in mice for prolonged and constant release of a hydrophilic drug. *Drug Delivery Trans Res*, 2(2), 106–111.

Shafei, AM, & Nekoo, S. R. (2013). Heat conduction of a hollow cylinder via generalized Hankel transform. *Intl Res J Appl Basic Sci*, 3, 758–769.

Siepmann, J., Ainaoui, A., Vergnaud, J. M., & Bodmeier, R. (1998). Calculation of the dimensions of drug-polymer devices based on diffusion parameters. *J Pharm Sci*, 87(7), 827–832.

Vergnaud, J. M., & Rosca, I.-D.. (2005). *Assessing bioavailability of drug delivery systems: mathematical and numerical treatment*. Boca Raton, FL: CRC: Taylor & Francis.

7

SUSTAINED DRUG RELEASE FROM A HOLLOW SPHERE IN THREE DIMENSIONS

7.1 METHOD OF GREEN'S FUNCTION IN 3-D SPHERICAL COORDINATES

7.1.1 The Laplace's Equation

The complete Laplace's equation in 3-D spherical coordinates reads

$$\frac{1}{r^2}\frac{\partial}{\partial r}\left(r^2\frac{\partial c}{\partial r}\right)+\frac{1}{r^2\sin\theta}\frac{\partial}{\partial\theta}\left(\sin\theta\frac{\partial c}{\partial r}\right)+\frac{1}{r^2\sin^2\theta}\frac{\partial^2 c}{\partial\phi^2}=0 \tag{7.1}$$

where r measures the radius; θ indicates the zenith angle or latitude; ϕ represents the azimuthal angle or longitude. Making the change of variable $x = \cos(\theta)$, Eq. (7.1) reduces to

$$\begin{aligned}&\frac{1}{r^2(x^2-1)}\left[\left(\frac{\partial^2}{\partial r^2}c\,(x,r,\phi)\right)r^2x^2-\left(\frac{\partial^2}{\partial x^2}c\,(x,r,\phi)\right)x^4+2\left(\frac{\partial}{\partial r}c\,(x,r,\phi)\right)rx^2\right.\\&-2\left(\frac{\partial}{\partial x}c\,(x,r,\phi)\right)x^3-\left(\frac{\partial^2}{\partial r^2}c\,(x,r,\phi)\right)r^2+2\left(\frac{\partial^2}{\partial x^2}c\,(x,r,\phi)\right)x^2-2\left(\frac{\partial}{\partial r}c\,(x,r,\phi)\right)r\\&\left.+2x\frac{\partial}{\partial x}c\,(x,r,\phi)-\frac{\partial^2}{\partial x^2}c(x,r,\phi)-\frac{\partial^2}{\partial\phi^2}c(x,r,\phi)\right]=0\end{aligned} \tag{7.2}$$

Closed-form Solutions for Drug Transport through Controlled-Release Devices in Two and Three Dimensions, First Edition. Laurent Simon and Juan Ospina.

We assume a solution of the form $c(x, r, \phi) = f(r)g(x)H(\phi)$ where the functions $f(r)$, $g(x)$ and $H(\varphi)$ satisfy, respectively, the following ordinary differential equations

$$\frac{d^2}{dr^2}f(r) = \frac{c_2 f(r)}{r^2} - 2\frac{\frac{d}{dr}f(r)}{r} \tag{7.3}$$

$$\frac{d^2}{dx^2}g(x) = \frac{c_2 g(x)}{x^2 - 1} - \frac{c_3 g(x)}{x^4 - 2x^2 + 1} - 2\frac{x\frac{d}{dx}g(x)}{x^2 - 1} \tag{7.4}$$

and

$$\frac{d^2}{d\phi^2}H(\phi) = c_3 H(\phi) \tag{7.5}$$

By solving Eqs. (7.3), (7.4) and (7.5), we get

$$c(x, r, \phi) = (C_1 e^{\sqrt{c_3}\phi} + C_2 e^{-\sqrt{c_3}\phi}) \times \left(r^{-\frac{1}{2}+\frac{1}{2}\sqrt{1+4c_2}} C_4 + r^{-\frac{1}{2}-\frac{1}{2}\sqrt{1+4c_2}} C_3\right)\left(C_5 P^{i\sqrt{c_3}}_{-\frac{1}{2}+\frac{1}{2}\sqrt{1+4c_2}}(x) + C_6 Q^{i\sqrt{c_3}}_{-\frac{1}{2}+\frac{1}{2}\sqrt{1+4c_2}}(x)\right) \tag{7.6}$$

which includes the **associated Legendre function of the first kind** P_n^m and the **associated Lengendre function of the second kind** Q_n^m. To obtain a periodic solution with respect to the variable ϕ, we require

$$\sqrt{c_3} = -im \tag{7.7}$$

where m is an integer. The following equation is derived from Eq. (7.7):

$$c_3 = -m^2 \tag{7.8}$$

leading to

$$c(x, r, \phi) = (C_1 e^{im\phi} + C_2 e^{-im\phi}) \times \left(C_3 r^{-\frac{1}{2}+\frac{1}{2}\sqrt{1+4c_2}} + C_4 r^{-\frac{1}{2}-\frac{1}{2}\sqrt{1+4c_2}}\right)\left(C_5 P^m_{-\frac{1}{2}+\frac{1}{2}\sqrt{1+4c_2}}(x) + C_6 Q^m_{-\frac{1}{2}+\frac{1}{2}\sqrt{1+4c_2}}(x)\right) \tag{7.9}$$

By definition of the associated Legendre function,

$$-\frac{1}{2} + \frac{1}{2}\sqrt{1 + 4c_2} = L \tag{7.10}$$

or

$$c_2 = L(1 + L) \tag{7.11}$$

where L is a positive integer. Equation (7.9) becomes

$$c(x, r, \phi) = (C_1 e^{im\phi} + C_2 e^{-im\phi})(C_3 r^L + C_4 r^{-1-L})(C_5 P_L^m(x) + C_6 Q_L^m(x)) \tag{7.12}$$

Given that the associated Legendre polynomials $P_L^m(x)$ are defined only when $-L \leq m \leq L$, Eq. (7.12) can be rewritten as

$$c(x,r,\phi) = e^{im\phi}(C_3 r^L + C_4 r^{-1-L})(C_5 P_L^m(x) + C_6 Q_L^m(x)) \tag{7.13}$$

or

$$c(x,r,\phi) = \sum_{L=0}^{\infty}\left[\sum_{m=-L}^{L} e^{im\phi}\left(A_{L,m}r^L + B_{L,m}r^{-1-L}\right)(C_{L,m}P_L^m(x) + E_{L,m}Q_L^m(x))\right] \tag{7.14}$$

after applying the **principle of superposition**. Equation (7.14) is rewritten as

$$c(x,r,\phi) = \sum_{L=0}^{\infty}\left[\sum_{m=-L}^{L} e^{im\phi}\left(A_{L,m}r^L + B_{L,m}r^{-1-L}\right)\left(C_{L,m}P_L^m(\cos(\theta)) + E_{L,m}Q_L^m(\cos(\theta))\right)\right] \tag{7.15}$$

after using $x = \cos(\theta)$. In many applications, the domain of interest includes $\theta = 0$ or $\theta = \pi$. The functions $Q_L^m(x)$ must be eliminated because of singularities at $x = 1$ and $x = -1$. Then Eq. (7.15) becomes

$$c(x,r,\phi) = \sum_{L=0}^{\infty}\left[\sum_{m=-L}^{L} e^{im\phi}\left(A_{L,m}r^L + B_{L,m}r^{-1-L}\right)P_L^m(\cos(\theta))\right] \tag{7.16}$$

The product $e^{im\phi}P_L^m(\cos(\theta))$ appears in the definition of the **spherical harmonics** $Y_L^m(\theta,\phi)$ (Byerly, 2003; Hobson, 1955):

$$Y_L^m(\theta,\phi) = \sqrt{\frac{(2L+1)(L-m)!}{4\pi(L+m)!}}P_L^m(\cos(\theta))e^{im\phi};\ -L \leq m \leq L \tag{7.17}$$

Therefore,

$$c(x,r,\phi) = \sum_{L=0}^{\infty}\left[\sum_{m=-L}^{L} Y_L^m(\theta,\phi)(A_{L,m}r^L + B_{L,m}r^{-1-L})\right] \tag{7.18}$$

When the domain of interest is the interior of a sphere, Eq. (7.18) is simplified to

$$c(x,r,\phi) = \sum_{L=0}^{\infty}\left[\sum_{m=-L}^{L} Y_L^m(\theta,\phi)A_{L,m}r^L\right] \tag{7.19}$$

If the domain is outside of the sphere, we have

$$c(x,r,\phi) = \sum_{L=0}^{\infty}\left[\sum_{m=-L}^{L} Y_L^m(\theta,\phi)B_{L,m}r^{-1-L}\right] \tag{7.20}$$

When the region investigated is confined between two concentric spheres, the solution is given by Eq. (7.18).

7.1.2 The Poisson's Equation

In section 7.1.1, the Laplace's equation was solved. Although the problem covers a number of drug delivery applications, it does not include models with a sink or a source term. The **Poisson's equation** (Arfken et al., 2013; McQuarrie, 2003) describes such situations:

$$\nabla^2 c = V \tag{7.21}$$

In order to solve Eq. (7.21), we define the corresponding Green's function (Cole, 2011):

$$\nabla^2 G = \delta \tag{7.22}$$

where G is the Green's function and δ represents the Dirac delta function. From Eqs. (7.21) and (7.22), we deduce that

$$G\nabla^2 c - c\nabla^2 G = GV - \delta c \tag{7.23}$$

or

$$\nabla \cdot (G\nabla c - c\nabla G) = GV - \delta c \tag{7.24}$$

using the principles of vector calculus. Integrating both sides of Eq. (7.24) over a volume of interest v yields

$$\iiint_v \nabla \cdot (G\nabla c - c\nabla G)dv = \iiint_v (GV - \delta c)dv \tag{7.25}$$

and $\iiint_v (.)$ is a volume integral. Application of the **divergence** or **Gauss' theorem** (Jenson & Jeffreys, 1977; Larson et al., 2008) to Eq. (7.25) results in

$$\oiint_S (G\nabla c - c\nabla G) \cdot d\vec{S} = \iiint_v (GV - \delta c)dv \tag{7.26}$$

where $\oiint_S (.)$ represents a surface integral and $\vec{S}$ is a vector. Using the properties of the Dirac delta function, Eq. (7.26) reduces to

$$c = \iiint_v GVdv - \oiint_S (G\nabla c - c\nabla G) \cdot d\vec{S} \tag{7.27}$$

Equation (7.27) provides the general solution for Eq. (7.21) in terms of the Green's function defined by Eq. (7.22).

Two specials cases of Eq. (7.27) are worth mentioning:

- ***The Green's function satisfies*** $G(S) = 0$. Equation (7.27) becomes

$$c = \iiint_v GVdv + \oiint_S (c\nabla G) \cdot d\vec{S} \tag{7.28}$$

- ***The Green's function satisfies*** $\nabla G \cdot d\vec{S} = 0$. Equation (7.27) then becomes

$$c = \iiint_v GVdv - \oiint_S (G\nabla c) \cdot d\vec{S} \tag{7.29}$$

In particular, when $c(S) = 0$,

$$c = \iiint_v GVdv \tag{7.30}$$

The explicit form of Eq. (7.30) is

$$c(\vec{r}) = \iiint_{v'} G(\vec{r}, \vec{r}')V(\vec{r}')dv' \tag{7.31}$$

In order to apply Eq. (7.31), it is necessary to know the exact form of the Green's function $G(\vec{r}, \vec{r}')$ in spherical coordinates.

To obtain an expression for the Green's function, we rewrite Eq. (7.22) as

$$\nabla^2 G(\vec{r}, \vec{r}') = \delta(\vec{r} - \vec{r}') \tag{7.32}$$

In spherical coordinates, the Dirac delta function $\delta(\vec{r} - \vec{r}')$ is given explicitly by

$$\delta(\vec{r} - \vec{r}') = \frac{\delta(r - r')\delta(\cos(\theta) - \cos(\theta'))\delta(\phi - \phi')}{r^2} \tag{7.33}$$

In addition, the completeness relation for the spherical harmonics is

$$\delta(\cos(\theta) - \cos(\theta'))\delta(\phi - \phi') = \sum_{L=0}^{\infty} \sum_{m=-L}^{L} Y_L^m(\theta, \phi) Y *_L^m (\theta', \phi') \tag{7.34}$$

where $Y *_L^m (\theta', \phi') = \sqrt{\frac{(2L+1)(L-m)!}{4\pi(L+m)!}} P_L^m(\cos(\theta'))e^{-im\phi'}$, resulting in

$$\delta(\vec{r} - \vec{r}') = \frac{\delta(r - r')\sum_{L=0}^{\infty} \sum_{m=-L}^{L} Y_L^m(\theta, \phi) Y *_L^m (\theta', \phi')}{r^2} \tag{7.35}$$

Now, Eq. (7.32) takes the form

$$\nabla^2 G(r, \theta, \phi, r', \theta', \phi') = \frac{\delta(r - r')\sum_{L=0}^{\infty} \sum_{m=-L}^{L} Y_L^m(\theta, \phi) Y *_L^m (\theta', \phi')}{r^2} \tag{7.36}$$

Since Eq. (7.36) is linear, we look for a solution of the form

$$G(r, \theta, \phi, r', \theta', \phi') = \sum_{L=0}^{\infty} \sum_{m=-L}^{L} f_{Lm}(r, r') Y_L^m(\theta, \phi) Y *_L^m (\theta', \phi') \tag{7.37}$$

where the function $f_{Lm}(r, r')$, denoted below as $f(r)$, satisfies the ordinary differential equation

$$\frac{2\frac{d}{dr}f(r) + r\frac{d^2}{dr^2}f(r)}{r} - \frac{L(L+1)f(r)}{r^2} = \frac{\delta(r-r')}{r^2} \tag{7.38}$$

The solution of Eq. (7.38) is given by

$$f(r) = C_2 r^L + C_1 r^{-L-1} - \frac{\text{Heaviside}(r' - r)(r'^{(-L-1)} r^L - r'^L r^{(-L-1)})}{2L+1} \tag{7.39}$$

See Chapter 1 for the definition of the *Heaviside* function. When $r < r'$, Eq. (7.39) reduces to

$$f(r) = C_2 r^L + C_1 r^{(-L-1)} - \frac{r'^{(-L-1)} r^L - r'^L r^{(-L-1)}}{2L+1} \tag{7.40}$$

The singular part of Eq. (7.40) at $r = 0$ must be eliminated according to

$$C_1 r^{(-L-1)} + \frac{r'^L r^{(-L-1)}}{2L+1} = 0 \tag{7.41}$$

Then, the parameter C_1 is obtained:

$$C_1 = -\frac{r'^L}{2L+1} \tag{7.42}$$

Therefore,

$$f(r) = C_2 r^L - \frac{r'^L r^{(-L-1)}}{2L+1} - \frac{\text{Heaviside}(r' - r)(r'^{(-L-1)} r^L - r'^L r^{(-L-1)})}{2L+1} \tag{7.43}$$

When $r > r'$, we have

$$f(r) = C_2 r^L - \frac{r'^L r^{(-L-1)}}{2L+1} \tag{7.44}$$

which gives

$$C_2 R^L - \frac{r'^L R^{(-L-1)}}{2L+1} = 0 \tag{7.45}$$

after applying the constraint $f(R) = 0$. This condition is appropriate for problems with a Dirichlet boundary condition at R, such as the one described in Section 7.2. In these cases, $G(S) = 0$, which implies that $G(R) = 0$ and, therefore, $f(R) = 0$.

The parameter C_2 is obtained from Eq. (7.45):

$$C_2 = \frac{r'^L R^{(-L-1)}}{(2L+1)R^L} \tag{7.46}$$

Replacing C_2 in Eq. (7.43), we get

$$f_{Lm}(r, r') =$$

$$f(r) = \frac{r^L r'^L R^{(-L-1)}}{(2L+1)R^L} - \frac{r'^L r^{(-L-1)}}{2L+1} - \frac{\text{Heaviside}(r' - r)(r'^{(-L-1)} r^L - r'^L r^{(-L-1)})}{2L+1} \tag{7.47}$$

Inserting this expression for $f_{Lm}(r, r')$ or $f(r)$ in Eq. (7.37) gives the explicit form of the Green's function in the interior of a sphere. The Green's function can then be used in Eq. (7.31) to yield the solution to Eq. (7.21).

7.2 MODEL FOR MOLECULAR TRANSPORT ACROSS THE WALL OF A HOLLOW SPHERE

Diffusion in two-dimensional spherical coordinates was introduced in Chapter 4. The analytical solution was given in terms of Legendre polynomials. In some cases, it is more convenient to apply the Green's function to solve Laplace's equation in three variables. Transport of a molecule through the stomach linings is expressed by a diffusion equation with an added source term Q (Fig. 7.1):

$$D\left[\frac{1}{r^2}\frac{\partial}{\partial r}\left(r^2\frac{\partial c}{\partial r}\right)+\frac{1}{r^2\sin\theta}\frac{\partial}{\partial\theta}\left(\sin\theta\frac{\partial c}{\partial r}\right)+\frac{1}{r^2\sin^2\theta}\frac{\partial^2 c}{\partial\varphi^2}\right]+Q=0 \tag{7.48}$$

where D is the diffusion coefficient in the walls of the stomach. Note that $0 \le \theta \le \pi$ and $0 \le \varphi \le 2\pi$. Materials are assumed to diffuse quickly inside the stomach making it unnecessary to set up a separate equation in this compartment. In this case, a substance, such as alcohol, enters the stomach at a rate of Q and is transported to the inner surface of the stomach lining. Diffusion then occurs through the walls of the stomach before the molecules are absorbed into the bloodstream. Note that absorption can also take place in the small intestine. The radial domain of the problem is $0 \le r \le R$ such that the origin is placed at the center of the sphere. The input rate, Q, is defined by the following equation:

$$Q(0 \le r \le r_1) = g(\theta, \phi);\ Q(r_1 < r < R) = 0 \tag{7.49}$$

where r_1 and R are the inner and outer radii of the spherically-shaped stomach. In most practical situations, g is chosen as a fixed value that is independent of direction. The dependence of g

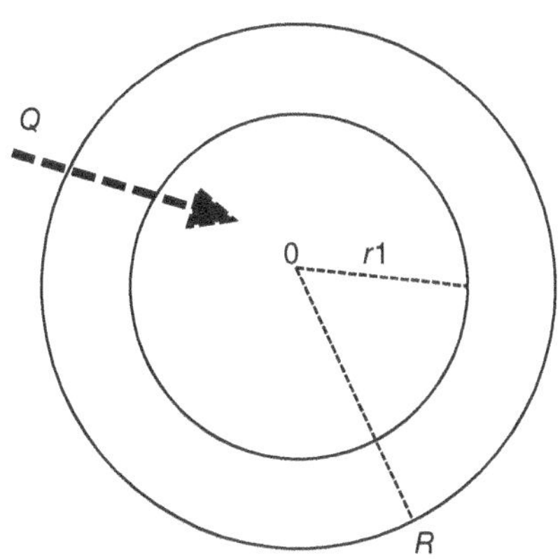

Figure 7.1 Diagram of the spherically-shaped stomach with an added source term Q.

on θ is to assure that the approach, outlined in this chapter, is applicable to more complex scenarios. A symmetry condition is imposed at $r = 0$:

$$\left.\frac{\partial c\,(r,\theta,\varphi)}{\partial r}\right|_{r=0} = 0 \tag{7.50}$$

while the blood flow is represented by a perfect sink:

$$c(R,\theta,\varphi) = 0 \tag{7.51}$$

A periodic boundary condition is imposed on φ:

$$c(R,\theta,\varphi) = c(R,\theta,\varphi + 2\pi) \tag{7.52}$$

In addition, the solution is finite at $\theta = 0$ and $\theta = \pi$:

$$c(R,0,\varphi) = \text{finite value} \tag{7.53}$$

and

$$c(R,\pi,\varphi) = \text{finite value} \tag{7.54}$$

Transport equations, with source terms, are used by researchers in a variety of fields. Photon transport in a turbid medium was investigated in Fantini et al. (1997). Some researchers provided an analytical solution for a multilayer heat conduction problem (Jain et al., 2010). Their system included heat sources in the concentric layers. Studies of photoactivatable green fluorescent protein diffusion following multiphoton photoconversion are also undertaken using a similar modeling approach (Calvert et al., 2007).

7.3 ANALYTICAL SOLUTION OF THE TRANSPORT MODEL IN 3-D SPHERICAL COORDINATES

The solution of the problem, defined by Eqs. (7.48)–(7.54), will be constructed using the Green's function method. Equation (7.48) is restated as follows:

$$\nabla^2 c = -\frac{Q}{D} \tag{7.55}$$

Note that Eq. (7.55) has a form similar to Eq. (7.21) with $V = -Q/D$. The solution of Eq. (7.55) is given by Eq. (7.31) according to

$$c(\vec{r}) = \iiint_{v'} G(\vec{r},\vec{r}\,')\frac{Q(\vec{r}\,')}{-D}dv' \tag{7.56}$$

Boundary conditions (7.50)–(7.54) are automatically satisfied. Replacing the expression for $G(r,\theta,\phi,r',\theta',\phi')$ from Eq. (7.37) into Eq. (7.56) gives

$$c(\vec{r}) = \iiint_{v'} \left(\sum_{L=0}^{\infty}\sum_{m=-L}^{L} f_{Lm}\,(r,r')\,Y_L^m(\theta,\phi)Y *_L^m\,(\theta',\phi')\right)\frac{Q(\vec{r}\,')}{-D}dv' \tag{7.57}$$

which leads to

$$c(r,\theta,\phi)$$

$$= -\frac{1}{D}\sum_{L=0}^{\infty}\sum_{m=-L}^{L} Y_L^m(\theta,\phi)\int_0^{r_1} f_{Lm}(r,r')r'^2dr'\int_0^{2\pi}\int_0^{\pi} g(\theta',\phi')Y*_L^m(\theta',\phi')\sin(\theta')d\theta'd\phi' \tag{7.58}$$

after introducing condition (7.49). Substitution of $f_{Lm}(r,r')$, from Eq. (7.47), into Eq. (7.58) leads to the complete solution. The concentration is plotted in Fig. 7.2 for the following source function and parameter values: $g(\theta,\phi) = \cos(\phi)^2$, $D = 1$, $r_1 = 0.2$ and $R = 1$. When $g(\theta,\phi) = \cos(\phi)^2$, Eq. (7.58) reduces to

$$c(r<r_1,\theta) = -\frac{2\pi}{D}\left(\begin{array}{c}\frac{1}{6}\frac{1}{\pi}\left(1/3r^2\left(\frac{r}{R}-1\right)+\frac{1}{6}\frac{3Rr^2-3Rr_1{}^2-2r^3+2r_1{}^3}{R}\right)+\\ \frac{1}{3}\frac{P_2(\cos(\theta))}{\pi}\left(\frac{1}{25}r^2\left(\frac{r^5}{R^5}-1\right)+\int_r^{R_1} r'^2\left(\frac{1}{5}\frac{r^2r'^2}{R^5}-\frac{1}{5}\frac{r^2}{r'^3}\right)\mathrm{d}r'\right)\end{array}\right) \tag{7.59}$$

and

$$c(r\geq r_1,\theta) = -\frac{2\pi}{D}\left(-\frac{1}{18\pi}\left(-\frac{r_1{}^3}{R}+\frac{r_1{}^3}{r}\right)-\frac{P_2(\cos(\theta))}{75\pi}\left(-\frac{r_1{}^5r^2}{R^5}+\frac{r_1{}^5}{r^3}\right)\right) \tag{7.60}$$

7.4 SUMMARY

This chapter introduces the reader to the use of the Green's function to study the Poisson's and Laplace's equations in 3-D spherical coordinates. The solution of the Laplace's equation was derived in terms of spherical harmonics. A critical step in solving the Poisson's equation was to apply the divergence or Gauss' theorem, which relates a volume integral to a surface

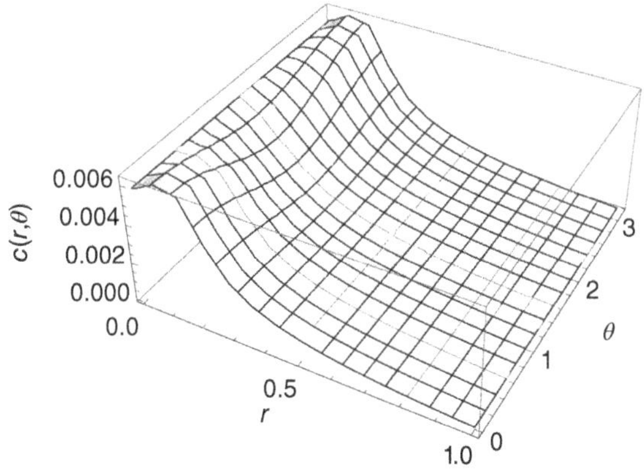

Figure 7.2 Three-dimensional plot of the concentration using the following source function and parameter values: $g(\theta,\phi) = \cos(\phi)^2$, $D = 1$, $r_1 = 0.2$ and $R = 1$.

integral. We applied the results to predict molecular transport through the walls of a hollow sphere. Other applications of the method, outlined in this chapter, involve photon transport in a turbid medium, multilayer heat conduction and protein diffusion.

7.5 APPENDIX: MAPLE, MATHEMATICA AND MAXIMA CODE LISTINGS

7.5.1 Maple Code: threespheresteady1.mws (Eq. (7.20))

```
> restart:with(VectorCalculus):with(PDETools):
>
> eq:=expand(Laplacian(c(r,theta,phi),'spherical'[r,theta,phi]))=0:
> x=cos(theta):
> tr := {theta = arccos(x)}:
eq1A:=simplify(dchange(tr,eq)):
> c(r,x,phi)=f(r)*g(x)*H(phi):
> eq2:=pdsolve(eq1A,HINT=f(r)*g(x)*H(phi)):
> eq3:=factor(build(eq2)):
>
 aux:=_C1*exp(_c[3]^(1/2)*phi)^2+_C2=exp(_c[3]^(1/2)*phi)*(_C1*exp
(_c[3]^(1/2)*phi)+_C2*exp(-_c[3]^(1/2)*phi)):
>
 auxA:=r^(1/2*(1+4*_c[2])^(1/2))*_C3+_C4*r^(-1/2*(1+4*_c[2])^(1/2))
=sqrt(r)*(r^(1/2*(1+4*_c[2])^(1/2)-1/2)*_C3+_C4*r^(-1/2*(1+4*_c[2])^
(1/2)-1/2)):
> eq3A:=subs(aux,auxA,eq3):
>
 eq6:=_c[3]^(1/2)=-m*I:
>
> eq7:=isolate(eq6,_c[3]):
> eq8:=simplify(subs(eq6,eq3A),power,symbolic):
>
>
>
>
>
>
>
>
 eq13:=-1/2+1/2*(1+4*_c[2])^(1/2)=L:
>
> eq14:=_c[2]=factor(solve(eq13,_c[2])):
>
> eq15:=simplify(simplify(subs(eq14,eq8),power,symbolic)):
> eq16:=subs(_C1=0, _C2=1,eq15):
>
> eq17:=subs(_C3=A[L,m],_C4=B[L,m],_C5=C[L,m],_C6=E[L,m],eq16):
>
```

```
 eq18:=C(x,r,phi)=Sum(Sum(rhs(eq17),m=-L..L),L=0..infinity):
> eq19:=c(r,theta,phi)=subs(x=cos(theta),rhs(eq18))  :
> eq20:=subs(E[L,m]=0,C[L,m]=1,eq19):
> aux:=exp(m*phi*I)=SphericalY(L,m,theta,phi)/LegendreP(L,m,cos
(theta)):
> eq21:=subs(aux,eq20):
> eq22:=subs(B[L,m]=0,eq21):
>
> eq23:=subs(A[L,m]=0,eq21);
```

$$eq23 := c(r,\ \theta,\ \phi) = \sum_{L=0}^{\infty}\left(\sum_{m=-L}^{L} \texttt{SphericalY}(L,\ m,\ \theta,\ \phi)\ B_{L,\ m} r^{-1-L}\right)$$

7.5.2 Maple Code: threespheresteady2.mws (Eq. (7.58)*)

* The symbol η is used for the diffusion coefficient; u is the concentration c.

```
> restart:with(VectorCalculus):
> eq:=factor(Laplacian(f(r),spherical[r,theta,phi]))-L*(L+1)*f(r)/r^2
=Dirac(r-'r'')/r^2;
>
>
>
```

$$eq := \frac{\left(\frac{d^2}{dr^2}f(r)\right)\ r+2\left(\frac{d}{dr}f(r)\right)}{r} - \frac{L\ (L+1)f(r)}{r^2} = \frac{\texttt{Dirac}(r-r')}{r^2}$$

```
> eq1:=dsolve(eq):
> eq2:=eq1 assuming r < 'r'':
> eq3:=subs(r^L=0,rhs(eq2))=0:
> eq4:=isolate(eq3,_C1):
> eq5:=subs(eq4,eq1):
> eq6:=eq5 assuming r > 'r'':
> eq7:=subs(r=R,rhs(eq6))=0:
> eq7A:=isolate(eq7,_C2):
> eq8:=subs(eq7A,eq5):
> eq9:=simplify(expand((eq8)),power,symbolic) assuming r < 'r'':
> eq10:=simplify(expand((eq8)),power,symbolic) assuming r > 'r'':
> eq11:=G(r,'r'',theta,Theta,phi,Phi)=Sum(Sum(SphericalY(L,m,theta,
phi)*SphericalY(L,m,Theta,Phi)*rhs(eq8),m=-L..L),L=0..infinity):
> eq12:=u(r,theta,phi)=-(1/eta)*Int(Int(Int(('r'')^2*g(Theta,Phi)*rhs
(eq11)*sin(Theta),'r''=0..R[1]),Theta=0..Pi),Phi=0..2*Pi):
>
> eq13:=u(r,theta,phi)=-(1/eta)*Sum(Sum(SphericalY(L,m,theta,phi)*Int
(Int(SphericalY(L,m,Theta,Phi)*g(Theta,Phi)*sin(Theta),Theta=0..Pi),
Phi=0..2*Pi)*Int((r^L/(2*L+1)*'r''^L*R^(-L-1)/(R^L)-1/(2*L+1)*'r''
^L*r^(-L-1)-Heaviside('r''-r)*('r''^(-L-1)*r^L-'r''^L*r^(-L-1))/(2*
L+1))*('r'')^2,'r''=epsilon..R[1]),m = -L .. L),L =0 .. infinity);
>
>
>
>
```

$$eq13 := u(r,\theta,\phi) = -\frac{1}{\eta}\left(\sum_{L=0}^{\infty}\left(\sum_{m=-L}^{L}\text{SphericalY}(L,m,\theta,\phi)\left(\int_0^{2\pi}\int_0^{\pi}\text{SphericalY}(L,m,\Theta,\Phi)\, g(\Theta,\Phi)\sin(\Theta)\,d\Theta\,d\Phi\right)\left(\int_{\epsilon}^{R_1}\left(\frac{r^L r'^L R^{-L-1}}{(2L+1)R^L}-\frac{r'^L r^{-L-1}}{2L+1}-\frac{\text{Heaviside}(r'-r)\,(r^L r'^{-L-1}-r^{-L-1}r'^L)}{2L+1}\right)r'^2\,dr'\right)\right)\right)$$

>

7.5.3 Mathematica Code: threespheresteady1.nb (Eq. (7.20))

eq = Laplacian$[C[r,\theta,\phi],\{r,\theta,\phi\}$,"Spherical"]//Expand

$$\frac{\mathrm{Csc}[\theta]^2C^{(0,0,2)}[r,\theta,\phi]}{r^2}+\frac{\mathrm{Cot}[\theta]C^{(0,1,0)}[r,\theta,\phi]}{r^2}+\frac{C^{(0,2,0)}[r,\theta,\phi]}{r^2}+\frac{2C^{(1,0,0)}[r,\theta,\phi]}{r}+C^{(2,0,0)}[r,\theta,\phi]$$

Mutiplicativeseparation:

eq2 = eq/.$C \to$ Function$[\{r,\theta,\phi\}, f[r]*g[\theta]*h[\phi]]$

$$\frac{2g[\theta]h[\phi]f'[r]}{r}+\frac{\mathrm{Cot}[\theta]f[r]h[\phi]g'[\theta]}{r^2}+g[\theta]h[\phi]f''[r]+\frac{f[r]h[\phi]g''[\theta]}{r^2}+\frac{\mathrm{Csc}[\theta]^2f[r]g[\theta]h''[\phi]}{r^2}$$

eq3 = Expand[eq2/$(f[r]*g[\theta]*h[\phi])$] == 0

$$\frac{2f'[r]}{rf[r]}+\frac{\mathrm{Cot}[\theta]g'[\theta]}{r^2g[\theta]}+\frac{f''[r]}{f[r]}+\frac{g''[\theta]}{r^2g[\theta]}+\frac{\mathrm{Csc}[\theta]^2h''[\phi]}{r^2h[\phi]}==0$$

$$eq3A=\frac{h''[\phi]}{h[\phi]}==-m\text{^}2$$

$$\frac{h''[\phi]}{h[\phi]}==-m^2$$

$$eq4=\frac{\mathrm{Cot}[\theta]g'[\theta]}{r^2g[\theta]}+\frac{g''[\theta]}{r^2g[\theta]}-(m\text{^}2)*\mathrm{Csc}[\theta]^2r\text{^}2==-\alpha/r\text{^}2$$

$$-\frac{m^2\mathrm{Csc}[\theta]^2}{r^2}+\frac{\mathrm{Cot}[\theta]g'[\theta]}{r^2g[\theta]}+\frac{g''[\theta]}{r^2g[\theta]}==-\frac{\alpha}{r^2}$$

$$eq4A=\frac{2f'[r]}{rf[r]}-\frac{\alpha}{r^2}+\frac{f''[r]}{f[r]}==0$$

$$-\frac{\alpha}{r^2}+\frac{2f'[r]}{rf[r]}+\frac{f''[r]}{f[r]}==0$$

eq5 = DSolve[eq4, g, θ]

$$\left\{\left\{g\to\text{Function}\left[\begin{array}{l}\{\theta\},C[1]\text{LegendreP}\left[\frac{1}{2}\left(-1+\sqrt{1+4\alpha}\right),m,\mathrm{Cos}[\theta]\right]\\+C[2]\text{LegendreQ}\left[\frac{1}{2}\left(-1+\sqrt{1+4\alpha}\right),m,\mathrm{Cos}[\theta]\right]\end{array}\right]\right\}\right\}$$

eq50 =

$$\left\{\left\{g\to\text{Function}\left[\{\theta\},C[1]\text{LegendreP}\left[\frac{1}{2}\left(-1+\sqrt{1+4\alpha}\right),m,\mathrm{Cos}[\theta]\right]+C[2]\text{LegendreQ}\left[\frac{1}{2}\left(-1+\sqrt{1+4\alpha}\right),m,\mathrm{Cos}[\theta]\right]\right]\right\}\right\}$$

$$\left\{\left\{g\to\text{Function}\left[\begin{array}{l}\{\theta\},C[1]\text{LegendreP}\left[\frac{1}{2}\left(-1+\sqrt{1+4\alpha}\right),m,\mathrm{Cos}[\theta]\right]\\+C[2]\text{LegendreQ}\left[\frac{1}{2}\left(-1+\sqrt{1+4\alpha}\right),m,\mathrm{Cos}[\theta]\right]\end{array}\right]\right\}\right\}$$

eq5A = (DSolve[eq4A, f, r]/.$C[1]-> A[1]$)/.$C[2]\to B[1]$

$$\left\{\left\{f\to\text{Function}\left[\{r\},r^{\frac{1}{2}i(i-\sqrt{-1-4\alpha})}A[1]+r^{\frac{1}{2}i(i+\sqrt{-1-4\alpha})}B[1]\right]\right\}\right\}$$

eq5B = (DSolve[eq3A, h,ϕ]/.$C[1]\to F[1]$)/.$C[2]\to F[2]$

$$\{\{h\to\text{Function}[\{\phi\},F[1]\mathrm{Cos}[m\phi]+F[2]Sin[m\phi]]\}\}$$

$$\text{eq6} = \left(r^{\frac{1}{2}i\left(i-\sqrt{-1-4\alpha}\right)}A[1] + r^{\frac{1}{2}i(i+\sqrt{-1-4\alpha})}B[1]\right)$$
$$\left(C[1]\,\texttt{LegendreP}\left[\tfrac{1}{2}\left(-1+\sqrt{1+4\alpha}\right), m, \text{Cos}[\theta]\right] + C[2]\,\texttt{LegendreQ}\left[\tfrac{1}{2}\left(-1+\sqrt{1+4\alpha}\right), m, \text{Cos}[\theta]\right]\right)$$
$$(F[1]\text{Exp}[i*m*\phi] + F[2]\text{Exp}[-i*m*\phi])$$

$$\left(r^{\frac{1}{2}i\left(i-\sqrt{-1-4\alpha}\right)}A[1] + r^{\frac{1}{2}i(i+\sqrt{-1-4\alpha})}B[1]\right)(e^{im\phi}F[1] + e^{-im\phi}F[2])$$
$$\left(\begin{array}{l} C[1]\,\texttt{LegendreP}\left[\frac{1}{2}\left(-1+\sqrt{1+4\alpha}\right), m, \text{Cos}[\theta]\right] \\ +C[2]\texttt{LegendreQ}\left[\frac{1}{2}\left(-1+\sqrt{1+4\alpha}\right), m, \text{Cos}[\theta]\right] \end{array}\right)$$

$$\text{eq6A} = \tfrac{1}{2}(-1+\sqrt{1+4\alpha}) == n$$

$$\tfrac{1}{2}(-1+\sqrt{1+4\alpha}) == n$$

$$\text{eq6B} = \texttt{Solve}[\text{eq6A}, \alpha]$$

$$\{\{\alpha \to n+n^2\}\}$$

$$\text{eq6C} = \text{eq6}/.\alpha \to n+n^2$$

$$\left(r^{\frac{1}{2}i\left(i-\sqrt{-1-4(n+n^2)}\right)}A[1] + r^{\frac{1}{2}i(i+\sqrt{-1-4(n+n^2)})}B[1]\right)$$
$$(e^{im\phi}F[1] + e^{-im\phi}F[2])\left(\begin{array}{l} C[1]\,\texttt{LegendreP}\left[\frac{1}{2}\left(-1+\sqrt{1+4\left(n+n^2\right)}\right), m, \text{Cos}[\theta]\right] \\ +C[2]\texttt{LegendreQ}\left[\frac{1}{2}\left(-1+\sqrt{1+4\left(n+n^2\right)}\right), m, \text{Cos}[\theta]\right] \end{array}\right)$$

$$\text{eq6D} = \texttt{Simplify}[\text{eq6C}, n > 0]$$

$$e^{-im\phi}r^n(A[1] + r^{-1-2n}B[1])(e^{2im\phi}F[1] + F[2])(C[1]\texttt{LegendreP}[n, m, \text{Cos}[\theta]] + C[2]\texttt{LegendreQ}[n, m, \text{Cos}[\theta]])$$

$$\text{eq6E} = ((((((\text{eq6D}/.A[1] \to A_{n,m})/.B[1] \to B_{n,m})/.C[1]->C[n,m])/.C[2] \to F[n,m])/.F[1] \to 1)/.F[2] \to 0)$$

$$e^{im\phi}r^n(C[n,m]\texttt{LegendreP}[n, m, \text{Cos}[\theta]] + F[n,m]\texttt{LegendreQ}[n, m, \text{Cos}[\theta]])(A_{n,m} + r^{-1-2n}B_{n,m})$$

$$\text{eq6F} = \texttt{Sum}[\texttt{Sum}[e^{im\phi}r^n(C[n,m]\texttt{LegendreP}[n, m, \text{Cos}[\theta]] + F[n,m]\texttt{LegendreQ}[n, m, \text{Cos}[\theta]])(A_{n,m} + r^{-1-2n}B_{n,m}), \{m, -n, n\}], \{n, 0, inf\}]$$

$$\sum_{n=0}^{inf}\left(\sum_{m=-n}^{n} e^{im\phi}r^n(C[n,m]\,\texttt{LegendreP}[n, m, \text{Cos}[\theta]] + F[n,m]\texttt{LegendreQ}[n, m, \text{Cos}[\theta]])(A_{n,m} + r^{-1-2n}B_{n,m})\right)$$

$$\text{eq6G} = C[r,\theta,\phi] == \text{eq6F}$$

$$C[r,\theta,\phi] == \sum_{n=0}^{inf}\left(\sum_{m=-n}^{n} e^{im\phi}r^n(C[n,m]\,\texttt{LegendreP}[n, m, \text{Cos}[\theta]] + F[n,m]\texttt{LegendreQ}[n, m, \text{Cos}[\theta]])(A_{n,m} + r^{-1-2n}B_{n,m})\right)$$

$$\text{eq6H} = C[r,\theta,\phi] == (\text{eq6F}/.F[n,m]->0)/.C[n,m] \to 1$$

$$C[r,\theta,\phi] == \sum_{n=0}^{inf}\left(\sum_{m=-n}^{n} e^{im\phi}r^n\texttt{LegendreP}[n, m, \text{Cos}[\theta]](A_{n,m} + r^{-1-2n}B_{n,m})\right)$$

eq6I =
eq6H/.LegendreP[n, m, Cos[θ]] →
$e^{-im\phi}$ * SphericalHarmonicY[n, m, Cos[θ], ϕ]

$$C[r,\theta,\phi] == \sum_{n=0}^{inf}\left(\sum_{m=-n}^{n} r^n \texttt{SphericalHarmonicY}[n,m,\text{Cos}[\theta],\phi](A_{n,m}+r^{-1-2n}B_{n,m})\right)$$

eq7 = eq6I/.$B_{n,m}$ → 0

$$C[r,\theta,\phi] == \sum_{n=0}^{inf}\left(\sum_{m=-n}^{n} r^n \texttt{SphericalHarmonicY}[n,m,\text{Cos}[\theta],\phi]A_{n,m}\right)$$

eq8 = eq6I/.$A_{n,m}$ → 0

$$C[r,\theta,\phi] == \sum_{n=0}^{inf}\left(\sum_{m=-n}^{n} r^{-1-n} \texttt{SphericalHarmonicY}[n,m,\text{Cos}[\theta],\phi]B_{n,m}\right)$$

7.5.4 Mathematica Code: threespheresteady2.nb (Eq. (7.58)*)

* The symbol η is used for the diffusion coefficient; C is the concentration c.

---

eq = (Laplacian[$f[r]$, {r, θ, ϕ},"Spherical"]//Expand)−
$L(L+1)f[r]/r$^2 == DiracDelta[$r-\rho$]/r^2

$$-\frac{L(1+L)f[r]}{r^2}+\frac{2f'[r]}{r}+f''[r]==\frac{\texttt{DiracDelta}[r-\rho]}{r^2}$$

eq5 = DSolve[eq,f, r]

$$\left\{\left\{f\to\texttt{Function}\left[\{r\},r^{\frac{1}{2}i(i-\sqrt{-1-4L-4L^2})}C[1]+r^{\frac{1}{2}i(i+\sqrt{-1-4L-4L^2})}C[2]-\frac{1}{\sqrt{-(1+2L)^2}}ir^{-\frac{1}{2}-\frac{1}{2}i\sqrt{-(1+2L)^2}}\rho^{-\frac{1}{2}-\frac{1}{2}i\sqrt{-(1+2L)^2}}(r^{i\sqrt{-(1+2L)^2}}-\rho^{i\sqrt{-(1+2L)^2}})\texttt{HeavisideTheta}[r-\rho]\right]\right\}\right\}$$

eq5A = $r^{\frac{1}{2}i(i-\sqrt{-1-4L-4L^2})}C[1]+r^{\frac{1}{2}i(i+\sqrt{-1-4L-4L^2})}C[2]-$
$\frac{1}{\sqrt{-(1+2L)^2}}ir^{-\frac{1}{2}-\frac{1}{2}i\sqrt{-(1+2L)^2}}\rho^{-\frac{1}{2}-\frac{1}{2}i\sqrt{-(1+2L)^2}}(r^{i\sqrt{-(1+2L)^2}}-\rho^{i\sqrt{-(1+2L)^2}})$
HeavisideTheta[$r-\rho$]

$r^{\frac{1}{2}i(i-\sqrt{-1-4L-4L^2})}C[1]+r^{\frac{1}{2}i(i+\sqrt{-1-4L-4L^2})}C[2]-$
$\frac{1}{\sqrt{-(1+2L)^2}}ir^{-\frac{1}{2}-\frac{1}{2}i\sqrt{-(1+2L)^2}}\rho^{-\frac{1}{2}-\frac{1}{2}i\sqrt{-(1+2L)^2}}(r^{i\sqrt{-(1+2L)^2}}-\rho^{i\sqrt{-(1+2L)^2}})$
HeavisideTheta[$r-\rho$]

eq50 = Simplify[eq5A, $L>0$]

$$\frac{1}{1+2L}r^{-1-L}\rho^{-1-L}((1+2L)\rho^{1+L}(r^{1+2L}C[1]+C[2])+(r^{1+2L}-\rho^{1+2L})\texttt{HeavisideTheta}[r-\rho])$$

eq5B = Simplify[eq50, $r<\rho$]

$$r^LC[1]+r^{-1-L}C[2]$$

eq5C = eq50/.$C[2]$ → 0

$$\frac{r^{-1-L}\rho^{-1-L}((1+2L)r^{1+2L}\rho^{1+L}C[1]+(r^{1+2L}-\rho^{1+2L})\texttt{HeavisideTheta}[r-\rho])}{1+2L}$$

eq6 = Simplify[eq5C, $r>\rho$]

$$\frac{-r^{-1-L}\rho^L+r^L(\rho^{-1-L}+C[1]+2LC[1])}{1+2L}$$

eq6A = (eq6/.$r\to R$) == 0

$$\frac{-R^{-1-L}\rho^L+R^L(\rho^{-1-L}+C[1]+2LC[1])}{1+2L}==0$$

eq6B = Solve[eq6A, $C[1]$]

$\left\{\left\{C[1] \to -\frac{R^{-1-2L}\rho^{-1-L}(R^{1+2L}-\rho^{1+2L})}{1+2L}\right\}\right\}$

eq6C = eq5C/.$C[1] \to -\frac{R^{-1-2L}\rho^{-1-L}(R^{1+2L}-\rho^{1+2L})}{1+2L}$

$\frac{r^{-1-L}\rho^{-1-L}(-r^{1+2L}R^{-1-2L}(R^{1+2L}-\rho^{1+2L})+(r^{1+2L}-\rho^{1+2L})\texttt{HeavisideTheta}[r-\rho])}{1+2L}$

eq6D = `Simplify`[eq6C, $r < \rho$]

$$\frac{\left(\frac{r}{\rho}\right)^L(-1+R^{-1-2L}\rho^{1+2L})}{\rho+2L\rho}$$

eq6E = `Simplify`[eq6C, $r > \rho$]

$-\frac{r^{-1-L}\rho^L - r^L R^{-1-2L}\rho^L}{1+2L}$

eq6F = $G[r,\rho,\theta,\Theta]$ =
`Sum[Sum[SphericalHarmonicY`$[L,m,\mathrm{Cos}[\theta],\phi] *$
`Conjugate[SphericalHarmonicY`$[L,m,\mathrm{Cos}[\Theta],\Phi]] *$ eq6C, $\{m,-L,L\}], \{L,0,inf\}]$

$$\sum_{L=0}^{inf}\left(\sum_{m=-L}^{L}\frac{1}{1+2L}r^{-1-L}\rho^{-1-L}\texttt{Conjugate}[\texttt{SphericalHarmonicY}[L,m,\mathrm{Cos}[\Theta],\Phi]]\right.$$
$$(-r^{1+2L}R^{-1-2L}(R^{1+2L}-\rho^{1+2L})+(r^{1+2L}-\rho^{1+2L})\texttt{HeavisideTheta}[r-\rho])$$
$$\left.\texttt{SphericalHarmonicY}[L,m,\mathrm{Cos}[\theta],\phi]\right)$$

eq6G = Int[eq6C $*\ \rho$^2, $\{\rho, 0, R_1\}$]

$$\mathrm{Int}\left[\frac{r^{-1-L}\rho^{1-L}(-r^{1+2L}R^{-1-2L}(R^{1+2L}-\rho^{1+2L})+(r^{1+2L}-\rho^{1+2L})\texttt{HeavisideTheta}[r-\rho])}{1+2L}, \{\rho,0,R_1\}\right]$$

eq6H =
$C[r,\theta,\phi]$ ==
$(-1/\eta)$ *
`Sum[`
`Sum[SphericalHarmonicY`$[L,m,\mathrm{Cos}[\theta],\phi] *$
`Integrate[Integrate[Conjugate[SphericalHarmonicY`$[L,m,\mathrm{Cos}[\Theta],\Phi]] * Sin[\Theta]$
$* g[\Theta,\Phi]$,
$\{\Theta,0,\pi\}], \{\Phi,0,2*\pi\}] *$ eq6G, $\{m,-L,L\}], \{L,0,inf\}]$

$C[r,\theta,\phi]$ ==

$$-\frac{1}{\eta}\left(\sum_{L=0}^{inf}\left(\sum_{m=-L}^{L}\mathrm{Int}\left[1/(1+2L)\,r^{-1-L}\rho^{1-L}\left(\begin{array}{c}-r^{1+2L}R^{-1-2L}(R^{1+2L}-\rho^{1+2L})\\+(r^{1+2L}-\rho^{1+2L})\texttt{HeavisideTheta}[r-\rho]\end{array}\right),\{\rho,0,R_1\}\right]\right.\right.$$
$$\left(\int_0^{2\pi}\left(\int_0^{\pi}\texttt{Conjugate}[\texttt{SphericalHarmonicY}[L,m,\mathrm{Cos}[\Theta],\Phi]]\,g[\Theta,\Phi]Sin[\Theta]\,d\Theta\right)\right.$$
$$\left.(d\Phi)\right)$$
$$\left.\left.\texttt{SphericalHarmonicY}[L,m,\mathrm{Cos}[\theta],\phi]\right)\right)$$

(* We can replace "Int" by "Integrate" to evaluate the integral *)

$C[r,\theta,\phi]$ ==

$$-\frac{1}{\eta}$$
$$\left(\sum_{L=0}^{inf}\left(\sum_{m=-L}^{L}\texttt{Integrate}\left[1/(1+2L)\,r^{-1-L}\rho^{1-L}\right.\right.\right.$$
$$(-r^{1+2L}R^{-1-2L}(R^{1+2L}-\rho^{1+2L})+(r^{1+2L}-\rho^{1+2L})\texttt{HeavisideTheta}[r-\rho]),$$
$$\left.\{\rho,0,R_1\}\right]$$
$$\left(\int_0^{2\pi}\left(\int_0^{\pi}\texttt{Conjugate}[\texttt{SphericalHarmonicY}[L,m,\mathrm{Cos}[\Theta],\Phi]]\,g[\Theta,\Phi]Sin[\Theta]d\Theta\right)d\Phi\right)$$
$$\left.\left.\texttt{SphericalHarmonicY}[L,m,\mathrm{Cos}[\theta],\phi]\right)\right);$$

7.5.5 wxMaxima Code: threespheresteady1.wxm (Eq. (7.20))

```
(%i1) load(vect)$
(%i2) scalefactors([[r*sin(theta)*cos(phi),r*sin(theta)*sin(phi),r*cos
(theta)],r,theta,phi]);
(%o2) done
(%i3) depends(C,[r,theta,phi]);
(%o3) [C(r, θ, ϕ)]
(%i4) eq0: express(laplacian(C));
```

(%o4) $$\frac{\frac{d}{d\,\theta}\left(|\sin(\theta)|\,\left(\frac{d}{d\,\theta}\,C\right)\right)+\frac{d}{d\,r}\left(r^2|\sin(\theta)|\left(\frac{d}{d\,r}\,C\right)\right)+\frac{d}{d\,\phi}\,\frac{\frac{d}{d\,\phi}\,C}{|\sin(\theta)|}}{r^2|\sin(\theta)|}$$

```
(%i5) eq0A: subst(abs(sin(phi))=sin(phi),eq0),simp, diff, expand$
Multiplicative separation of variables: wxMaxima was not able to solve
the equations.The Maple solution was used in these cases.
(%i6) eq8: C=f(r)*g(theta)*h(phi)$
(%i7) eq9: (subst(eq8,eq0A))$
(%i8) eq10: eq9,simp, diff$
(%i9) eq11: eq10/(f(r)*g(theta)*h(phi)), simp, expand$
(%i10) eq12A: 'diff(h(phi),phi,2)/h(phi)=-m^2$
(%i11) eq12B: 'diff(h(phi),phi,2)=-m^2*h(phi)$
(%i12) eq12: 'diff(g(theta),theta,2)/(r^2*g(theta))+(cos(theta)*('diff
(g(theta),theta,1)))/(r^2*g(theta)*sin(theta))-m^2/(r^2*sin(theta)
^2)=-alpha/r^2$
(%i14) eq12A: desolve(eq12B,h(phi))$
Is m zero or nonzero? nonzero;
(%i15) eq13: g(theta)=_C3*Legendre_p(-1/2+1/2*(1+4*_c[1])^(1/2),m,cos
(theta))+_C4*Legendre_q(-1/2+1/2*(1+4*_c[1])^(1/2),m,cos(theta))$
(%i16) eq14: sqrt(4*_c[1]+1)/2-1/2=n$
(%i17) assume_pos: n$
(%i18) eq14A: solve(eq14,_c[1])$
(%i19) eq14B: subst(eq14A,eq13)$
(%i20) eq14C: sqrt(4*(n^2+n)+1)/2-1/2 = n$
(%i21) eq14C1: sqrt(4*(n^2+n)+1)/2= n+1/2$
(%i22) eq14D: subst(eq14C,eq14B)$
(%i23) eq15: subst(_C4=B[n,m],subst(_C3=A[n,m],eq14D))$
(%i24) eq16: 'diff(f(r),r,2)/f(r)+(2*('diff(f(r),r,1)))/(r*f(r))
-alpha/r^2$
(%i25) eq16A: eq16*f(r), expand, simp$
(%i26) eq17: desolve(eq16A, f(r))$
(%i27) eq18: f(r)=(_C1*r^(1/2*(1+4*_c[1])^(1/2)-1/2)+_C2*r^(-1/2-1/2*
(1+4*_c[1])^(1/2)))$
(%i28) eq19: subst(eq14A,eq18)$
(%i29) eq20: subst(eq14C1,eq19)$
(%i30) eq21: subst(_C2=E[n,m],subst(_C1=C[n,m],eq20))$
(%i31) eq22: C(r,theta,phi)=rhs(eq21)*rhs(eq15)*(F[n,m]*e^(i*m*phi)+G
[n,m]*e^(-i*m*phi))$
(%i32) eq23: C(r,theta,phi)=sum (sum(rhs(eq22),m,-n,n), n, 0, inf)$
(%i33) eq23A: subst(F[n,m]=1,subst(G[n,m]=0,subst(A[n,m]=1,subst
(B[n,m]=0,eq23))))$
```

```
(%i36) eq23B: subst(Legendre_p(n,m,cos(theta))=Y(n,m,cos(theta),phi)/
e^(i*m*phi),eq23A);
```

$$(\%o36)\quad C(r,\ \theta,\ \phi) = \sum_{n=0}^{\infty}\sum_{m=-n}^{n} y(n,\ m,\ \cos(\theta),\ \phi)\ (C_{n,\ m}r^{n} + E_{n,\ m}r^{-n-1})$$

```
(%i35) eq23C: subst(E[n,m]=0,eq23B);
```

$$(\%o35)\quad C(r,\ \theta,\ \phi) = \sum_{n=0}^{\infty}\left(\sum_{m=-n}^{n} C_{n,\ m}y(n,\ m,\ \cos(\theta),\ \phi)\right) r^{n}$$

```
(%i37) eq23C: subst(C[n,m]=0,eq23B);
```

$$(\%o37)\quad C(r,\ \theta,\ \phi) = \sum_{n=0}^{\infty}\left(\sum_{m=-n}^{n} E_{n,\ m}y(n,\ m,\ \cos(\theta),\ \phi)\right) r^{-n-1}$$

```
The function "Legendre_p" should be implemented in wxMaxima; "Y" is
used here to match the notation in the text.
```

PROBLEMS

7.1. Develop the concentration profile for the model in Section 7.2 when $g(\theta, \phi) = 1$.

7.2. Derive the following expressions for the concentration profile obtained in Section 7.3:

$$c(r < r_1, \theta) = -\frac{2\pi}{D}\left(\begin{array}{l} \frac{1}{6}\frac{1}{\pi}\left(1/3r^2\left(\frac{r}{R}-1\right)+\frac{1}{6}\frac{3Rr^2-3Rr_1{}^2-2r^3+2r_1{}^3}{R}\right) \\ +\frac{1}{3}\frac{P_2(\cos(\theta))}{\pi}\left(\frac{1}{25}r^2\left(\frac{r^5}{R^5}-1\right)+\int_r^{R_1} r'^2\left(\frac{1}{5}\frac{r^2r'^2}{R^5}-\frac{1}{5}\frac{r^2}{r'^3}\right)\mathrm{d}r'\right)\end{array}\right) \tag{1}$$

and

$$c(r \geq r_1, \theta) = -\frac{2\pi}{D}\left(-\frac{1}{18\pi}\left(-\frac{r_1{}^3}{R}+\frac{r_1{}^3}{r}\right)-\frac{P_2(\cos(\theta))}{75\pi}\left(-\frac{r_1{}^5r^2}{R^5}+\frac{r_1{}^5}{r^3}\right)\right) \tag{2}$$

7.3. Derive the concentration for the model in Section 7.2 when $g(\theta, \phi) = g(\theta)$, i.e., is only a function of θ.

7.4. Develop an expression for the flux through the outer surface:

$$J(\theta) = -D\left.\frac{\partial c(r, \theta, \varphi)}{\partial r}\right|_{r=R} \tag{1}$$

using the concentration in Problem 7.2.

7.5. Plot the flux obtained in Problem 7.4 for the following parameter values: $D = 1$, $r_1 = 0.2$ and $R = 1$.

REFERENCES

Arfken GB, Weber H-J, Harris FE. *Mathematical methods for physicists : a comprehensive guide*. 7th ed. Amsterdam: Elsevier; 2013.

Byerly WE. *An elementary treatise on Fourier's series and spherical, cylindrical, and ellipsoidal harmonics, with applications to problems in mathematical physics*. Mineola, NY: Dover Publications; 2003.

Calvert PD, Peet JA, Bragin A, Schiesser WE, Pugh EN Jr. Fluorescence relaxation in 3D from diffraction-limited sources of PAGFP or sinks of EGFP created by multiphoton photoconversion. *J Microsc* 2007;225(Pt 1):49–71.

Cole KD. *Heat conduction using Green's functions*. 2nd ed. Boca Raton: CRC Press; 2011.

Fantini S, Franceschini MA, Gratton E. Effective source term in the diffusion equation for photon transport in turbid media. *Appl Opt* 1997;36(1):156–163.

Hobson, E. W. (1955). *The theory of spherical and ellipsoidal harmonics*. New York,: Chelsea Pub. Co.

Jain PK, Singh S, Rizwan U. An exact analytical solution for two-dimensional, unsteady, multilayer heat conduction in spherical coordinates. *Int J Heat Mass Transf* 2010;53(9–10):2133–2142.

Jenson VG, Jeffreys GV. *Mathematical methods in chemical engineering*. 2nd ed. London: Academic Press; 1977.

Larson R, Hostetler RP, Edwards BH. *Essential calculus : early transcendental functions*. Boston: Houghton Mifflin; 2008.

McQuarrie DA. *Mathematical methods for scientists and engineers*. Sausalito, CA: University Science Books; 2003.

8

ANALYSIS OF TRANSIENT GROWTH FACTOR TRANSPORT THROUGH DOUBLE-LAYERED SCAFFOLDS

8.1 LAPLACE AND FOURIER-BESSEL-BASED METHODS IN 2-D CYLINDRICAL COORDINATES

A reaction–diffusion equation in 2-D cylindrical coordinates is of the form

$$\frac{\partial c(r,z,t)}{\partial t} = \frac{\eta\left(\frac{\partial}{\partial r}c(r,z,t) + r\frac{\partial^2}{\partial r^2}c(r,z,t) + r\frac{\partial^2}{\partial z^2}c(r,z,t)\right)}{r} + kc(r,z,t) \tag{8.1}$$

where η is the diffusivity and k is a rate constant. The initial condition is

$$c(r,z,0) = c_0 \tag{8.2}$$

To illustrate the solution procedure, we assume that the process occurs in a cylindrical domain defined by $0 < z < H$ and $0 < r < R$ with the following boundary conditions:

$$c(R,z,t) = 0 \tag{8.3}$$

$$c(r,0,t) = 0 \tag{8.4}$$

$$c(r,H,t) = 0 \tag{8.5}$$

The Laplace transform of Eq. (8.1) is given by

$$sC(r,z) - c_0 = \eta\frac{\partial^2}{\partial r^2}C(r,z) + \eta\frac{\partial^2}{\partial z^2}C(r,z) + \frac{\eta\frac{\partial}{\partial r}C(r,z)}{r} + kC(r,z) \tag{8.6}$$

Closed-form Solutions for Drug Transport through Controlled-Release Devices in Two and Three Dimensions, First Edition. Laurent Simon and Juan Ospina.

with

$$C(r,z) = \int_0^\infty c(r,z,t)\mathrm{e}^{-st}\mathrm{d}t \tag{8.7}$$

A solution to Eq. (8.6) of the form $C(r,z) = H(r) + F(r,z)$ is assumed, where $H(r)$ satisfies

$$sH(r) - c_0 = \eta\frac{\mathrm{d}^2}{\mathrm{d}r^2}H(r) + \frac{\eta\frac{\mathrm{d}}{\mathrm{d}r}H(r)}{r} + kH(r) \tag{8.8}$$

and $F(r,z)$ obeys the equation

$$-kF(r,z) + sF(r,z) - \frac{\eta}{r}\frac{\partial F}{\partial r}(r,z) - \eta\frac{\partial^2 F}{\partial r^2}(r,z) - \eta\frac{\partial^2 F}{\partial z^2}(r,z) = 0 \tag{8.9}$$

The solution to Eq. (8.8) is

$$H(r) = C_2 J_0\left(r\sqrt{\frac{-s+k}{\eta}}\right) + C_1 Y_0\left(r\sqrt{\frac{-s+k}{\eta}}\right) - \frac{c_0}{-s+k} \tag{8.10}$$

To avoid the logarithmic singularity at $r = 0$, C_1 is set to zero, which gives

$$H(r) = C_2 J_0\left(r\sqrt{\frac{-s+k}{\eta}}\right) - \frac{c_0}{-s+k} \tag{8.11}$$

Based on the boundary condition (8.3), we require that $H(R) = 0$. The following equation was obtained:

$$C_2 J_0\left(R\sqrt{\frac{-s+k}{\eta}}\right) - \frac{c_0}{-s+k} = 0 \tag{8.12}$$

From Eq. (8.12), it easy to verify that

$$C_2 = \frac{c_0}{(-s+k)\left(J_0\left(R\sqrt{\frac{-s+k}{\eta}}\right)\right)} \tag{8.13}$$

The substitution of C_2 into Eq. (8.11) gives

$$H(r) = \frac{c_0 J_0\left(r\sqrt{\frac{-s+k}{\eta}}\right)}{(-s+k)J_0\left(R\sqrt{\frac{-s+k}{\eta}}\right)} - \frac{c_0}{-s+k} \tag{8.14}$$

To solve Eq. (8.9), function F is written as $F(r,z) = f(r)g(z)$ where $f(r)$ and $g(z)$ satisfy the following equations

$$\frac{\mathrm{d}^2}{\mathrm{d}r^2}f(r) = c_1 f(r) - \frac{1}{r}\frac{\mathrm{d}}{\mathrm{d}r}f(r) \tag{8.15}$$

and

$$\frac{\mathrm{d}^2}{\mathrm{d}r^2}g(r) = -c_1 g(z) - \frac{g(z)(-s+k)}{\eta} \tag{8.16}$$

It follows from the solution of Eqs. (8.15) and (8.16) that

$$F(r,z) = \left(C_3 \sin\left(z\sqrt{\frac{c_1\eta - s + k}{\eta}} \right) + C_4 \cos\left(z\sqrt{\frac{c_1\eta - s + k}{\eta}} \right) \right) \times (C_1 J_0(r\sqrt{-c_1}) + C_2 Y_0(r\sqrt{-c_1})) \tag{8.17}$$

Again, to prevent a logarithmic singularity, the constant C_2 is zero and without loss of generality, we set $C_1 = 1$, which leads to

$$F(r,z) = J_0(r\sqrt{-c_1}) \left(C_3 \sin\left(z\sqrt{\frac{c_1\eta - s + k}{\eta}} \right) + C_4 \cos\left(z\sqrt{\frac{c_1\eta - s + k}{\eta}} \right) \right) \tag{8.18}$$

Now, the equality $F(R,z) = 0$ holds according to Eq. (8.3). Hence

$$J_0(R\sqrt{-c_1}) \left(C_3 \sin\left(z\sqrt{\frac{c_1\eta - s + k}{\eta}} \right) + C_4 \cos\left(z\sqrt{\frac{c_1\eta - s + k}{\eta}} \right) \right) = 0 \tag{8.19}$$

Therefore,

$$R\sqrt{-c_1} = \alpha_{0,n} \tag{8.20}$$

or equivalently:

$$c_1 = -\frac{\alpha_{0,n}^2}{R^2} \tag{8.21}$$

where $\alpha_{0,n}$ are the zeros of J_0, which leads to

$$F(r,z) = J_0\left(\frac{\alpha_{0,n} r}{R}\right) \times \left(C_3 \sin\left(\frac{z}{R}\sqrt{\frac{-\alpha_{0,n}{}^2\eta - sR^2 + kR^2}{\eta}} \right) + C_4 \cos\left(\frac{z}{R}\sqrt{\frac{-\alpha_{0,n}{}^2\eta - sR^2 + kR^2}{\eta}} \right) \right) \tag{8.22}$$

Application of the **principle of superposition** to Eqs. (8.22) and (8.14) results in

$$C(r,z) = \frac{c_0}{-s+k} \frac{J_0\left(r\sqrt{\frac{-s+k}{\eta}}\right)}{J_0\left(R\sqrt{\frac{-s+k}{\eta}}\right)} - \frac{c_0}{-s+k} + \sum_{n=1}^{\infty} J_0\left(\frac{\alpha_{0,n} r}{R}\right) \times \left(A_n \sin\left(\frac{z}{R}\sqrt{\frac{R^2k - R^2s - \eta\alpha_{0,n}{}^2}{\eta}} \right) + B_n \cos\left(\frac{z}{R}\sqrt{\frac{R^2k - R^2s - \eta\alpha_{0,n}{}^2}{\eta}} \right) \right) \tag{8.23}$$

Furthermore, the boundary condition (8.4) gives

$$\frac{c_0}{-s+k}\frac{J_0\left(r\sqrt{\frac{-s+k}{\eta}}\right)}{J_0\left(R\sqrt{\frac{-s+k}{\eta}}\right)} - \frac{c_0}{-s+k} + \sum_{n=1}^{\infty} J_0\left(\frac{\alpha_{0,n}r}{R}\right)B_n = 0 \tag{8.24}$$

Multiplying both sides of Eq. (8.24) by $rJ_0\left(\frac{\alpha_{0,m}r}{R}\right)$ and integrating from $r=0$ to $r=R$ yields

$$-\frac{J_1(\alpha_{0,m})R^4c_0}{\alpha_{0,m}(R^2k-R^2s-\eta\alpha_{0,m}{}^2)} + \sum_{n=1}^{\infty}\left(B_n\int_0^R rJ_0\left(\frac{\alpha_{0,m}r}{R}\right)J_0\left(\frac{\alpha_{0,n}r}{R}\right)\mathrm{d}r\right) = 0 \tag{8.25}$$

With the following relationship

$$\int_0^R rJ_0\left(\frac{\alpha_{0,m}r}{R}\right)J_0\left(\frac{\alpha_{0,n}r}{R}\right)\mathrm{d}r = \frac{1}{2}R^2(J_1(\alpha_{0,n}))^2\delta_{n,m} \tag{8.26}$$

Eq. (8.25) becomes

$$-\frac{J_1(\alpha_{0,m})R^4c_0}{\alpha_{0,m}(R^2k-R^2s-\eta\alpha_{0,m}{}^2)} + \frac{1}{2}R^2(J_1(\alpha_{0,n}))^2\delta_{n,m} = 0 \tag{8.27}$$

After using the boundary condition (8.5) in Eq. (8.23), the following equation is obtained:

$$\frac{c_0}{-s+k}\frac{J_0\left(r\sqrt{\frac{-s+k}{\eta}}\right)}{J_0\left(R\sqrt{\frac{-s+k}{\eta}}\right)} - \frac{c_0}{-s+k} + \sum_{n=1}^{\infty} J_0\left(\frac{\alpha_{0,n}r}{R}\right)\times$$
$$\left(A_n\sin\left(\frac{H}{R}\sqrt{\frac{R^2k-R^2s-\eta\alpha_{0,n}{}^2}{\eta}}\right)+B_n\cos\left(\frac{H}{R}\sqrt{\frac{R^2k-R^2s-\eta\alpha_{0,n}{}^2}{\eta}}\right)\right)=0 \tag{8.28}$$

Multiplying both sides of Eq. (8.28) by $rJ_0\left(\frac{\alpha_{0,m}r}{R}\right)$ and integrating from $r=0$ to $r=R$, we deduce that

$$-\frac{J_1(\alpha_{0,m})R^4c_0}{\alpha_{0,m}(R^2k-R^2s-\eta\alpha_{0,m}{}^2)}$$
$$+\sum_{n=1}^{\infty}\left(\left(A_n\sin\left(\frac{H}{R}\sqrt{\frac{R^2k-R^2s-\eta\alpha_{0,n}{}^2}{\eta}}\right)+B_n\cos\left(\frac{H}{R}\sqrt{\frac{R^2k-R^2s-\eta\alpha_{0,n}{}^2}{\eta}}\right)\right)\times \int_0^R rJ_0\left(\frac{\alpha_{0,m}r}{R}\right)J_0\left(\frac{\alpha_{0,n}r}{R}\right)\mathrm{d}r\right)=0 \tag{8.29}$$

By applying Eq. (8.26), Eq. (8.29) reduces to

$$-\frac{J_1(\alpha_{0,m})R^4c_0}{\alpha_{0,m}(R^2k-R^2s-\eta\alpha_{0,m}{}^2)}+\sum_{n=1}^{\infty}\left(\left(A_n\sin\left(\frac{H}{R}\sqrt{\frac{R^2k-R^2s-\eta\alpha_{0,n}{}^2}{\eta}}\right)+B_n\cos\left(\frac{H}{R}\sqrt{\frac{R^2k-R^2s-\eta\alpha_{0,n}{}^2}{\eta}}\right)\right)\times\frac{1}{2}R^2(J_1(\alpha_{0,n}))^2\delta_{n,m}\right)=0 \tag{8.30}$$

The parameters A_m and B_m are calculated from Eqs. (8.27) and (8.30):

$$A_n=-\frac{2R^2c_0}{J_1(\alpha_{0,n})\alpha_{0,n}(R^2k-R^2s-\eta\alpha_{0,n}{}^2)}\left(\frac{\cos\left(\frac{H}{R}\sqrt{\frac{R^2k-R^2s-\eta\alpha_{0,n}{}^2}{\eta}}\right)-1}{\sin\left(\frac{H}{R}\sqrt{\frac{R^2k-R^2s-\eta\alpha_{0,n}{}^2}{\eta}}\right)}\right) \tag{8.31}$$

and

$$B_n=\frac{2R^2c_0}{J_1(\alpha_{0,n})\alpha_{0,n}(R^2k-R^2s-\eta\alpha_{0,n}{}^2)} \tag{8.32}$$

Then, Eq. (8.23) becomes

$$C(r,z)=\frac{c_0}{-s+k}\frac{J_0\left(r\sqrt{\frac{-s+k}{\eta}}\right)}{J_0\left(R\sqrt{\frac{-s+k}{\eta}}\right)}-\frac{c_0}{-s+k}+\sum_{n=1}^{\infty}J_0\left(\frac{\alpha_{0,n}r}{R}\right)\times\left(-\frac{2R^2c_0\sin\left(\frac{z}{R}\sqrt{\frac{R^2k-R^2s-\eta\alpha_{0,n}{}^2}{\eta}}\right)}{J_1(\alpha_{0,n})\alpha_{0,n}(R^2k-R^2s-\eta\alpha_{0,n}{}^2)}\left(\frac{\cos\left(\frac{H}{R}\sqrt{\frac{R^2k-R^2s-\eta\alpha_{0,n}{}^2}{\eta}}\right)-1}{\sin\left(\frac{H}{R}\sqrt{\frac{R^2k-R^2s-\eta\alpha_{0,n}{}^2}{\eta}}\right)}\right)\times+\frac{2R^2c_0\cos\left(\frac{z}{R}\sqrt{\frac{R^2k-R^2s-\eta\alpha_{0,n}{}^2}{\eta}}\right)}{J_1(\alpha_{0,n})\alpha_{0,n}(R^2k-R^2s-\eta\alpha_{0,n}{}^2)}\right) \tag{8.33}$$

after substituting A_n and B_n. Equation (8.33) gives the solution in the Laplace domain. To obtain the solution in the time domain, we must compute the **inverse Laplace transform** (Graf, 2004; Loney, 2007). This computation is performed using the **Bromwich integral** (Debnath & Bhatta, 2007; Thomson, 1950) and the **residue theorem** (Loney, 2007). The solution is written, formally, as the following Bromwich integral:

$$c(r,z,t)=\frac{1}{2\pi i}\int C(r,z)e^{st}ds \tag{8.34}$$

and the Bromwich integral is computed using the residue theorem as

$$c(r,z,t)=\sum_{pole}Residue(C(r,z)e^{st})|_{s=pole} \tag{8.35}$$

Substituting $C(r, z)$, from Eq. (8.33), into Eq. (8.34), we have

$$c(r,z,t) = \frac{1}{2\pi i} \times \int \left[\frac{c_0}{-s+k} \frac{J_0\left(r\sqrt{\frac{-s+k}{\eta}}\right)}{J_0\left(R\sqrt{\frac{-s+k}{\eta}}\right)} - \frac{c_0}{-s+k} + \sum_{n=1}^{\infty} J_0\left(\frac{\alpha_{0,n} r}{R}\right) \times \left(-\frac{2R^2 c_0 \sin\left(\frac{z}{R}\sqrt{\frac{R^2 k - R^2 s - \eta\alpha_{0,n}{}^2}{\eta}}\right)}{J_1(\alpha_{0,n})\alpha_{0,n}(R^2 k - R^2 s - \eta\alpha_{0,n}{}^2)} \left(\frac{\cos\left(\frac{H}{R}\sqrt{\frac{R^2 k - R^2 s - \eta\alpha_{0,n}{}^2}{\eta}}\right) - 1}{\sin\left(\frac{H}{R}\sqrt{\frac{R^2 k - R^2 s - \eta\alpha_{0,n}{}^2}{\eta}}\right)} \right) \times + \frac{2R^2 c_0 \cos\left(\frac{z}{R}\sqrt{\frac{R^2 k - R^2 s - \eta\alpha_{0,n}{}^2}{\eta}}\right)}{J_1(\alpha_{0,n})\alpha_{0,n}(R^2 k - R^2 s - \eta\alpha_{0,n}{}^2)} \right) \right] e^{st} ds \tag{8.36}$$

Applying the residue theorem to the Bromwich integrals in Eq. (8.36) gives the following expression:

$$c(r,z,t) = \sum_{n=1}^{\infty} \left(\sum_{m=1}^{\infty} -\frac{4c_0((-1)^m - 1)}{J_1(\alpha_{0,n})\alpha_{0,n}\pi m} e^{\frac{(-\pi^2 R^2 \eta m^2 + H^2 R^2 k - H^2 \eta \alpha_{0,n}{}^2)t}{H^2 R^2}} \sin\left(\frac{\pi m z}{H}\right) J_0\left(\frac{\alpha_{0,n} r}{R}\right) \right) \tag{8.37}$$

Note that the solution is given in terms of the ***Fourier-Bessel series*** (Zauderer, 2006).

8.2 GOVERNING EQUATIONS FOR TRANSPORT THROUGH DOUBLE-LAYERED SCAFFOLDS

The focus is on the system defined in Chapter 3: a scaffold with two connected compartments, layers 1 and 2, packed on top of each other. A similar design was able to sequentially deliver VEGF and PDGF resulting in spatially-organized tissues (Chen et al., 2007b). The two reaction–diffusion equations, which explain transport across the device, are (Chen et al., 2007a, b):

$$\frac{\partial c_1}{\partial t} = \eta \left(\frac{1}{r}\frac{\partial}{\partial r}\left(r\frac{\partial c_1}{\partial r}\right) + \frac{\partial^2 c_1}{\partial z^2} \right) + k_r(t)c_{10} - k_c c_1 \tag{8.38}$$

$$\frac{\partial c_2}{\partial t} = \eta \left(\frac{1}{r}\frac{\partial}{\partial r}\left(r\frac{\partial c_2}{\partial r}\right) + \frac{\partial^2 c_2}{\partial z^2} \right) + k_r(t)c_{20} - k_c c_2 \tag{8.39}$$

where $c_1(r,\ z)$ is the concentration of VEGF or PDGF in the cylindrical region bounded by $0 < r < R$ and $0 < z < b$ (i.e., layer 1); $c_2(r,\ z)$ is its concentration in the space formed by $0 < r < R$ and $-a < z < 0$ (i.e., layer 2); η is the diffusion coefficient in the polymer; k_2 and k_c are release and degradation rates, respectively; c_{10} and c_{20} represent the initial drug concentrations.

The release-rate constant was a function of time:

$$k_r(t \le T_1) = k_{r1}$$
$$k_r(t > T_1) = k_{r2} \tag{8.40}$$

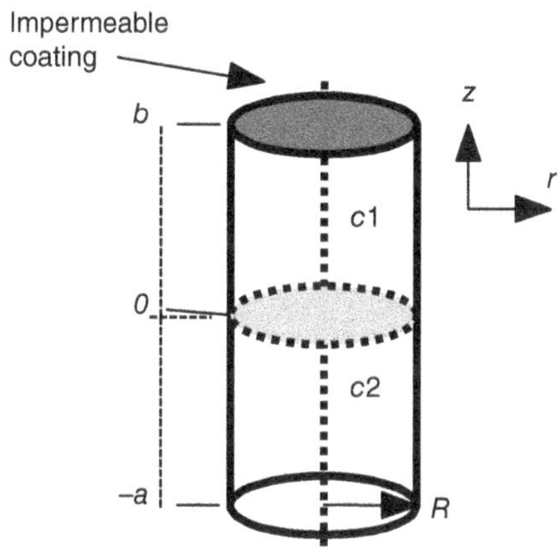

Figure 8.1 Schematic of the double-layered scaffold system with a no-flux boundary condition at $z = b$.

Except for Eq. (8.47), the boundary conditions were described in Chapter 3 (Fig. 8.1):

$$\left.\frac{\partial c_1(r,z,t)}{\partial r}\right|_{r=0} = 0 \tag{8.41}$$

$$\left.\frac{\partial c_2(r,z,t)}{\partial r}\right|_{r=0} = 0 \tag{8.42}$$

$$c_1(R,z,t) = 0 \tag{8.43}$$

$$c_2(R,z,t) = 0 \tag{8.44}$$

$$\left.\frac{\partial c_1(r,z,t)}{\partial z}\right|_{z=0} = \left.\frac{\partial c_2(r,z,t)}{\partial z}\right|_{z=0} \tag{8.45}$$

$$c_1(r,0,t) = c_2(r,0,t) \tag{8.46}$$

$$\left.\frac{\partial c_1(r,z,t)}{\partial z}\right|_{z=b} = 0 \tag{8.47}$$

$$c_2(r,-a,t) = 0 \tag{8.48}$$

At the beginning of the experiment, the concentrations in the two layers are

$$c_1(r,z,0) = c_{10} \tag{8.49}$$

$$c_2(r,z,0) = c_{20} \tag{8.50}$$

The fluxes can be computed by the following expressions:

$$J_1 = -2\eta\pi R\left(\int_0^b \left.\frac{\partial c_1(r,z,t)}{\partial r}\right|_{r=R} \mathrm{d}z\right) \tag{8.51}$$

$$J_2 = -2\eta\pi R\left(\int_{-a}^0 \left.\frac{\partial c_2(r,z,t)}{\partial r}\right|_{r=R} \mathrm{d}z\right) \tag{8.52}$$

$$J_3 = 2\eta\pi \left(\int_0^R \left(\frac{\partial c_2 (r, z, t)}{\partial z} \bigg|_{z=-a} \right) r\mathrm{d}r \right) \tag{8.53}$$

and

$$J_4 = -2\eta\pi \left(\int_0^R \left(\frac{\partial c_1 (r, z, t)}{\partial z} \bigg|_{z=b} \right) r\mathrm{d}r \right) \tag{8.54}$$

Finally, the total flux is

$$J = J_1 + J_2 + J_3 + J_4 \tag{8.55}$$

8.3 CONCENTRATION PROFILE OF VASCULAR ENDOTHELIAL GROWTH FACTOR IN TWO LAYERS

In the Laplace domain, Eq.(8.38), with the conditions (8.40) and (8.49), takes the form

$$\begin{aligned} sC_1(r,z) - c_{10} &= \eta \frac{\partial^2}{\partial r^2} C_1(r,z) + \eta \frac{\partial^2}{\partial z^2} C_1(r,z) + \frac{\eta \frac{\partial}{\partial r} C_1(r,z)}{r} \\ &+ c_{10} \left(-\frac{k_1 \left(\mathrm{e}^{-sT_1} - 1 \right)}{s} + \frac{k_2}{s\mathrm{e}^{sT_1}} \right) - k_C C_1(r,z) \end{aligned} \tag{8.56}$$

In the Laplace domain, Eq.(8.39), with conditions (8.40) and (8.50), becomes

$$\begin{aligned} sC_2(r,z) - c_{20} &= \eta \frac{\partial^2}{\partial r^2} C_2(r,z) + \eta \frac{\partial^2}{\partial z^2} C_2(r,z) + \frac{\eta \frac{\partial}{\partial r} C_2(r,z)}{r} \\ &+ c_{20} \left(-\frac{k_1 \left(\mathrm{e}^{-sT_1} - 1 \right)}{s} + \frac{k_2}{s\mathrm{e}^{sT_1}} \right) - k_C C_2(r,z) \end{aligned} \tag{8.57}$$

A solution to Eq. (8.56) of the form $C_1(r,z) = H_1(r) + F_1(r,z)$ is first assumed, where

$$sH_1(r) - c_{10} = \eta \frac{\mathrm{d}^2}{\mathrm{d}r^2} H_1(r) + \frac{\eta \frac{\mathrm{d}}{\mathrm{d}r} H_1(r)}{r} + c_{10} \left(-\frac{k_1 \left(\mathrm{e}^{-sT_1} - 1 \right)}{s} + \frac{k_2}{s\mathrm{e}^{sT_1}} \right) - k_C H_1(r) \tag{8.58}$$

and

$$sF_1(r,z) - \frac{\eta \frac{\partial}{\partial r} F_1(r,z)}{r} - \eta \frac{\partial^2}{\partial r^2} F_1(r,z) - \eta \frac{\partial^2}{\partial z^2} F_1(r,z) - k_C F_1(r,z) = 0 \tag{8.59}$$

The general solution of Eq. (8.58) is given by

$$H_1(r) = C_2 J_0 \left(\sqrt{\frac{-s - k_C}{\eta}} r \right) + C_1 Y_0 \left(\sqrt{\frac{-s - k_C}{\eta}} r \right) + \frac{((s + k_1)\mathrm{e}^{sT_1} - k_1 + k_2)\mathrm{e}^{-sT_1} c_{10}}{s(s + k_C)} \tag{8.60}$$

To avoid logarithmic singularity at $r = 0$ and satisfy the boundary condition (8.41), C_1, is set to zero, which yields

$$H_1(r) = C_2 J_0\left(\sqrt{\frac{-s-k_C}{\eta}}r\right) + \frac{((s+k_1)e^{sT_1} - k_1 + k_2)e^{-sT_1}c_{10}}{s(s+k_C)} \tag{8.61}$$

and

$$C_2 J_0\left(\sqrt{\frac{-s-k_C}{\eta}}R\right) + \frac{((s+k_1)e^{sT_1} - k_1 + k_2)e^{-sT_1}c_{10}}{s(s+k_C)} = 0 \tag{8.62}$$

after enforcing the boundary condition (8.43). As a result, C_2 is

$$C_2 = -\frac{((s+k_1)e^{sT_1} - k_1 + k_2)e^{-sT_1}c_{10}}{s(s+k_C)J_0\left(R\sqrt{\frac{-s-k_C}{\eta}}\right)} \tag{8.63}$$

Equation (8.61) becomes

$$H_1(r) = -\frac{((s+k_1)e^{sT_1} - k_1 + k_2)e^{-sT_1}c_{10}J_0\left(\sqrt{\frac{-s-k_C}{\eta}}r\right)}{s(s+k_C)J_0\left(R\sqrt{\frac{-s-k_C}{\eta}}\right)} + \frac{((s+k_1)e^{sT_1} - k_1 + k_2)e^{-sT_1}c_{10}}{s(s+k_C)} \tag{8.64}$$

Now, for a solution to Eq. (8.59), we proceed as before and write $F_1(r,z) = f_1(r)g_1(z)$ where

$$\frac{d^2}{dr^2}f_1(r) = c_1 f_1(r) - \frac{\frac{d}{dr}f_1(r)}{r} \tag{8.65}$$

and

$$\frac{d^2}{dz^2}g_1(z) = -c_1 g_1(z) + \frac{g_1(z)(s+k_C)}{\eta} \tag{8.66}$$

The solutions of Eq. (8.65) and (8.66) lead to

$$F_1(r,z) = (C_1 J_0(r\sqrt{-c_1}) + C_2 Y_0(r\sqrt{-c_1})) \times\left(C_3 \sin\left(z\sqrt{\frac{\eta c_1 - s - k_C}{\eta}}\right) + C_4 \cos\left(z\sqrt{\frac{\eta c_1 - s - k_C}{\eta}}\right)\right) \tag{8.67}$$

Again, to avoid the logarithmic singularity at $r = 0$ and to satisfy the boundary condition (8.41), C_2 is equal to zero. Without loss of generality, we make $C_1 = 1$. Hence,

$$F_1(r,z) = J_0(r\sqrt{-c_1})\left(C_3 \sin\left(z\sqrt{\frac{\eta c_1 - s - k_C}{\eta}}\right) + C_4 \cos\left(z\sqrt{\frac{\eta c_1 - s - k_C}{\eta}}\right)\right) \tag{8.68}$$

The boundary condition (8.47) applied to Eq. (8.68) yields

$$J_0(r\sqrt{-c_1})\times \left(C_3\sqrt{\frac{\eta c_1 - s - k_C}{\eta}}\cos\left(b\sqrt{\frac{\eta c_1 - s - k_C}{\eta}}\right) - C_4\sqrt{\frac{\eta c_1 - s - k_C}{\eta}}\sin\left(b\sqrt{\frac{\eta c_1 - s - k_C}{\eta}}\right)\right) = 0 \tag{8.69}$$

The parameter C_4 is calculated from Eq. (8.69):

$$C_4 = \frac{C_3\cos\left(b\sqrt{\frac{\eta c_1 - s - k_C}{\eta}}\right)}{\sin\left(b\sqrt{\frac{\eta c_1 - s - k_C}{\eta}}\right)} \tag{8.70}$$

Replacing C_4 in Eq. (8.68) produces

$$F_1(r,z) = \frac{\left[J_0\left(r\sqrt{-c_1}\right)C_3\begin{pmatrix}\sin\left(z\sqrt{\frac{\eta c_1 - s - k_C}{\eta}}\right)\sin\left(b\sqrt{\frac{\eta c_1 - s - k_C}{\eta}}\right) + \\ \cos\left(z\sqrt{\frac{\eta c_1 - s - k_C}{\eta}}\right)\cos\left(b\sqrt{\frac{\eta c_1 - s - k_C}{\eta}}\right)\end{pmatrix}\right]}{\sin\left(b\sqrt{\frac{\eta c_1 - s - k_C}{\eta}}\right)} \tag{8.71}$$

Applying the boundary condition (8.43) to Eq. (8.71) yields

$$\frac{J_0(R\sqrt{-c_1})C_3\begin{pmatrix}\sin\left(z\sqrt{\frac{\eta c_1 - s - k_C}{\eta}}\right)\sin\left(b\sqrt{\frac{\eta c_1 - s - k_C}{\eta}}\right) + \\ \cos\left(z\sqrt{\frac{\eta c_1 - s - k_C}{\eta}}\right)\cos\left(b\sqrt{\frac{\eta c_1 - s - k_C}{\eta}}\right)\end{pmatrix}}{\sin\left(b\sqrt{\frac{\eta c_1 - s - k_C}{\eta}}\right)} = 0 \tag{8.72}$$

From Eq. (8.72), we can show that

$$c_1 = -\frac{{\alpha_{0,n}}^2}{R^2} \tag{8.73}$$

which gives

$$F_1(r,z) = \frac{-iC_3J_0\left(\frac{\alpha_{0,n}r}{R}\right)\begin{pmatrix}-\sinh\left(\frac{z}{R}\sqrt{\frac{R^2s+R^2k_C+\eta{\alpha_{0,n}}^2}{\eta}}\right)\sinh\left(\frac{b}{R}\sqrt{\frac{R^2s+R^2k_C+\eta{\alpha_{0,n}}^2}{\eta}}\right) + \\ \cosh\left(\frac{z}{R}\sqrt{\frac{R^2s+R^2k_C+\eta{\alpha_{0,n}}^2}{\eta}}\right)\cosh\left(\frac{b}{R}\sqrt{\frac{R^2s+R^2k_C+\eta{\alpha_{0,n}}^2}{\eta}}\right)\end{pmatrix}}{\sinh\left(\frac{b}{R}\sqrt{\frac{R^2s+R^2k_C+\eta{\alpha_{0,n}}^2}{\eta}}\right)} \tag{8.74}$$

after substitution in Eq. (8.71). The **principle of superposition** applied to Eqs. (8.64) and (8.74) results in

$$C_1(r,z) = -\frac{((s+k_1)e^{sT_1}-k_1+k_2)e^{-sT_1}c_{10}J_0\left(r\sqrt{\frac{-s-k_C}{\eta}}\right)}{s(s+k_C)J_0\left(R\sqrt{\frac{-s-k_C}{\eta}}\right)} + \frac{((s+k_1)e^{sT_1}-k_1+k_2)e^{-sT_1}c_{10}}{s(s+k_C)}$$
$$+\sum_{n=1}^{\infty}\left[\frac{\begin{array}{c}-iA_nJ_0\left(\frac{\alpha_{0,n}r}{R}\right)\times\\ \left(\begin{array}{c}-\sinh\left(\frac{z}{R}\sqrt{\frac{R^2s+R^2k_C+\eta\alpha_{0,n}{}^2}{\eta}}\right)\sinh\left(\frac{b}{R}\sqrt{\frac{R^2s+R^2k_C+\eta\alpha_{0,n}{}^2}{\eta}}\right)+\\ \cosh\left(\frac{z}{R}\sqrt{\frac{R^2s+R^2k_C+\eta\alpha_{0,n}{}^2}{\eta}}\right)\cosh\left(\frac{b}{R}\sqrt{\frac{R^2s+R^2k_C+\eta\alpha_{0,n}{}^2}{\eta}}\right)\end{array}\right)\end{array}}{\left(\sinh\left(\frac{b}{R}\sqrt{\frac{R^2s+R^2k_C+\eta\alpha_{0,n}{}^2}{\eta}}\right)\right)}\right] \tag{8.75}$$

Using a procedure similar to the one outlined above, the solution of Eq. (8.57), with the boundary conditions (8.42), (8.44) and (8.45), is given by

$$C_2(r,z) = -\frac{((s+k_1)e^{sT_1}-k_1+k_2)e^{-sT_1}c_{20}J_0\left(r\sqrt{\frac{-s-k_C}{\eta}}\right)}{s(s+k_C)J_0\left(R\sqrt{\frac{-s-k_C}{\eta}}\right)}$$
$$+\frac{((s+k_1)e^{sT_1}-k_1+k_2)e^{-sT_1}c_{20}}{s(s+k_C)}$$
$$+\sum_{n=1}^{\infty}J_0\left(\frac{\alpha_{0,n}r}{R}\right)$$
$$\times\left(iA_n\sinh\left(\frac{z}{R}\sqrt{\frac{R^2s+R^2k_C+\eta\alpha_{0,n}{}^2}{\eta}}\right)+B_n\cosh\left(\frac{z}{R}\sqrt{\frac{R^2s+R^2k_C+\eta\alpha_{0,n}{}^2}{\eta}}\right)\right) \tag{8.76}$$

Applying the boundary condition (8.46) to Eqs. (8.75) and (8.76) gives

$$-\frac{((s+k_1)e^{sT_1}-k_1+k_2)e^{-sT_1}c_{10}J_0\left(r\sqrt{\frac{-s-k_C}{\eta}}\right)}{s(s+k_C)J_0\left(R\sqrt{\frac{-s-k_C}{\eta}}\right)} + \frac{((s+k_1)e^{sT_1}-k_1+k_2)e^{-sT_1}c_{10}}{s(s+k_C)}+$$

$$\sum_{n=1}^{\infty}\left[\frac{-iA_nJ_0\left(\frac{\alpha_{0,n}r}{R}\right)\left(\cosh\left(\frac{b}{R}\sqrt{\frac{R^2s+R^2k_C+\eta\alpha_{0,n}{}^2}{\eta}}\right)\right)}{\left(\sinh\left(\frac{b}{R}\sqrt{\frac{R^2s+R^2k_C+\eta\alpha_{0,n}{}^2}{\eta}}\right)\right)}\right]$$

$$=\sum_{n=1}^{\infty}B_nJ_0\left(\frac{\alpha_{0,n}r}{R}\right)$$

$$-\frac{((s+k_1)e^{sT_1}-k_1+k_2)e^{-sT_1}c_{20}J_0\left(r\sqrt{\frac{-s-k_C}{\eta}}\right)}{s(s+k_C)J_0\left(R\sqrt{\frac{-s-k_C}{\eta}}\right)}+\frac{((s+k_1)e^{sT_1}-k_1+k_2)e^{-sT_1}c_{20}}{s(s+k_C)} \tag{8.77}$$

Multiplying both sides of Eq. (8.77) by $rJ_0\left(\frac{\alpha_{0,m}r}{R}\right)$, integrating from $r=0$ to $r=R$; and applying Eq. (8.26), we obtain

$$\frac{(se^{sT_1}+k_1e^{sT_1}-k_1+k_2)J_1(\alpha_{0,m})R^4c_{10}}{(R^2s+R^2k_C+\eta\alpha_{0,m}{}^2)\alpha_{0,m}e^{sT_1}s}$$
$$-\frac{\frac{1}{2}iA_n\cosh\left(\frac{b}{R}\sqrt{\frac{R^2s+R^2k_C+\eta\alpha_{0,n}{}^2}{\eta}}\right)R^2\left(J_1\left(\frac{\alpha_{0,n}r}{R}\right)\right)^2}{\sinh\left(\frac{b}{R}\sqrt{\frac{R^2s+R^2k_C+\eta\alpha_{0,n}{}^2}{\eta}}\right)}$$
$$=\frac{(se^{sT_1}+k_1e^{sT_1}-k_1+k_2)J_1(\alpha_{0,m})R^4c_{20}}{(R^2s+R^2k_C+\eta\alpha_{0,m}{}^2)\alpha_{0,m}e^{sT_1}s}+\frac{1}{2}B_nR^2\left(J_1\left(\frac{\alpha_{0,n}r}{R}\right)\right)^2 \tag{8.78}$$

Finally, applying the boundary condition (8.48) to Eqs. (8.75) and (8.76) gives

$$-\frac{((s+k_1)e^{sT_1}-k_1+k_2)e^{-sT_1}c_{20}J_0\left(r\sqrt{\frac{-s-k_C}{\eta}}\right)}{s(s+k_C)J_0\left(\sqrt{\frac{-s-k_C}{\eta}}R\right)}+\frac{((s+k_1)e^{sT_1}-k_1+k_2)e^{-sT_1}c_{20}}{s(s+k_C)}$$
$$+\sum_{n=1}^{\infty}\left(\begin{array}{l}-iA_n\sinh\left(\frac{a}{R}\sqrt{\frac{R^2s+R^2k_C+\eta\alpha_{0,n}{}^2}{\eta}}\right)\\+B_n\cosh\left(\frac{a}{R}\sqrt{\frac{R^2s+R^2k_C+\eta\alpha_{0,n}{}^2}{\eta}}\right)\end{array}\right)J_0\left(\frac{\alpha_{0,n}r}{R}\right)=0 \tag{8.79}$$

Multiplying both sides of Eq. (8.79) by $rJ_0\left(\frac{\alpha_{0,m}r}{R}\right)$, integrating from $r = 0$ to $r = R$ and applying Eq. (8.26), we obtain

$$\frac{(se^{sT_1} + k_1e^{sT_1} - k_1 + k_2)J_1(\alpha_{0,m})R^4c_{20}}{(R^2s + R^2k_C + \eta\alpha_{0,m}{}^2)\alpha_{0,m}e^{sT_1}s} + \left(\begin{array}{l} -\frac{1}{2}iA_n \sinh\left(\frac{a}{R}\sqrt{\frac{R^2s+R^2k_C+\eta\alpha_{0,n}{}^2}{\eta}}\right) \\ +\frac{1}{2}B_n \cosh\left(\frac{a}{R}\sqrt{\frac{R^2s+R^2k_C+\eta\alpha_{0,n}{}^2}{\eta}}\right)\end{array}\right)R^2(J_1(\alpha_{0,n}))^2 = 0 \tag{8.80}$$

Solving Eqs. (8.78) and (8.80), with respect to A_n and B_n, and replacing the results in Eqs. (8.75) and (8.76), we obtain the solution in the Laplace domain. The final expressions are very large and will be not presented here. Finally, after performing the inverse Laplace transform using the Bromwich integral and the residue theorem, the corresponding solutions in the time domain are derived. These expressions are not presented because of space limitation.

Several techniques are available in the literature to invert a function $\overline{\Psi}(s)$ from the Laplace to the time domain $\Psi(t)$. Three of these methods are outlined here. The residue theorem yields (Loney, 2007)

$$\Psi(t) = \sum_{m=1}^{\infty}\left(\frac{P\left(s_m\right)e^{s_m t}}{\left.\frac{dQ(s)}{ds}\right|_{s=s_m}}\right) \tag{8.81}$$

where $P(s)$ and $Q(s)$ are such that $\overline{\Psi}(s) = P(s)/Q(s)$ and s_n are the roots of $Q(s) = 0$. The Zakian method can also be implemented (Rice & Do, 1995) :

$$\Psi(t) = \frac{2}{t}\sum_{m=1}^{5} \mathrm{Re}(K_m\overline{\Psi}(a_m/t)) \tag{8.82}$$

where K_m and a_m are complex numbers defined in the Zakian algorithm. The package "*FixedTalbotNumericalLaplaceInversion.m*" (Abate & Valkó, 2004) performs a numerical Laplace inversion in Mathematica (Wolfram Research, Inc.).

8.4 SUMMARY

A Laplace and Fourier-Bessel-based method was introduced to solve a reaction–diffusion problem in 2-D cylindrical coordinates. The model contained a time-dependent rate coefficient. The boundary conditions were those applied in Section 3 with the exception of a no-flux boundary condition enforced at the top of the cylindrical device. A closed-form solution was first developed in terms of the Laplace variable. Time-domain solutions were computed after inversion of the Laplace transforms via the Bromwich integral and the residue theorem. Numerical tools are available in the literature to carry out the inversion.

8.5 APPENDIX: MAPLE CODE LISTINGS

8.5.1 Maple Code: twocylindertransient.mws (Eq. (8.37))

```
> restart:with(VectorCalculus):with(inttrans):
> eq1:=diff(c(r,z,t),t)=eta*Laplacian(c(r,z,t),cylindrical[r,theta,z])
      +k*c(r,z,t);
>
```

$$eq1 := \frac{\partial}{\partial t} c(r, z, t) = \frac{\eta\left(\frac{\partial}{\partial t} c(r, z, t) + r\left(\frac{\partial^2}{\partial r^2} c(r, z, t)\right) + r\left(\frac{\partial^2}{\partial z^2} c(r, z, t)\right)\right)}{r} + k c(r, z, t)$$

```
>
>
> eq4:=c(r,z,0)=c[0]:
> eq4A:= c(R,z,t)=0:
> eq4B:=c(r,0,t)=0:
> eq4C:=c(r,H,t)=0:
> eq6:=laplace(eq1,t,s):
> eq6A:=subs(eq4,eq6):
>
>
> with(PDETools):
> eq10:=subs(laplace(c(r,z,t),t,s)=C(r,z),eq6A
):
> C(r,z)=Int(c(r,z,t)*exp(-s*t),t=0..infinity):
>
> eq11:=C(r,z)=H(r):
> eq11A:=eval(subs(eq11,eq10)):
> eq11B:=dsolve(eq11A):
>
> eq11B1:=subs(_C1=0,eq11B):
> eq11B2:=subs(r=R,rhs(eq11B1))=0:
> eq11B3:=isolate(eq11B2,_C2):
> eq11B4:=subs(eq11B3,eq11B1):
>
```

We look for a solution of the following form:

```
> eq11C:=C(r,z)=rhs(eq11B4)+F(r,z):
> eq11D:=eval(subs(eq11C,eq10)):
> eq11E:=expand(simplify(lhs(eq11D)-rhs(eq11D)))=0:
>
> eq13:=pdsolve(eq11E,HINT=f(r)*g(z)):
> eq13A:=factor(build(eq13)):
> eq13B:=subs(_C2=0,_C1=1,eq13A):
> eq15:=subs(r=R,rhs(eq13B))=0:
> eq15A:=(-_c[1])^(1/2)*R=alpha[0,n]:
> eq15B:=isolate(eq15A,_c[1]):
> eq16:=simplify(subs(eq15B,eq13B),power,symbolic):
> eq17:=C(r,z)=rhs(eq11B4)+Sum(subs(_C3=A[n],_C4=B[n],rhs(eq16)),
n=1..infinity):
>
```

```
> eq18:=eval(subs(z=0,rhs(eq17))=0):
>
> eq18A:=factor(subs(BesselJ(0,alpha[0,m])=0,expand(int(rhs(eq11B4)*r*
BesselJ(0,alpha[0,m]/R*r),r=0..R))))+Sum(B[n]*Int(r*BesselJ(0,
alpha[0,m]/R*r)*BesselJ(0,alpha[0,n]/R*r),r=0..R),
n =1 .. infinity)=0:
> eq18B:=Int(r*BesselJ(0,alpha[0,m]/R*r)*BesselJ(0,alpha[0,n]/R*r),
r =0 .. R)=int(r*BesselJ(0,alpha[0,m]/R*r)*BesselJ(0,alpha[0,n]/
R*r),r =0 .. R):
>
eq18C := subs({BesselJ(0,alpha[0,m]) = 0,BesselJ(0,alpha[0,n]) = 0},eq18B):
> eq18D:=Int(r*BesselJ(0,alpha[0,n]/R*r)*BesselJ(0,alpha[0,n]/R*r),
r =0 .. R)=int(r*BesselJ(0,alpha[0,n]/R*r)^2,r =0 .. R):
> eq18E:=subs(BesselJ(0,alpha[0,n])=0,eq18D):
> eq18F:=lhs(eq18C)=rhs(eq18E)*delta[n,m]:
> eq18G:=subs(eq18F,eq18A):
> eq18H:=-c[0]*R^4*BesselJ(1,alpha[0,m])/alpha[0,m]/(-s*R^2+
k*R^2-alpha[0,m]^2*eta)+(1/2*B[m]*R^2*BesselJ(1,alpha[0,m])^2)=0:
>
> eq19:=subs(z=H,rhs(eq17))=0:
> eq19A:=factor(subs(BesselJ(0,alpha[0,m])=0,expand(int(rhs(eq11B4)*r*
BesselJ(0,alpha[0,m]/R*r),r=0..R))))+Sum(expand(Int(r*BesselJ(0,
alpha[0,m]/R*r)*BesselJ(0,alpha[0,n]/R*r)*eval(subs(z=H,A[n]*
sin(1/eta^(1/2)*(-alpha[0,n]^2*eta-s*R^2+k*R^2)^(1/2)/R*H)+B[n]*
cos(1/eta^(1/2)*(-alpha[0,n]^2*eta-s*R^2+k*R^2)^(1/2)/R*H))),
r=0..R)),n=1..infinity)=0:
> eq19B:=subs(eq18F,eq19A):
> eq19C:=-c[0]*R^4*BesselJ(1,alpha[0,m])/alpha[0,m]/(-s*R^2+k*R^2-
alpha[0,m]^2*eta)+(1/2*(A[m]*sin(1/eta^(1/2)*(-alpha[0,m]^2*eta-
s*R^2+k*R^2)^(1/2)/R*H)+B[m]*cos(1/eta^(1/2)*(-alpha[0,m]^2*eta-
s*R^2+k*R^2)^(1/2)/R*H))*R^2*BesselJ(1,alpha[0,m])^2) =0:
>
eq20 := subs(m = n,simplify(solve({eq18H,eq19C},{A[m],B[m]}))):
> eq21:=subs(eq20,eq17):
> c(r,z,t)=(1/(2*Pi*i))*Int(C(r,z)*exp(s*t),s):
> c(r,z,t)=Sum(Eval(Residue(C(r,z)*exp(s*t)),s=pole),pole):
> c(r,z,t) = (1/(2*Pi*i))*Int((BesselJ(0,((-s+k)/eta)^(1/2)*r)*c[0]/
(-s+k)/BesselJ(0,((-s+k)/eta)^(1/2)*R)-c[0]/(-s+k))*exp(s*t),s)+
(1/(2*Pi*i))*Sum(Int(((-2*c[0]*R^2*(-1+cos(1/eta^(1/2)*
(-alpha[0,n]^2*eta-s*R^2+k*R^2)^(1/2)/R*H))/BesselJ(1,alpha[0,n])/
alpha[0,n]/sin(1/eta^(1/2)*(-alpha[0,n]^2*eta-s*R^2+k*R^2)^(1/2)/
R*H)/(-alpha[0,n]^2*eta-s*R^2+k*R^2)*sin(1/eta^(1/2)*
(-alpha[0,n]^2*eta-s*R^2+k*R^2)^(1/2)/R*z)+2*c[0]*R^2/
BesselJ(1,alpha[0,n])/alpha[0,n]/(-alpha[0,n]^2*eta-s*R^2+k*R^2)*
cos(1/eta^(1/2)*(-alpha[0,n]^2*eta-s*R^2+k*R^2)^(1/2)/R*z)))*
exp(s*t),s)*BesselJ(0,alpha[0,n]/R*r),n =1 .. infinity):
> eq21A:=factor(-2*c[0]*R^2*(-1+cos(1/eta^(1/2)*(-alpha[0,n]^2*eta-
s*R^2+k*R^2)^(1/2)/R*H))/BesselJ(1,alpha[0,n])/alpha[0,n]/
sin(1/eta^(1/2)*(-alpha[0,n]^2*eta-s*R^2+k*R^2)^(1/2)/R*H)/
```

```
(-alpha[0,n]^2*eta-s*R^2+k*R^2)*sin(1/eta^(1/2)*(-alpha[0,n]^2*
eta-s*R^2+k*R^2)^(1/2)/R*z)+2*c[0]*R^2/BesselJ(1,alpha[0,n])/
alpha[0,n]/(-alpha[0,n]^2*eta-s*R^2+k*R^2)*cos(1/eta^(1/2)*
(-alpha[0,n]^2*eta-s*R^2+k*R^2)^(1/2)/R*z)):
> eq21B:=denom(eq21A):
> eq21C:=-alpha[0,n]^2*eta-s*R^2+k*R^2=0:
> eq21D:=expand(isolate(eq21C,s)):
> eq21E:=frontend(coeff,[convert(series(eq21A*exp(s*t),eq21D,3),
polynom),s+1/R^2*alpha[0,n]^2*eta-k,-1]):
> eq21F:= sin(1/eta^(1/2)*(-alpha[0,n]^2*eta-s*R^2+k*R^2)^(1/2)/R*H)=0:
> eq21G:=(1/eta^(1/2)*(-alpha[0,n]^2*eta-s*R^2+k*R^2)^(1/2)/R*H)=m*Pi:
> eq21H:=isolate(eq21G,s):
> eq21J:=numer(eq21A)*exp(s*t)/diff(denom(eq21A),s):
> eq21K:=factor(simplify(subs(eq21H,eq21J),power,symbolic))
assuming m::integer:
> eq21L:=simplify(eq11B4):
> eq21M:=residue(rhs(eq21L)*exp(s*t),s=k):
> eq21N:=BesselJ(0,((-s+k)/eta)^(1/2)*R)=0:
> eq21O:=((-s+k)/eta)^(1/2)*R=alpha[0,n]:
> eq21P:=isolate(eq21O,s):
> eq21Q:=numer(rhs(eq21L))*exp(s*t)/diff(denom(rhs(eq21L)),s):
> eq21R:=subs(BesselJ(0,alpha[0,n])=0,simplify(subs(eq21P,eq21Q),
power,symbolic)):
> eq21S:=c(r,z,t)=simplify(Sum(eq21R,n=1..infinity)+
(Sum(eq21E*BesselJ(0,alpha[0,n]/R*r),n=1..infinity)))+
Sum(Sum(eq21K*BesselJ(0,alpha[0,n]/R*r),m=1..infinity),
n=1..infinity);
```

$$\mathrm{eq21S} := c(r,\ z,\ t) = \sum_{n=1}^{\infty}\left(\sum_{n=1}^{\infty}\left(-\frac{1}{\mathrm{BesselJ}(1,\ \alpha_{0,\ n})\ \alpha_{0,\ n}\pi\ m}\left(4\ c_0\ \mathrm{e}^{\frac{\left(-\pi^2 R^2 \eta\ m^2+H^2R^2k-H^2\eta\ \alpha_{0,n}^2\right)\ t}{H^2\ R^2}}\ \sin\left(\frac{\pi m z}{H}\right)\right.\right.\right.$$
$$\left.\left.\left.((-1)^m-1)\ \mathrm{BesselJ}\left(0,\ \frac{\alpha_{0,\ n}r}{R}\right)\right.\right.\right.$$

```
>
```

8.5.2 Maple Code: twocylindtransientcase.mws (Eqs. (8.78) and (8.80))

```
> restart:with(VectorCalculus):with(inttrans):
> eq1:=diff(c[1](r,z,t),t)=eta*Laplacian(c[1](r,z,t),
cylindrical[r,theta,z])+k(t)*c[10]-k[C]*c[1](r,z,t);
>
```

$$\mathrm{eq1} := \frac{\partial}{\partial t}\ c_1(r,\ z,\ t) = \frac{\eta\ \left(\frac{\partial}{\partial r}\ c_1(r,\ z,\ t)+r\left(\frac{\partial^2}{\partial r^2}\ c_1(r,\ z,\ t)\right)+r\ \left(\frac{\partial^2}{\partial z^2}\ c_1(r,\ z,\ t)\right)\right)}{r}$$
$$+\ k(t)c_{10} - k_C c_1(r,\ z,\ t)$$

```
> eq2:=diff(c[2](r,z,t),t)=eta*Laplacian(c[2](r,z,t),
cylindrical[r,theta,z])+k(t)*c[20]-k[C]*c[2](r,z,t);
>
```

$$eq2 := \frac{\partial}{\partial t} c_2(r, z, t) = \frac{\eta \left(\frac{\partial}{\partial r} c_2(r, z, t) + r\left(\frac{\partial^2}{\partial r^2} c_2(r, z, t) \right) + r \left(\frac{\partial^2}{\partial z^2} c_2(r, z, t) \right) \right)}{r} + k(t) c_{20} - k_C c_2(r, z, t)$$

```
> eq3:=k(t)=piecewise(t<=T[1],k[1],k[2]):
>
> eq4:=c[1](r,z,0)=c[10]:
> eq4A:=c[2](r,z,0)=c[20]:
> eq5:=subs(eq3,eq1):
> eq5A:=subs(eq3,eq2):
> eq6:=laplace(eq5,t,s):
```

Using the initial conditions:

```
> eq6A:=subs(eq4,eq6):
> eq7:=laplace(eq5A,t,s):
> eq7A:=subs(eq4A,eq7):
> eq8:=laplace(PIECEWISE([k[1], t <= T[1]],[k[2], otherwise]),t,s)=
int(k[1]*exp(-s*t),t=0..T[1]) +int(k[2]*exp(-s*t),
t=T[1]..infinity) assuming T[1]>0 and s>0:
```

$$s\,laplace(c_1(r, z, t), t, s) - c_{10} = \eta \left(\frac{\partial}{\partial r} \left(\frac{\partial}{\partial r} laplace(c_1(r, z, t), t, s) \right) \right) + \eta \left(\frac{\partial}{\partial z} \left(\frac{\partial}{\partial z} laplace(c_1(r, z, t), t, s) \right) \right) + \frac{\eta \left(\frac{\partial}{\partial r} laplace(c_1(r, z, t), t, s) \right)}{r} + c_{10} \left(-\frac{k_1 \left(e^{-s\,T_1} - 1 \right)}{s} + \frac{k_2}{s\,e^{s T_1}} \right) - k_C\, laplace(c_1(r, z, t), t, s)$$

```
> eq9:=subs(eq8,eq6A):
> eq9A:=subs(eq8,eq7A):
> with(PDETools):
> eq10:=subs(laplace(c[1](r,z,t),t,s)=C[1](r,z),eq9):
> eq10A:=subs(laplace(c[2](r,z,t),t,s)=C[2](r,z),eq9A):
>
> eq11:=C[1](r,z)=H[1](r):
> eq11A:=eval(subs(eq11,eq10)):
> eq11B:=dsolve(eq11A):
>
> eq11B1:=subs(_C1=0,eq11B):
> eq11B2:=subs(r=R,rhs(eq11B1))=0:
> eq11B3:=isolate(eq11B2,_C2):
> eq11B4:=subs(eq11B3,eq11B1):
```

We look for a solution of the following form:

```
> eq11C:=C[1](r,z)=rhs(eq11B4)+F[1](r,z):
> eq11D:=eval(subs(eq11C,eq10)):
> eq11E:=expand(simplify(lhs(eq11D)-rhs(eq11D)))=0:
> eq12:=C[2](r,z)=H[2](r):
> eq12A:=eval(subs(eq12,eq10A)):
>
> eq12B:=dsolve(eq12A):
> eq12B1:=subs(_C1=0,eq12B):
> eq12B2:=subs(r=R,rhs(eq12B1))=0:
> eq12B3:=isolate(eq12B2,_C2):
> eq12B4:=subs(eq12B3,eq12B1):
```

We look for a solution of the form:

```
> eq12C:=C[2](r,z)=rhs(eq12B4)+F[2](r,z):
> eq12D:=eval(subs(eq12C,eq10A)):
> eq12E:=expand(simplify(lhs(eq12D)-rhs(eq12D)))=0:
> eq13:=pdsolve(eq11E,HINT=f[1](r)*g[1](z)):
> eq13A:=factor(build(eq13)):
> eq13B:=subs(_C2=0,_C1=1,eq13A):
> Eval(diff(c[1](r,z,t),z),z=b)=0:
> eq13C:=subs(z=b,diff(rhs(eq13B),z))=0:
> eq13D:=isolate(eq13C,_C4):
> eq13E:=factor(subs(eq13D,eq13B)):
> eq14:=pdsolve(eq12E,HINT=f[2](r)*g[2](z)):
>
eq14A := subs({_C1 = E1,_C2 = E2,_C3 = E3,_C4 = E4},factor(build(eq14))):
> eq14B:=subs(E2=0,E1=1,eq14A):
> eq14C:=eval(subs(z=0,diff(rhs(eq14B),z)=diff(rhs(eq13E),z))):
> eq14D:=isolate(eq14C,E3):
> eq14E:=subs(eq14D,eq14B):
> eq15:=subs(r=R,rhs(eq13E))=0:
> eq15A:=(-_c[1])^(1/2)*R=alpha[0,n]:
> eq15B:=isolate(eq15A,_c[1]):
> eq16:=simplify(subs(eq15B,eq13E),power,symbolic):
>
> eq16A:=simplify(subs(eq15B,eq14E),power,symbolic):
```

We use the principle of superposition:

```
> eq17:=C[1](r,z)=rhs(eq11B4)+Sum(subs(_C3=A[n],rhs(eq16)),
n=1..infinity):
>
        eq17A := C[2](r,z) = rhs(eq12B4)+
Sum(subs({_C3 = A[n],E4 = B[n]},rhs(eq16A)),n = 1..infinity) :
> eq18:=eval(subs(z=0,rhs(eq17))=subs(z=0,rhs(eq17A))):
> eq18A:=factor(subs(BesselJ(0,alpha[0,m])=0,expand(int(rhs(eq11B4)*r*
BesselJ(0,alpha[0,m]/R*r),r=0..R))))+Sum(expand(Int(r*BesselJ(0,
alpha[0,m]/R*r)*eval(subs(z=0,_C3=A[n],rhs(eq16))),r=0..R)),
n=1..infinity)=factor(subs(BesselJ(0,alpha[0,m])=0,
expand(int(rhs(eq12B4)*r*BesselJ(0,alpha[0,m]/R*r),r=0..R))))+
Sum(expand(Int(r*BesselJ(0,alpha[0,m]/R*r)*eval(subs(z=0,_C3=A[n],
E4=B[n],rhs(eq16A))),r=0..R)),n=1..infinity):
> eq18B:=Int(r*BesselJ(0,alpha[0,m]/R*r)*BesselJ(0,alpha[0,n]/R*r),
r =0 .. R)=int(r*BesselJ(0,alpha[0,m]/R*r)*BesselJ(0,alpha[0,n]/
R*r),r =0 .. R):
>
eq18C := subs({BesselJ(0,alpha[0,m]) = 0,BesselJ(0,alpha[0,n]) = 0},eq18B):
> eq18D:=Int(r*BesselJ(0,alpha[0,n]/R*r)*BesselJ(0,alpha[0,n]/R*r),
r =0 .. R)=int(r*BesselJ(0,alpha[0,n]/R*r)^2,r =0 .. R):
> eq18E:=subs(BesselJ(0,alpha[0,n])=0,eq18D):
> eq18F:=lhs(eq18C)=rhs(eq18E)*delta[n,m]:
> eq18G:=subs(eq18F,eq18A):
```

```
> eq18H:=(exp(s*T[1])*s+exp(s*T[1])*k[1]+k[2]-k[1])*c[10]*R^4*
BesselJ(1,alpha[0,m])/(s*R^2+k[C]*R^2+alpha[0,m]^2*eta)/
alpha[0,m]/exp(s*T[1])/s+(-1/2*I*A[n]*cosh(1/eta^(1/2)*
(eta*alpha[0,n]^2+s*R^2+k[C]*R^2)^(1/2)/R*b)/sinh(1/eta^(1/2)*
(eta*alpha[0,n]^2+s*R^2+k[C]*R^2)^(1/2)/R*b)*R^2*
BesselJ(1,alpha[0,n])^2) = (exp(s*T[1])*s+exp(s*T[1])*k[1]+
k[2]-k[1])*c[20]*R^4*BesselJ(1,alpha[0,m])/(s*R^2+k[C]*R^2+
alpha[0,m]^2*eta)/alpha[0,m]/exp(s*T[1])/s+(1/2*B[n]*R^2*
BesselJ(1,alpha[0,n])^2);
>
```

$$eq18H := \frac{(s\, e^{s\,T_1}+k_1\, e^{s\,T_1}-k_1+k_2)\ \mathtt{BesselJ}(1,\ \alpha_{0,\ m})\ R^4 c_{10}}{(R^2 s+R^2 k_C+\eta\ \alpha_{0,\ m}^2)\ \alpha_{0,\ m}\, e^{s\ T_1}\, s} - \frac{\frac{1}{2}\ IA_n \cosh\left(\frac{\sqrt{R^2 s+R^2 k_C+\eta\ \alpha_{0,\ n}^2}\ b}{\sqrt{\eta}\, R}\right) R^2 \mathtt{BesselJ}(1,\ \alpha_{0,\ n})^2}{\sinh\left(\frac{\sqrt{R^2 s+R^2 k_C+\eta\ \alpha_{0,\ n}^2}\ b}{\sqrt{\eta}\ R}\right)}$$

$$= \frac{(s\, e^{s\,T_1}+k_1\, e^{s\,T_1}-k_1+k_2)\ \mathtt{BesselJ}(1,\ \alpha_{0,\ m}) R^4 c_{20}}{(R^2 s+R^2 k_C+\eta\ \alpha_{0,\ m}^2)\ \alpha_{0,\ m}\, e^{s T_1}\, s} + \frac{1}{2} B_n R^2 \mathtt{BesselJ}(1,\ \alpha_{0,\ n})^2$$

```
> eq19:=subs(z=-a,rhs(eq17A)):
> -BesselJ(0,((-s-k[C])/eta)^(1/2)*r)*((s+k[1])*exp(s*T[1])+k[2]-k[1])*
c[20]*exp(-s*T[1])/s/(s+k[C])/BesselJ(0,((-s-k[C])/eta)^(1/2)*R)+
((s+k[1])*exp(s*T[1])+k[2]-k[1])*c[20]*exp(-s*T[1])/s/(s+k[C])+
Sum((A[n]*sinh(-1/eta^(1/2)*(eta*alpha[0,n]^2+s*R^2+k[C]*
R^2)^(1/2)/R*a)*I+B[n]*cosh(-1/eta^(1/2)*(eta*alpha[0,n]^2+s*
R^2+k[C]*R^2)^(1/2)/R*a))*BesselJ(0,alpha[0,n]/R*r),
n =1 .. infinity)=0:
> eq19A:=factor(subs(BesselJ(0,alpha[0,m])=0,expand(int(rhs(eq12B4)*r*
BesselJ(0,alpha[0,m]/R*r),r=0..R))))+Sum(expand(Int(r*
BesselJ(0,alpha[0,m]/R*r)*eval(subs(z=-a,_C3=A[n],E4=B[n],
rhs(eq16A))),r=0..R)),n=1..infinity)=0:
> eq19B:=subs(eq18F,eq19A):
> eq19C:=(exp(s*T[1])*s+exp(s*T[1])*k[1]+k[2]-k[1])*c[20]*R^4*
BesselJ(1,alpha[0,m])/(s*R^2+k[C]*R^2+alpha[0,m]^2*eta)/
alpha[0,m]/exp(s*T[1])/s+(1/2*(-I*A[n]*sinh(1/eta^(1/2)*
(eta*alpha[0,n]^2+s*R^2+k[C]*R^2)^(1/2)/R*a)+B[n]*
cosh(1/eta^(1/2)*(eta*alpha[0,n]^2+s*R^2+k[C]*R^2)^(1/2)/R*a))*
R^2*BesselJ(1,alpha[0,n])^2) =0;
```

$$eq19C := \frac{(s e^{s\ T_1}+k_1\, e^{s\ T_1}-k_1+k_2)\ \mathtt{BesselJ}(1,\ \alpha_{0,\ m})\ R^4 c_{20}}{(R^2 s+R^2 k_C+\eta\ \alpha_{0,\ m}^2)\ \alpha_{0,\ m}\, e^{s\ T_1}\, s} + \left(-\frac{1}{2}\ IA_n \sinh\left(\frac{\sqrt{R^2 s+R^2 k_C+\eta\ \alpha_{0,\ n}^2}\ a}{\sqrt{\eta}\ R}\right) + \frac{1}{2}\ B_n \cosh\left(\frac{\sqrt{R^2 s+R^2 k_C+\eta\ \alpha_{0,\ n}^2}\ a}{\sqrt{\eta}\ R}\right)\right) R^2 \mathtt{BesselJ}(1,\ \alpha_{0,\ n})^2 = 0$$

Equations eq18H and eq19C are solved for A[n] and B[n]:

```
> eq20 := simplify(solve({eq18H,eq19C},{A[n],B[n]})):
> eq21:=subs(eq20,eq17):
> eq21A:=subs(eq20,eq17A):
> eq26:=eq11B4;
> eq26A:=-BesselJ(0,((-s-k[C])/eta)^(1/2)*r)*c[10]*exp(-s*T[1])*
((s+k[1])*exp(s*T[1])+k[2]-k[1])/s/(s+k[C])/
BesselJ(0,((-s-k[C])/eta)^(1/2)*R);
> eq27:=eq12B4;
> eq27A:=-BesselJ(0,((-s-k[C])/eta)^(1/2)*r)*c[20]*((s+k[1])*
exp(s*T[1])+k[2]-k[1])*exp(-s*T[1])/s/(s+k[C])/
BesselJ(0,((-s-k[C])/eta)^(1/2)*R);
```

PROBLEMS

8.1. Plot the total flux $J(t)$ using the following parameter values: $\eta = 0.06048$ cm^2/day, $R = 0.235$ cm, $a = b = 0.15$ cm, $k_r(t \le T_1 = 2 \text{ days}) = 0.19872$ (1/day), $k_c = 19.9584$ (1/day), $c_{10} = 57638.7$ ng/ml, $k_r(t > T_1 = 2 \text{ days}) = 0.040608$ (1/day), $c_{20} = 115277$ ng/ml. Note: Use $N = 10$ where N is the number of terms in the series solution. The simulation time is 3 days.

8.2. Find the concentrations $c_1(r, z, t)$ and $c_2(r, z, t)$ using the parameters in Problem 8.1. Note: Use $N = 10$ where N is the number of terms in the series solution.

8.3. Plot the concentrations $c_1(r, z, t)$ and $c_2(r, z, t)$ determined in Problem 8.2 Use a simulation time of 40 days.
Note: Use $N = 10$ where N is the number of terms in the series solution.

8.4. Derive the concentration profiles $c_1(r, z, t)$ and $c_2(r, z, t)$ using the boundary conditions given in Section 8.3 with the exception that $c_1(r, b, t) = 0$ instead of $\left.\frac{\partial c_1(r,z,t)}{\partial z}\right|_{z=b} = 0$.

8.5. Derive the flux $J(t)/J(\infty)$ using the boundary conditions given in Section 8.3 with the exception that $c_1(r, b, t) = 0$ instead of $\left.\frac{\partial c_1(r,z,t)}{\partial z}\right|_{z=b} = 0$.

REFERENCES

Abate, J., & Valkó, P. P. (2004). Multi-precision Laplace transform inversion. *Int J Num Methods Eng*, 60(5), 979–993.

Chen, R. R., Silva, E. A., Yuen, W. W., Brock, A. A., Fischbach, C., Lin, A. S., . . . Mooney, D. J. (2007a). Integrated approach to designing growth factor delivery systems. *FASEB J*, 21(14), 3896–3903.

Chen, R. R., Silva, E. A., Yuen, W. W., & Mooney, D. J. (2007b). Spatio-temporal VEGF and PDGF delivery patterns blood vessel formation and maturation. *Pharm Res*, 24(2), 258–264.

Debnath, L, & Bhatta, D. (2007). *Integral transforms and their applications* (2nd ed.). Boca Raton: Chapman & Hall/CRC.

Graf, U. (2004). *Applied Laplace transforms and z-transforms for scientists and engineers : a computational approach using a Mathematica package*. Basel: Birkhäuser.

Loney, N. W. (2007). *Applied mathematical methods for chemical engineers* (2nd ed.). Boca Raton, FL: CRC/Taylor & Francis.

Rice, R. G., & Do, D. D. (1995). *Applied mathematics and modeling for chemical engineers*. New York: Wiley.

Thomson, W T. (1950). *Laplace transformation; theory and engineering applications*. New York,: Prentice-Hall.

Zauderer, E. (2006). *Partial differential equations of applied mathematics* (3rd ed.). Hoboken, N.J.: Wiley-Interscience.

9

MOLECULAR DIFFUSION THROUGH THE STOMACH LINING AND INTO THE BLOODSTREAM

9.1 LAPLACE TRANSFORMS, LEGENDRE FUNCTIONS AND SPHERICAL HARMONICS

The 3D-diffusion equation in spherical coordinates is

$$\frac{\partial}{\partial t}C(r,\theta,\phi,t) = \eta\left(\begin{array}{l} 2\frac{\frac{\partial}{\partial r}C(r,\theta,\phi,t)}{r} + \frac{\partial^2}{\partial r^2}C(r,\theta,\phi,t) + \frac{\cos(\theta)\frac{\partial}{\partial\theta}C(r,\theta,\phi,t)}{r^2\sin(\theta)} \\ + \frac{\frac{\partial^2}{\partial\theta^2}C(r,\theta,\phi,t)}{r^2} + \frac{\frac{\partial^2}{\partial\phi^2}C(r,\theta,\phi,t)}{r^2(\sin(\theta))^2} \end{array}\right) \tag{9.1}$$

The Laplace transform of Eq. (9.1) is given by

$$\begin{aligned} sC(r,\theta,\phi) - C(r,\theta,\phi,0) = {} & 2\frac{\eta\frac{\partial}{\partial r}C(r,\theta,\phi)}{r} + \eta\frac{\partial^2}{\partial r^2}C(r,\theta,\phi) + \frac{\eta\cos(\theta)\frac{\partial}{\partial\theta}C(r,\theta,\phi)}{r^2\sin(\theta)} \\ & + \frac{\eta\frac{\partial^2}{\partial\theta^2}C(r,\theta,\phi)}{r^2} + \frac{\eta\frac{\partial^2}{\partial\phi^2}C(r,\theta,\phi)}{r^2(\sin(\theta))^2} \end{aligned} \tag{9.2}$$

where

$$C(r,\theta,\phi) = \int_0^\infty C(r,\theta,\phi,t)\mathrm{e}^{-st}\mathrm{d}t \tag{9.3}$$

After considering the initial condition:

$$C(r,\theta,\phi,0) = 0 \tag{9.4}$$

Closed-form Solutions for Drug Transport through Controlled-Release Devices in Two and Three Dimensions, First Edition. Laurent Simon and Juan Ospina.

Eq. (9.2) becomes

$$sC(r,\theta,\phi) = 2\frac{\eta\frac{\partial}{\partial r}C(r,\theta,\phi)}{r} + \eta\frac{\partial^2}{\partial r^2}C(r,\theta,\phi) + \frac{\eta\cos(\theta)\frac{\partial}{\partial\theta}C(r,\theta,\phi)}{r^2\sin(\theta)} + \frac{\eta\frac{\partial^2}{\partial\theta^2}C(r,\theta,\phi)}{r^2} + \frac{\eta\frac{\partial^2}{\partial\phi^2}C(r,\theta,\phi)}{r^2(\sin(\theta))^2} \tag{9.5}$$

We make the following change of variables $x = \cos(\theta)$ and obtain

$$sC(x,r,\phi) = \frac{\eta}{r^2(x^2-1)}\begin{pmatrix} -2\left(\frac{\partial}{\partial r}C(x,r,\phi)\right)r + 2\left(\frac{\partial}{\partial r}C(x,r,\phi)\right)rx^2 - \left(\frac{\partial^2}{\partial r^2}C(x,r,\phi)\right)r^2 \\ +\left(\frac{\partial^2}{\partial r^2}C(x,r,\phi)\right)r^2x^2 + 2\left(\frac{\partial}{\partial x}C(x,r,\phi)\right)x - 2\left(\frac{\partial}{\partial x}C(x,r,\phi)\right)x^3 \\ -\frac{\partial^2}{\partial x^2}C(x,r,\phi) + 2\left(\frac{\partial^2}{\partial x^2}C(x,r,\phi)\right)x^2 - \left(\frac{\partial^2}{\partial x^2}C(x,r,\phi)\right)x^4 \\ -\frac{\partial^2}{\partial\phi^2}C(x,r,\phi) \end{pmatrix} \tag{9.6}$$

To solve Eq. (9.6), a solution of the form $C(x,r,\phi) = f(x)g(r)H(\phi)$ is assumed where the new functions satisfy the following equations:

$$\frac{d^2}{dr^2}f(r) = -\frac{c_2 f(r)}{r^2\eta} + \frac{sf(r)r - 2\eta\frac{d}{dr}f(r)}{\eta r} \tag{9.7}$$

$$\frac{d^2}{dx^2}g(x) = -\frac{c_2 g(x)}{(x^2-1)\eta} - \frac{c_3 g(x)}{x^4 - 2x^2 + 1} - 2\frac{x\frac{d}{dx}g(x)}{x^2-1} \tag{9.8}$$

and

$$\frac{d^2}{d\phi^2}H(\phi) = c_3 H(\phi) \tag{9.9}$$

After solving the system formed by Eqs. (9.7), (9.8) and (9.9), $C(x,r,\phi)$ becomes

$$C(x,r,\phi) = \frac{C_1 e^{\phi\sqrt{c_3}} + C_2 e^{-\phi\sqrt{c_3}}}{\sqrt{r}}\left(C_5 P^{i\sqrt{c_3}}_{\frac{1}{2}\frac{\sqrt{\eta-4c_2}-\sqrt{\eta}}{\sqrt{\eta}}}(x) + C_6 Q^{i\sqrt{c_3}}_{\frac{1}{2}\frac{\sqrt{\eta-4c_2}-\sqrt{\eta}}{\sqrt{\eta}}}(x)\right) \times\left(C_3 J_{\frac{1}{2}\sqrt{\frac{\eta-4c_2}{\eta}}}\left(r\sqrt{-\frac{s}{\eta}}\right) + C_4 Y_{\frac{1}{2}\sqrt{\frac{\eta-4c_2}{\eta}}}\left(r\sqrt{-\frac{s}{\eta}}\right)\right) \tag{9.10}$$

which includes the **associated Legendre function of the first kind** P_n^m and the **associated Lengendre function of the second kind** Q_n^m (see chapter 7); J and Y are Bessel functions.

The periodicity in ϕ is enforced by setting

$$\sqrt{c_3} = -mi \tag{9.11}$$

As a result, Eq. (9.10) is rewritten as

$$C(x,r,\phi)=\frac{C_1\mathrm{e}^{-im\phi}+C_2\mathrm{e}^{im\phi}}{\sqrt{r}}\left(C_5P^m_{\frac{1}{2}\frac{\sqrt{\eta-4c_2}-\sqrt{\eta}}{\sqrt{\eta}}}(x)+C_6Q^m_{\frac{1}{2}\frac{\sqrt{\eta-4c_2}-\sqrt{\eta}}{\sqrt{\eta}}}(x)\right)$$
$$\times\left(C_3J_{\frac{1}{2}\sqrt{\frac{\eta-4c_2}{\eta}}}\left(i\times r\sqrt{\frac{s}{\eta}}\right)+C_4Y_{\frac{1}{2}\sqrt{\frac{\eta-4c_2}{\eta}}}\left(i\times r\sqrt{\frac{s}{\eta}}\right)\right) \tag{9.12}$$

As stated in Chapter 7, the Legendre functions are defined only when

$$\frac{1}{2}\frac{\sqrt{\eta-4c_2}-\sqrt{\eta}}{\sqrt{\eta}}=L \tag{9.13}$$

where $L=0,1,2,3,\ \ldots$ In this case, we have

$$C(x,r,\phi)=\frac{C_1\mathrm{e}^{-im\phi}+C_2\mathrm{e}^{im\phi}}{\sqrt{r}}(C_5P^m_L(x)+C_6Q^m_L(x))$$
$$\times\left(C_3J_{L+\frac{1}{2}}\left(i\times r\sqrt{\frac{s}{\eta}}\right)+C_4Y_{L+\frac{1}{2}}\left(i\times r\sqrt{\frac{s}{\eta}}\right)\right) \tag{9.14}$$

Without loss of generality, Eq. (9.14) is written as (see Chapter 7)

$$C(r,\theta,\phi)=\frac{\mathrm{e}^{im\phi}}{\sqrt{r}}(C_5P^m_L(\cos(\theta))+C_6Q^m_L(\cos(\theta)))$$
$$\times\left(C_3J_{L+\frac{1}{2}}\left(i\times r\sqrt{\frac{s}{\eta}}\right)+C_4Y_{L+\frac{1}{2}}\left(i\times r\sqrt{\frac{s}{\eta}}\right)\right) \tag{9.15}$$

Note that C_1 is set equal to zero because, when applying the **principle of superposition**, the sums $\sum_{m=-L}^{L}\mathrm{e}^{im\phi}$ and $\sum_{m=-L}^{L}\mathrm{e}^{-im\phi}$ are equal. The functions $Q^m_L(\cos(\theta))$ must be eliminated because of singularity at $\theta=0$ or $\theta=\pi$. Therefore,

$$C(r,\theta,\phi)=\frac{\mathrm{e}^{im\phi}}{\sqrt{r}}P^m_L(\cos(\theta))\left(C_3J_{L+\frac{1}{2}}\left(i\times r\sqrt{\frac{s}{\eta}}\right)+C_4Y_{L+\frac{1}{2}}\left(i\times r\sqrt{\frac{s}{\eta}}\right)\right) \tag{9.16}$$

In terms of **spherical harmonics**, Eq. (9.16) is

$$C(r,\theta,\phi)=\frac{Y^m_L(\theta,\phi)}{\sqrt{r}}\left(C_3J_{L+\frac{1}{2}}\left(i\times r\sqrt{\frac{s}{\eta}}\right)+C_4Y_{L+\frac{1}{2}}\left(i\times r\sqrt{\frac{s}{\eta}}\right)\right) \tag{9.17}$$

The integration constants C_3 and C_4 must be determined using the appropriate boundary conditions and, in general, they are functions of s. Application of the principle of superposition leads to

$$C(r,\theta,\phi)=\sum_{L=0}^{\infty}\left[\sum_{m=-L}^{L}\frac{Y^m_L(\theta,\phi)}{\sqrt{r}}\left(A_{L,m}J_{L+\frac{1}{2}}\left(i\times r\sqrt{\frac{s}{\eta}}\right)+B_{L,m}Y_{L+\frac{1}{2}}\left(i\times r\sqrt{\frac{s}{\eta}}\right)\right)\right] \tag{9.18}$$

The inverse Laplace transform of Eq. (9.18) is formally computed using the following Bromwich integral:

$$C(r,\theta,\phi,t) = \frac{1}{2}\frac{1}{\pi i}\sum_{L=0}^{\infty}\left(\sum_{m=-L}^{L}\left(\frac{Y_L^m(\theta,\phi)}{\sqrt{r}}\left(\begin{array}{l}\int J_{L+\frac{1}{2}}\left(i\times r\sqrt{\frac{s}{\eta}}\right)A_{L,m}(s)e^{st}ds+\\ \int B_{L,m}(s)Y_{L+\frac{1}{2}}\left(i\times r\sqrt{\frac{s}{\eta}}\right)e^{st}ds\end{array}\right)\right)\right) \tag{9.19}$$

The integrals in Eq. (9.19) can be computed using the residue theorem and the final solution is given in terms of a double series involving **spherical Bessel functions** (i.e., $\sqrt{\frac{\pi}{2}}\frac{J_{n+1/2}(z)}{\sqrt{z}}$ and $\sqrt{\frac{\pi}{2}}\frac{Y_{n+1/2}(z)}{\sqrt{z}}$) and **spherical harmonics**.

9.2 SPHERICAL DIFFUSION IN THREE DIMENSIONS

Transport of a molecule through the stomach lining was analyzed at steady state. A source term Q was added to account for the rate at which a molecule enters the stomach (Fig. 7.1). In this chapter, the focus is on the system dynamic behavior:

$$\frac{\partial c}{\partial t} = \eta\left[\frac{1}{r^2}\frac{\partial}{\partial r}\left(r^2\frac{\partial c}{\partial r}\right) + \frac{1}{r^2\sin\theta}\frac{\partial}{\partial\theta}\left(\sin\theta\frac{\partial c}{\partial r}\right) + \frac{1}{r^2\sin^2\theta}\frac{\partial^2 c}{\partial\phi^2}\right] + Q \tag{9.20}$$

The constant η is the diffusion coefficient in the walls of the stomach. The angles are defined in the ranges: $0 \le \theta \le \pi$ and $0 \le \phi \le 2\pi$. The input rate Q is described by the following equation:

$$Q(0 \le r \le r_1) = g(\theta,\ \phi,\ t);\quad Q(r_1 < r < R) = 0 \tag{9.21}$$

where r_1 and R are the inner and outer radii of the spherically-shaped stomach. The boundary conditions are

$$\left.\frac{\partial c(r,\ \theta,\ \phi,\ t)}{\partial r}\right|_{r=0} = 0 \tag{9.22}$$

$$c(R,\ \theta,\ \phi,\ t) = 0 \tag{9.23}$$

$$c(R,\ \theta,\ \phi,\ t) = c(R,\ \theta,\ \phi + 2\pi,\ t) \tag{9.24}$$

$$c(R,\ 0,\ \phi) = \text{finite value} \tag{9.25}$$

and

$$c(R,\ \pi,\ \phi) = \text{finite value} \tag{9.26}$$

Initially, the system is free of the diffusant:

$$c(r,\theta,\phi,0) = 0 \tag{9.27}$$

Examples of problems modeled by transport equations with source terms are photon transport (Fantini et al., 1997), multilayer heat conduction (Jain et al., 2010) and protein diffusion (Calvert et al., 2007).

9.3 ANALYTICAL SOLUTION OF THE TRANSIENT TRANSPORT MODEL IN 3-D SPHERICAL COORDINATES

The Laplace transform of Eq. (9.20) is

$$sC(r,\theta,\phi) = 2\frac{\eta\frac{\partial}{\partial r}C(r,\theta,\phi)}{r} + \eta\frac{\partial^2}{\partial r^2}C(r,\theta,\phi) + \frac{\eta\cos(\theta)\frac{\partial}{\partial\theta}C(r,\theta,\phi)}{r^2\sin(\theta)}$$
$$+\frac{\eta\frac{\partial^2}{\partial\theta^2}C(r,\theta,\phi)}{r^2} + \frac{\eta\frac{\partial^2}{\partial\phi^2}C(r,\theta,\phi)}{r^2(\sin(\theta))^2} + Q(\theta,\phi,s) \tag{9.28}$$

after using Eqs. (9.3) and (9.27), where

$$Q(\theta,\phi,s) = \int_0^\infty Q(r,\theta,\phi,t)\mathrm{e}^{-st}\mathrm{d}t \tag{9.29}$$

Note that Eq. (9.28) has the form

$$\nabla^2 C = \frac{s}{\eta}C - \frac{Q}{\eta} \tag{9.30}$$

The solution of Eq. (9.30) is developed using the **Green's function** defined by

$$\nabla^2 G = \frac{s}{\eta}G + \delta \tag{9.31}$$

Then, after applying the boundary condition (9.23), the solution of Eq. (9.30) is given by (7.31) (i.e., $C(\vec{r}) = \iiint_{v'} G(\vec{r},\vec{r}\,')V(\vec{r}\,')dv'$) with

$$V = -\frac{Q}{\eta} \tag{9.32}$$

To determine the explicit form of the Green's function given by Eq. (9.31), we assume a solution to Eq. (9.31) of the form given in Eq. (7.37) where the function $f_{Lm}(r,r')$, denoted below as $f(r)$, satisfies the ordinary differential equation:

$$\frac{2\frac{\mathrm{d}}{\mathrm{d}r}f(r) + r\frac{\mathrm{d}^2}{\mathrm{d}r^2}f(r)}{r} - \frac{L(L+1)f(r)}{r^2} = \frac{\delta(r-r')}{r^2} + \frac{sf(r)}{\eta} \tag{9.33}$$

which can be solved to get

$$f(r) = \frac{C_2}{\sqrt{r}}J_{L+\frac{1}{2}}\left(r\sqrt{-\frac{s}{\eta}}\right) + \frac{C_1}{\sqrt{r}}Y_{L+\frac{1}{2}}\left(r\sqrt{-\frac{s}{\eta}}\right) - \frac{1}{2}\frac{\pi(\mathrm{Heaviside}(\sqrt{r}-\sqrt{r'})+1)}{\sqrt{r'}\sqrt{r}}$$
$$\times\left(Y_{L+\frac{1}{2}}\left(r'\sqrt{-\frac{s}{\eta}}\right)J_{L+\frac{1}{2}}\left(r\sqrt{-\frac{s}{\eta}}\right) - J_{L+\frac{1}{2}}\left(r'\sqrt{-\frac{s}{\eta}}\right)Y_{L+\frac{1}{2}}\left(r\sqrt{-\frac{s}{\eta}}\right)\right) \tag{9.34}$$

See Chapter 1 for the definition of the **Heaviside function**. When $r < r'$, Eq. (9.34) is reduced to

$$f(r) = \frac{C_2}{\sqrt{r}} J_{L+\frac{1}{2}}\left(r\sqrt{-\frac{s}{\eta}}\right) + \frac{C_1}{\sqrt{r}} Y_{L+\frac{1}{2}}\left(r\sqrt{-\frac{s}{\eta}}\right) - \frac{\pi}{2\sqrt{r'}\sqrt{r}} \times \left(Y_{L+\frac{1}{2}}\left(r'\sqrt{-\frac{s}{\eta}}\right) J_{L+\frac{1}{2}}\left(r\sqrt{-\frac{s}{\eta}}\right) - J_{L+\frac{1}{2}}\left(r'\sqrt{-\frac{s}{\eta}}\right) Y_{L+\frac{1}{2}}\left(r\sqrt{-\frac{s}{\eta}}\right)\right) \tag{9.35}$$

To avoid singularity at $r = 0$, the terms containing the function $Y_{L+\frac{1}{2}}\left(r\sqrt{-\frac{s}{\eta}}\right)$ must be eliminated according to

$$\frac{C_1}{\sqrt{r}} + \frac{\pi}{2\sqrt{r'}\sqrt{r}} J_{L+\frac{1}{2}}\left(r'\sqrt{-\frac{s}{\eta}}\right) = 0 \tag{9.36}$$

which leads to

$$C_1 = -\frac{\pi}{2\sqrt{r'}} J_{L+\frac{1}{2}}\left(r'\sqrt{-\frac{s}{\eta}}\right) \tag{9.37}$$

Replacing C_1 in Eq. (9.34), we obtain

$$f(r) = \frac{C_2}{\sqrt{r}} J_{L+\frac{1}{2}}\left(r\sqrt{-\frac{s}{\eta}}\right) - \frac{\pi}{2\sqrt{r}\sqrt{r'}} J_{L+\frac{1}{2}}\left(r'\sqrt{-\frac{s}{\eta}}\right) Y_{L+\frac{1}{2}}\left(r\sqrt{-\frac{s}{\eta}}\right) - \frac{1}{2}\frac{\pi(\text{Heaviside}(\sqrt{r}-\sqrt{r'})+1)}{\sqrt{r'}\sqrt{r}} \times \left(Y_{L+\frac{1}{2}}\left(r'\sqrt{-\frac{s}{\eta}}\right) J_{L+\frac{1}{2}}\left(r\sqrt{-\frac{s}{\eta}}\right) - J_{L+\frac{1}{2}}\left(r'\sqrt{-\frac{s}{\eta}}\right) Y_{L+\frac{1}{2}}\left(r\sqrt{-\frac{s}{\eta}}\right)\right) \tag{9.38}$$

Now, when $r > r'$, Eq. (9.38) becomes

$$f(r) = \frac{C_2}{\sqrt{r}} J_{L+\frac{1}{2}}\left(r\sqrt{-\frac{s}{\eta}}\right) - \frac{\pi}{2\sqrt{r}\sqrt{r'}} J_{L+\frac{1}{2}}\left(r'\sqrt{-\frac{s}{\eta}}\right) Y_{L+\frac{1}{2}}\left(r\sqrt{-\frac{s}{\eta}}\right) - \frac{\pi}{\sqrt{r'}\sqrt{r}}\left(Y_{L+\frac{1}{2}}\left(r'\sqrt{-\frac{s}{\eta}}\right) J_{L+\frac{1}{2}}\left(r\sqrt{-\frac{s}{\eta}}\right) - J_{L+\frac{1}{2}}\left(r'\sqrt{-\frac{s}{\eta}}\right) Y_{L+\frac{1}{2}}\left(r\sqrt{-\frac{s}{\eta}}\right)\right) \tag{9.39}$$

Application of the boundary condition $f(R) = 0$ results in

$$\frac{C_2}{\sqrt{R}} J_{L+\frac{1}{2}}\left(R\sqrt{-\frac{s}{\eta}}\right) - \frac{\pi}{2\sqrt{r}\sqrt{r'}} J_{L+\frac{1}{2}}\left(r'\sqrt{-\frac{s}{\eta}}\right) Y_{L+\frac{1}{2}}\left(R\sqrt{-\frac{s}{\eta}}\right) - \frac{\pi}{\sqrt{r'}\sqrt{R}}\left(Y_{L+\frac{1}{2}}\left(r'\sqrt{-\frac{s}{\eta}}\right) J_{L+\frac{1}{2}}\left(R\sqrt{-\frac{s}{\eta}}\right) - J_{L+\frac{1}{2}}\left(r'\sqrt{-\frac{s}{\eta}}\right) Y_{L+\frac{1}{2}}\left(R\sqrt{-\frac{s}{\eta}}\right)\right) = 0 \tag{9.40}$$

Solving Eq. (9.40) with respect to C_2, we get

$$C_2 = \frac{\sqrt{R}}{J_{L+\frac{1}{2}}\left(R\sqrt{-\frac{s}{\eta}}\right)} \times \left[\begin{array}{l} \frac{\pi}{2\sqrt{r}\sqrt{r'}} J_{L+\frac{1}{2}}\left(r'\sqrt{-\frac{s}{\eta}}\right) Y_{L+\frac{1}{2}}\left(R\sqrt{-\frac{s}{\eta}}\right) \\ +\frac{\pi}{\sqrt{r'}\sqrt{R}}\left(Y_{L+\frac{1}{2}}\left(r'\sqrt{-\frac{s}{\eta}}\right) J_{L+\frac{1}{2}}\left(R\sqrt{-\frac{s}{\eta}}\right) - J_{L+\frac{1}{2}}\left(r'\sqrt{-\frac{s}{\eta}}\right) Y_{L+\frac{1}{2}}\left(R\sqrt{-\frac{s}{\eta}}\right)\right) \end{array}\right] \tag{9.41}$$

As a result, Eq. (9.38) can be written as

$$\begin{aligned} f(r) &= f_{Lm}(r, r') \\ &= \left(\begin{array}{l} \frac{\sqrt{R} J_{L+\frac{1}{2}}\left(r\sqrt{-\frac{s}{\eta}}\right)}{J_{L+\frac{1}{2}}\sqrt{r}\left(R\sqrt{-\frac{s}{\eta}}\right)} \\ \times\left[\begin{array}{l} \frac{\pi}{2\sqrt{r}\sqrt{r'}} J_{L+\frac{1}{2}}\left(r'\sqrt{-\frac{s}{\eta}}\right) Y_{L+\frac{1}{2}}\left(R\sqrt{-\frac{s}{\eta}}\right) \\ +\frac{\pi}{\sqrt{r'}\sqrt{R}}\left(Y_{L+\frac{1}{2}}\left(r'\sqrt{-\frac{s}{\eta}}\right) J_{L+\frac{1}{2}}\left(R\sqrt{-\frac{s}{\eta}}\right) - J_{L+\frac{1}{2}}\left(r'\sqrt{-\frac{s}{\eta}}\right) Y_{L+\frac{1}{2}}\left(R\sqrt{-\frac{s}{\eta}}\right)\right) \end{array}\right] \end{array}\right) \\ &\quad - \frac{\pi}{2\sqrt{r}\sqrt{r'}} J_{L+\frac{1}{2}}\left(r'\sqrt{-\frac{s}{\eta}}\right) Y_{L+\frac{1}{2}}\left(r\sqrt{-\frac{s}{\eta}}\right) - \frac{1}{2}\frac{\pi(\mathrm{Heaviside}(\sqrt{r}-\sqrt{r'})+1)}{\sqrt{r'}\sqrt{r}} \\ &\quad \times\left(Y_{L+\frac{1}{2}}\left(r'\sqrt{-\frac{s}{\eta}}\right) J_{L+\frac{1}{2}}\left(r\sqrt{-\frac{s}{\eta}}\right) - J_{L+\frac{1}{2}}\left(r'\sqrt{-\frac{s}{\eta}}\right) Y_{L+\frac{1}{2}}\left(r\sqrt{-\frac{s}{\eta}}\right)\right) \end{aligned} \tag{9.42}$$

The explicit form of the Green's function is derived after inserting $f(r)$ (denoted $f_{Lm}(r, r')$) into Eq. (7.37):

$$\begin{aligned} & G(r, \theta, \phi, r', \theta', \phi') \\ &= \sum_{L=0}^{\infty}\sum_{m=-L}^{L} Y_L^m(\theta, \phi) Y *_L^m (\theta', \phi') \\ &\times\left\{\begin{array}{l} \left(\begin{array}{l} \frac{\sqrt{R} J_{L+\frac{1}{2}}\left(r\sqrt{-\frac{s}{\eta}}\right)}{J_{L+\frac{1}{2}}\sqrt{r}\left(R\sqrt{-\frac{s}{\eta}}\right)} \\ \times\left[\begin{array}{l} \frac{\pi}{2\sqrt{r}\sqrt{r'}} J_{L+\frac{1}{2}}\left(r'\sqrt{-\frac{s}{\eta}}\right) Y_{L+\frac{1}{2}}\left(R\sqrt{-\frac{s}{\eta}}\right) \\ +\frac{\pi}{\sqrt{r'}\sqrt{R}}\left(Y_{L+\frac{1}{2}}\left(r'\sqrt{-\frac{s}{\eta}}\right) J_{L+\frac{1}{2}}\left(R\sqrt{-\frac{s}{\eta}}\right) - J_{L+\frac{1}{2}}\left(r'\sqrt{-\frac{s}{\eta}}\right) Y_{L+\frac{1}{2}}\left(R\sqrt{-\frac{s}{\eta}}\right)\right) \end{array}\right] \end{array}\right) \\ \quad - \frac{\pi}{2\sqrt{r}\sqrt{r'}} J_{L+\frac{1}{2}}\left(r'\sqrt{-\frac{s}{\eta}}\right) Y_{L+\frac{1}{2}}\left(r\sqrt{-\frac{s}{\eta}}\right) - \frac{1}{2}\frac{\pi(\mathrm{Heaviside}(\sqrt{r}-\sqrt{r'})+1)}{\sqrt{r'}\sqrt{r}} \\ \quad \times\left(Y_{L+\frac{1}{2}}\left(r'\sqrt{-\frac{s}{\eta}}\right) J_{L+\frac{1}{2}}\left(r\sqrt{-\frac{s}{\eta}}\right) - J_{L+\frac{1}{2}}\left(r'\sqrt{-\frac{s}{\eta}}\right) Y_{L+\frac{1}{2}}\left(r\sqrt{-\frac{s}{\eta}}\right)\right) \end{array}\right\} \end{aligned} \tag{9.43}$$

Using Eqs. (9.43) and (7.31), the solution of the system formed by Eqs. (9.20)–(9.27) in the Laplace domain is formally given by

$$C(r,\theta,\phi) = -\frac{1}{\eta}\sum_{L=0}^{\infty}\left(\begin{array}{c}\sum_{m=-L}^{L}\int_0^{2\pi}\int_0^{\pi} g\left(\theta',\phi',s\right)\sin(\theta') \\ \times\int_{\varepsilon}^{R_1} G(r,\theta,\phi,r',\theta',\phi')r'^2 dr' d\theta' d\phi'\end{array}\right) \tag{9.44}$$

or

$$C(r,\theta,\phi) = -\frac{1}{\eta}\sum_{L=0}^{\infty}\left(\begin{array}{c}\sum_{m=-L}^{L} Y_L^m(\theta,\phi)\int_0^{2\pi}\int_0^{\pi} Y*_L^m(\theta',\phi')g(\theta',\phi',s)\sin(\theta')d\theta' d\phi' \\ \times\int_{\varepsilon}^{R_1} f_{Lm}(r,r')r'^2 dr'\end{array}\right) \tag{9.45}$$

The solution in the time domain is derived by taking the inverse Laplace transform of $C(r,\theta,\phi)$:

$$c(r,\theta,\phi,t) = \frac{1}{2\pi i}\int e^{st} C(r,\theta,\phi)ds \tag{9.46}$$

Calculation of the concentration may be computationally intensive based on the integrals involved in Eq. (9.45).

Consider the model described above with $Q(0 \leq r \leq 0.2) = 1$, $Q(0.2 < r < 1) = 0$ and $\eta = 1$. In Mathematica, the concentration in the Laplace domain is written as

```
Ca2[r_, theta_, phi_, s_] :
= -(Sum[Sum[SphericalHarmonicY[L, m, theta, phi] *
(NIntegrate[NIntegrate
[Conjugate[SphericalHarmonicY[L, m, Theta, Phi]] *
g[Theta, Phi, s] * Sin[Theta], {Theta, 0, Pi}],
{Phi, 0, 2 * Pi}]) *
(NIntegrate
[(BesselJ[L + 1/2, Sqrt[-s/eta] * r] *
((1/2) * BesselY[L + 1/2, Sqrt[-s/eta] * R] * Pi *
BesselJ[L + 1/2, rp * Sqrt[-s/eta]]/(Sqrt[R] * Sqrt[rp])+
Pi * (BesselY[L + 1/2, rp * Sqrt[-s/eta]] *
BesselJ[L + 1/2, Sqrt[-s/eta] * R]-
BesselJ[L + 1/2, rp * Sqrt[-s/eta]] *
BesselY[L + 1/2, Sqrt[-s/eta] * R])/
(Sqrt[R] * Sqrt[rp])) *
Sqrt[R]/(Sqrt[r] * BesselJ[L + 1/2, Sqrt[-s/eta] * R])-
```

$$(1/2) * \text{BesselY}[L + 1/2, \text{Sqrt}[-s/\text{eta}] * r] * \text{Pi} *$$
$$\text{BesselJ}[L + 1/2, \text{rp} * \text{Sqrt}[-s/\text{eta}]]/(\text{Sqrt}[r] * \text{Sqrt}[\text{rp}])-$$
$$(1/2) * \text{Pi} * (\text{HeavisideTheta}[\text{Sqrt}[r] - \text{Sqrt}[\text{rp}]] + 1) *$$
$$(\text{BesselY}[L + 1/2, \text{rp} * \text{Sqrt}[-s/\text{eta}]] *$$
$$\text{BesselJ}[L + 1/2, \text{Sqrt}[-s/\text{eta}] * r]-$$
$$\text{BesselJ}[L + 1/2, \text{rp} * \text{Sqrt}[-s/\text{eta}]] *$$
$$\text{BesselY}[L + 1/2, \text{Sqrt}[-s/\text{eta}] * r])/(\text{Sqrt}[r] * \text{Sqrt}[\text{rp}])) *$$
$$\text{rp}\hat{}2, \{\text{rp}, \text{epsilon}, \text{R1}\}]), \{m, -L, L\}], \{L, 0, \text{infinity}\}])/$$
$$\text{eta} \tag{9.47}$$

Note that *rp* is used instead of r'. In addition, Eq. (9.47) is implemented using $g(\theta', \phi', s) = 1/s$, $\varepsilon = 0$ and $L = 2$. The integrals are evaluated with the command "NIntegrate". The inverse Laplace transform is computed using the Zakian method (Rice & Do, 1995):

$$\Psi(t) = \frac{2}{t}\sum_{in=1}^{5} \text{Re}(K_{in}\overline{\Psi}(a_{in}/t)) \tag{9.48}$$

where K_{in} and a_{in} are complex numbers defined in the Zakian algorithm:

$$a_{in} = \{12.83767675 + 1.666063445i, 12.22613209 + 5.012718792i,$$
$$10.93430308 + 8.409673116i, 8.776434715 + 11.92185389i,$$
$$5.225453361 + 15.72952905i\}; \tag{9.49}$$

and

$$K_{in} = \{-36902.0821 + 196990.4257i, 61277.02524 - 95408.62551i,$$
$$-28916.56288 + 18169.18531i, +4655.361138 - 1.901528642i,$$
$$-118.7414011 - 141.3036911i\}; \tag{9.50}$$

The concentration in Eq. (9.46) is

$$\text{Caza}[\text{r\_}, \text{theta\_}, \text{phi\_}, \text{t\_}] := \frac{2}{t}\sum_{i=1}^{5} \text{Re}\left[\text{Kin}\,[[i]]\,\text{Ca2}\left[r, \text{theta}, \text{phi}, \frac{\text{ain}\,[[i]]}{t}\right]\right]; \tag{9.51}$$

by using Mathematica. Based on this approach, $c\left(\frac{1}{2}, \frac{\pi}{2}, \frac{\pi}{3}, 5\right)$ is estimated as 0.00269. It is worth noting that the model with $Q(0 \le r \le 0.2) = 1; Q(0.2_1 < r < 1) = 0$ is a one-dimensional problem and is independent of θ and ϕ. A solution procedure for the 1-D model, written in Maple, is outlined in Appendix 9.5.2. This technique yields $c\left(r = \frac{1}{2}, t = 5\right) = 0.00267$.

9.4 SUMMARY

The transient diffusion of a drug in a spherical geometry was analyzed. The model contained a source term, Q, which varied in time and space. This non-homogeneous form of the Laplace's equation, called the Poisson's equation, was solved after explicit evaluation of the Green's function. The concentration was obtained as a function of the Laplace transform variable. An inversion procedure was necessary to invert the expression in the time domain. Implementation of the final result required the application of numerical integration using an advanced computational package, such as Mathematica. Depending on the shape of Q, transport may occur along a single direction. In such cases, techniques, available for one-dimensional partial differential equations, may be adequate to solve the problem.

9.5 APPENDIX: MAPLE CODE LISTINGS

9.5.1 Maple Code: threesphertransient.mws

```
> restart:with(VectorCalculus):with(inttrans):with(PDETools):
> Eq:=diff(C(r,theta,phi,t),t)=eta*expand(Laplacian(C(r,theta,phi,t),
    spherical[r,theta,phi]));
```

$$Eq := \frac{\partial}{\partial t} C(r, \theta, \phi, t) = \eta \left(\frac{2\left(\frac{\partial}{\partial r} C(r, \theta, \phi, t)\right)}{r} + \frac{\partial^2}{\partial r^2} C(r, \theta, \phi, t) + \frac{\cos(\theta)\left(\frac{\partial}{\partial \theta} C(r, \theta, \phi, t)\right)}{r^2 \sin(\theta)} + \frac{\frac{\partial^2}{\partial \theta^2} C(r, \theta, \phi, t)}{r^2} + \frac{\frac{\partial^2}{\partial \phi^2} C(r, \theta, \phi, t)}{r^2 \sin(\theta)^2} \right)$$

```
> Eq1:=laplace(Eq,t,s):
> Eq2:=subs(laplace(C(r,theta,phi,t),t,s)=C(r,theta,phi),Eq1):
> Eq3:=C(r,theta,phi)=Int(C(r,theta,phi,t)*exp(-s*t),t=0..infinity):
> Eq4:=C(r,theta,phi,0)=0:
> Eq5:=subs(Eq4,Eq2):
> x=cos(theta):
> tr := {theta = arccos(x)}:
Eq6:=simplify(dchange(tr,Eq5)):
> C(r,x,phi)=f(r)*g(x)*H(phi):
> Eq7:=pdsolve(Eq6,HINT=f(r)*g(x)*H(phi)):
> Eq8:=factor(build(Eq7)):
> aux:=_C1*exp(_c[3]^(1/2)*phi)^2+_C2=exp(_c[3]^(1/2)*phi)*
    (_C1*exp(_c[3]^(1/2)*phi)+_C2*exp(-_c[3]^(1/2)*phi)):
> Eq9:=subs(aux,Eq8):
> Eq10:=_c[3]^(1/2)=-m*I:
>
> Eq11:=isolate(Eq10,_c[3]^(1/2)):
> Eq12:=simplify(subs(Eq11,subs(Eq10,Eq9)),power,symbolic):
> Eq13:=1/2*((eta-4*_c[2])^(1/2)-eta^(1/2))/eta^(1/2)=L:
>
> Eq14:=_c[2]=factor(solve(Eq13,_c[2])):
>
> Eq15:=simplify(simplify(subs(Eq14,Eq12),power,symbolic)):
```

```
> Eq16:=C(r,theta,phi)=subs(x=cos(theta),rhs(subs(_C1=0, _C2=1,Eq15))):
> Eq17:=subs(_C6=0,_C5=1,Eq16):
> aux1:=exp(m*phi*I)=Y[L,m](theta,phi)/LegendreP(L,m,cos(theta)):
> Eq18:=subs(aux1,Eq17):
> Eq19:=C(r,theta,phi)=Sum(Sum(subs(_C3=A[L,m](s),_C4=B[L.m](s),
    rhs(Eq18)),m=-L..L),L=0..infinity):
> Eq20:=C(r,theta,phi,t)=(1/(2*pi*i))*Sum(Sum(Y[L,m](theta,phi)*
    (Int(BesselJ(L+1/2,s^(1/2)/eta^(1/2)*r*I)*A[L,m](s)*exp(s*t),s)+
    Int(B[L*m](s)*BesselY(L+1/2,s^(1/2)/eta^(1/2)*r*I)*exp(s*t),s))/
    r^(1/2),m = -L .. L),L =0 .. infinity):
> Eq21:=diff(C(r,theta,phi,t),t)=eta*expand(Laplacian(C(r,theta,phi,t),
    spherical[r,theta,phi]))+Q(theta,phi,t):
> Eq22:=laplace(Eq21,t,s):
> Eq23:=subs(laplace(Q(theta,phi,t),t,s)=Q(theta,phi,s),
    subs(C(r,theta,phi,0)=0,subs(laplace(C(r,theta,phi,t),t,s)=
    C(r,theta,phi),Eq22))):
> Eq24:=Q(theta,phi,s)=Int(Q(r,theta,phi,t)*exp(-s*t),t=0..infinity):
> V = -Q/eta:
> eq:=factor(Laplacian(f(r),spherical[r,theta,phi]))-L*(L+1)*f(r)/r^2=
    Dirac(r-'r'')/r^2+s*f(r)/eta:
>
> eq1:=eval(subs(Int=int,dsolve(eq))) assuming r>0 and L>0 and 'r''>0 :
>
> eq2:=eq1 assuming r < 'r'':
> eq2A:=Heaviside(r^(1/2)-'r''^(1/2)) assuming sqrt(r)<sqrt('r''):
> eq2B:=subs(Heaviside(r^(1/2)-'r''^(1/2))=0,eq2):
> eq2C:=collect(rhs(eq2B),BesselY(L+1/2,(-1/eta*s)^(1/2)*r)):
> eq3:=coeff(eq2C,BesselY(L+1/2,(-s/eta)^(1/2)*r),1)=0:
> eq4:=isolate(eq3,_C1):
> eq5:=subs(eq4,eq1):
> eq6:=eq5 assuming r > 'r'':
> eq6A:=subs(Heaviside(r^(1/2)-'r''^(1/2))=1,eq6):
>
> eq7:=subs(r=R,rhs(eq6A))=0:
> eq7A:=isolate(eq7,_C2):
> eq8:=subs(eq7A,eq5):
> eq11:=G(r,'r'',theta,Theta,phi,Phi)=Sum(Sum(SphericalY(L,m,theta,
    phi)*conjugate(SphericalY(L,m,Theta,Phi))*rhs(eq8),m=-L..L),
    L=0..infinity):
> eq12:=c(r,theta,phi,s)=-(1/eta)*Int(Int(Int(('r'')^2*g(Theta,Phi,s)*
    rhs(eq11)*sin(Theta),'r''=0..R[1]),Theta=0..Pi),Phi=0..2*Pi):
>
> eq13:=C(r,theta,phi)=-(1/eta)*Sum(Sum(SphericalY(L,m,theta,phi)*
    Int(Int(conjugate(SphericalY(L,m,Theta,Phi))*g(Theta,Phi,s)*
    sin(Theta),Theta=0..Pi),Phi=0..2*Pi)*Int(rhs(eq8)*('r'')^2,
    'r''=epsilon..R[1]),m = -L .. L),L =0 .. infinity):
> eq15:=C(r,theta,phi,t)=(1/(2*Pi*i))*Int(exp(s*t)*C(r,theta,phi),s):
> eq16:=subs(eq13,eq15);
```

$$eq16 := C(r,\ \theta,\ \phi,\ t) = \frac{1}{2}\ \frac{1}{\pi i}\ \Bigg(\int \Bigg(-\frac{1}{\eta}\ (e^{st}\Bigg(\sum_{L=0}^{\infty}\Bigg(\sum_{m=-L}^{L} \texttt{SphericalY}(L,\ m,\ \theta,\ \phi)\ \Bigg(\int_0^{2\pi}\int_0^{\pi} \overline{\texttt{SphericalY}(L,\ m,\ \Theta,\ \Phi)}\ g(\Theta,\ \Phi,\ s)\ \sin(\Theta)\ d\Theta\ d\Phi) \Bigg(\int_{\varepsilon}^{R_1}\Bigg(\frac{1}{\sqrt{r}\ \texttt{BesselJ}\left(L+\frac{1}{2},\ \sqrt{-\frac{s}{\eta}}\ R\right)}\ \Bigg(\texttt{BesselJ}\left(L+\frac{1}{2},\ \sqrt{-\frac{s}{\eta}}\ r\right)\ \Bigg(\frac{1}{2}\ \frac{\texttt{BesselY}\left(L+\frac{1}{2},\ \sqrt{-\frac{s}{\eta}}\ R\right)\ \pi\ \texttt{BesselJ}\left(L+\frac{1}{2},\ r'\sqrt{-\frac{s}{\eta}}\right)}{\sqrt{R}\sqrt{r'}}$$
$$+ \frac{1}{\sqrt{r'}\sqrt{R}}\ \Bigg(\pi\ \Bigg(\texttt{BesselY}\left(L+\frac{1}{2},\ r'\sqrt{-\frac{s}{\eta}}\right)\ \texttt{BesselJ}\left(L+\frac{1}{2},\ \sqrt{-\frac{s}{\eta}}\ R\right) - \texttt{BesselJ}\left(L+\frac{1}{2},\ r'\sqrt{-\frac{s}{\eta}}\right)\ \texttt{BesselY}\ \left(L+\frac{1}{2},\ \sqrt{-\frac{s}{\eta}}\ R\right)\Bigg)\Bigg)\Bigg)\ \sqrt{R}\Bigg)$$
$$-\frac{1}{2}\ \frac{\texttt{BesselY}\left(L+\frac{1}{2},\ \sqrt{-\frac{s}{\eta}}\ r\right)\ \pi\ \texttt{BesselJ}\left(L+\frac{1}{2},\ r'\sqrt{-\frac{s}{\eta}}\right)}{\sqrt{r}\ \sqrt{r'}}$$
$$-\frac{1}{2}\ \frac{1}{\sqrt{r'}\ \sqrt{r}}\ \Bigg(\pi\Bigg(\texttt{BesselY}\left(L+\frac{1}{2},\ r'\sqrt{-\frac{s}{\eta}}\right)\ \texttt{BesselJ}\left(L+\frac{1}{2},\ \sqrt{-\frac{s}{\eta}}\ r\right) - \texttt{BesselJ}\left(L+\frac{1}{2},\ r'\sqrt{-\frac{s}{\eta}}\right)\ \texttt{BesselY}\ \left(L+\frac{1}{2},\ \sqrt{-\frac{s}{\eta}}\ r\right)\Bigg)\ (\texttt{Heaviside}\left(\sqrt{r}-\sqrt{r'}\right)$$
$$+1)))\ r'^2\ dr'\Bigg)\Bigg)\Bigg)\Bigg)\Bigg)\ ds\Bigg)$$

```
>
```

9.5.2 Maple Code: chap9onesphertransient.mws

```
> restart:with(VectorCalculus):with(inttrans):with(PDETools):
> Eq:=diff(C(r,t),t)=eta*expand(Laplacian(C(r,t),spherical[r,theta,
    phi]))+1-Heaviside(r-R[1]);
```

$$Eq\ := \frac{\partial}{\partial t}\ C(r,\ t) = \eta\ \left(\frac{2\left(\frac{\partial}{\partial r}\ C(r,\ t)\right)}{r} + \frac{\partial^2}{\partial r^2}\ C(r,\ t)\right) + 1 - \texttt{Heaviside}(r - R_1)$$

```
> Eq1:=laplace(Eq,t,s):
> Eq2:=subs(laplace(C(r,t),t,s)=C(r),Eq1):
>
> Eq4:=C(r,0)=0;
```

$$Eq4\ := C(r,\ 0) = 0$$

```
> Eq5:=subs(Eq4,Eq2):
>
> Eq7:=dsolve(Eq5):
>
> eq2:=Eq7 assuming r < R[1]:
>
> eq5:=subs(_C1=0,Eq7):
> eq6:=eq5 assuming r >R[1]:
>
> eq7:=subs(r=R,rhs(eq6))=0:
```

```
> eq7A:=isolate(eq7,_C2):
> eq8:=subs(eq7A,eq5):
>
> eq11:=simplify(eq8) assuming r < R[1]:
> eq12:=coeff(convert((simplify(series(rhs(eq11),s=0,3))),polynom),
    s,-1):
>
> eq13:=sinh((s/eta)^(1/2)*R)=0:
>
> eq15:=(s/eta)^(1/2)*R=n*Pi*I:
> eq16:=isolate(eq15,s):
> eq17:=numer(rhs(eq11))*exp(s*t)/diff(denom(rhs(eq11)),s):
> eq18:=factor(simplify(subs(eq16,eq17),power,symbolic))
      assuming n::integer:
> eq19:=C(r,t)=eq12+Sum(eq18,n=1..infinity):
> eq11A:=simplify(eq8) assuming r > R[1]:
> eq12A:=coeff(convert((simplify(series(rhs(eq11A),s=0,3))),polynom),
    s,-1):
>
> eq13A:=sinh((s/eta)^(1/2)*R)=0:
> eq15A:=(s/eta)^(1/2)*R=n*Pi*I:
> eq16A:=isolate(eq15A,s):
> eq17A:=numer(rhs(eq11A))*exp(s*t)/diff(denom(rhs(eq11A)),s):
> eq18A:=factor(simplify(subs(eq16A,eq17A),power,symbolic)) assum-
ing n::integer:
> eq19A:=C(r,t)=eq12A+Sum(eq18A,n=1..infinity):
> eq20:=C(r,t)=piecewise(0>r and r< R[1],rhs(eq19),r>R[1],rhs(eq19A)):
```

PROBLEMS

9.1. Show that the steady-state concentration profile for the system described in Section 9.2 is given by

$$c(r,\theta,\phi) = -\frac{1}{\eta}\sum_{L=0}^{\infty}\left[\begin{array}{l}\displaystyle\sum_{m=-L}^{L} Y_L^m(\theta,\phi)\int_0^{r_1} f_{Lm}(r,r')r'^2dr' \\ \displaystyle\times\int_0^{2\pi}\int_0^{\pi} g(\theta',\phi')Y*_L^m(\theta',\phi')\sin(\theta')d\theta'd\phi'\end{array}\right] \tag{1}$$

9.2. Estimate $c\left(0.7,\frac{\pi}{2},\frac{\pi}{3},8\right)$ for the following system

$$\frac{\partial c}{\partial t} = \left[\frac{1}{r^2}\frac{\partial}{\partial r}\left(r^2\frac{\partial c}{\partial r}\right) + \frac{1}{r^2\sin\theta}\frac{\partial}{\partial\theta}\left(\sin\theta\frac{\partial c}{\partial r}\right) + \frac{1}{r^2\sin^2\theta}\frac{\partial^2 c}{\partial\phi^2}\right] + Q \tag{1}$$

$$Q(0 \le r \le 0.2) = 1; Q(0.2 < r < 1) = 0 \tag{2}$$

$$\left.\frac{\partial c\,(r,\theta,\phi,t)}{\partial r}\right|_{r=0} = 0 \tag{3}$$

$$c(R,\theta,\phi,t) = 0 \tag{4}$$

$$c(R,\theta,\phi,t) = c(R,\theta,\phi+2\pi,t) \tag{5}$$

$$c(R,0,\phi) = \text{finite value} \tag{6}$$

and

$$c(R,\pi,\phi) = \text{finite value} \tag{7}$$

The initial condition is

$$c(r,\theta,\phi,0) = 0 \tag{8}$$

Compare the result with the 1-D solution provided in Appendix 9.5.2.

9.3. Develop an expression for the average concentration for the system in Section 9.2.

9.4. Derive the concentration $c(r,\theta,\phi,t)$ for the following system:

$$\frac{\partial c}{\partial t} = \eta\left[\frac{1}{r^2}\frac{\partial}{\partial r}\left(r^2\frac{\partial c}{\partial r}\right) + \frac{1}{r^2\ \sin\ \theta}\ \frac{\partial}{\partial \theta}\left(\sin\ \theta\frac{\partial c}{\partial r}\right) + \frac{1}{r^2\ \sin^2\ \theta}\ \frac{\partial^2 c}{\partial \phi^2}\right] \tag{1}$$

$$\left.\frac{\partial c\,(r,\theta,\phi,t)}{\partial r}\right|_{r=0} = 0 \tag{2}$$

$$c(R,\theta,\phi,t) = 0 \tag{3}$$

$$c(R,\theta,\phi,t) = c(R,\theta,\phi+2\pi,t) \tag{4}$$

$$c(R,0,\phi) = \text{finite value} \tag{5}$$

and

$$c(R,\pi,\phi) = \text{finite value} \tag{6}$$

The initial condition is

$$c(r,\theta,\phi,0) = c_0 \tag{7}$$

9.5. Derive the concentration $c(r,\theta,\phi,t)$ for the following system:

$$\frac{\partial c}{\partial t} = \eta\left[\frac{1}{r^2}\frac{\partial}{\partial r}\left(r^2\frac{\partial c}{\partial r}\right) + \frac{1}{r^2 \sin\theta}\frac{\partial}{\partial \theta}\left(\sin\theta\frac{\partial c}{\partial r}\right) + \frac{1}{r^2\sin^2\theta}\frac{\partial^2 c}{\partial \phi^2}\right] + Q \tag{1}$$

$$Q(0 \le r \le r_1) = g(\theta,\phi,t);\ Q(r_1 < r < R) = 0 \tag{2}$$

$$\left.\frac{\partial c\,(r,\theta,\phi,t)}{\partial r}\right|_{r=0} = 0 \tag{3}$$

$$c(R, \theta, \phi, t) = 0 \tag{4}$$

$$c(R, \theta, \phi, t) = c(R, \theta, \phi + 2\pi, t) \tag{5}$$

$$c(R, 0, \phi) = \text{finite value} \tag{6}$$

and

$$c(R, \pi, \phi) = \text{finite value} \tag{7}$$

The initial condition is

$$c(r, \theta, \phi, 0) = c_0 \tag{8}$$

REFERENCES

Calvert, P. D., Peet, J. A., Bragin, A., Schiesser, W. E., & Pugh, E. N., Jr. (2007). Fluorescence relaxation in 3D from diffraction-limited sources of PAGFP or sinks of EGFP created by multiphoton photoconversion. *J Microsc*, 225(Pt 1), 49–71.

Fantini, S., Franceschini, M. A., & Gratton, E. (1997). Effective source term in the diffusion equation for photon transport in turbid media. *Appl Opt*, 36(1), 156–163.

Jain, P. K., Singh, S, & Rizwan, U (2010). An exact analytical solution for two-dimensional, unsteady, multilayer heat conduction in spherical coordinates. *Int J HeatMass Transfer*, 53(9–10), 2133–2142.

Rice, R. G., & Do, D. D. (1995). *Applied mathematics and modeling for chemical engineers*. New York: Wiley.

10

DIFFUSION-CONTROLLED LIGAND BINDING TO RECEPTORS ON CELL SURFACES

10.1 WEBER'S INTEGRAL

A classical problem that emerges in Mathematical Physics and Electrostatics is the computation of the electrostatic potential generated by an electrified circular disc (Stone & Goldbart, 2009). The disc is situated in the xy plane with its center at the origin and its axis coincides with the z-axis. Let a be the radius of the disc and Q the total electric charge upon it. We denote by V the electric potential at any point of the space caused by the charge on the disc. In the free space beyond the disc, the electric potential satisfies the Laplace's equation written in cylindrical coordinates with axial symmetry:

$$\frac{\frac{\partial}{\partial r}V(r,z)}{r} + \frac{\partial^2}{\partial r^2}V(r,z) + \frac{\partial^2}{\partial z^2}V(r,z) = 0 \tag{10.1}$$

A solution of the form

$$V(r,z) = h(r) + q(z) + f(r)g(z) \tag{10.2}$$

is proposed where

$$\frac{\mathrm{d}^2}{\mathrm{d}r^2}f(r) = c_1 f(r) - \frac{\frac{\mathrm{d}}{\mathrm{d}r}f(r)}{r} \tag{10.3}$$

$$\frac{\mathrm{d}^2}{\mathrm{d}z^2}g(z) = -c_1 g(z) \tag{10.4}$$

$$\frac{\frac{\mathrm{d}}{\mathrm{d}r}h(r)}{r} + \frac{\mathrm{d}^2}{\mathrm{d}r^2}h(r) = 0 \tag{10.5}$$

Closed-form Solutions for Drug Transport through Controlled-Release Devices in Two and Three Dimensions, First Edition. Laurent Simon and Juan Ospina.

and

$$\frac{d^2}{dz^2}q(z) = 0 \tag{10.6}$$

Solving Eqs. (10.3) and (10.4) leads to

$$f(r)g(z) = \left(C_3 \sin\left(z\sqrt{c_1}\right) + C_4 \cos\left(z\sqrt{c_1}\right)\right)\left(C_1 J_0\left(r\sqrt{-c_1}\right) + C_2 Y_0\left(r\sqrt{-c_1}\right)\right) \tag{10.7}$$

To guarantee a solution without singularities at $r = 0$, we use $C_2 = 0$. The constant C_1 is set equal to 1 without loss of generality. Then, Eq. (10.7) is reduced to

$$f(r)g(z) = J_0\left(r\sqrt{-c_1}\right)\left(C_3 \sin\left(z\sqrt{c_1}\right) + C_4 \cos\left(z\sqrt{c_1}\right)\right) \tag{10.8}$$

Given that the constant c_1 is a free parameter without any kind of restriction, the following equation holds:

$$\sqrt{-c_1} = m \tag{10.9}$$

where m is an arbitrary positive real number. Equation (10.8) becomes

$$f(r)g(z) = J_0(mr)(C_3 i \sinh(mz) + C_4 \cosh(mz)) \tag{10.10}$$

or

$$f(r)g(z) = J_0(mr)\left(iC_3\left(\frac{1}{2}e^{mz} - \frac{1}{2}e^{-mz}\right) + C_4\left(\frac{1}{2}e^{mz} + \frac{1}{2}e^{-mz}\right)\right) \tag{10.11}$$

Since $f(r)g(z)$ is expected to be finite at large z, Eq. (10.11) is written as

$$J_0(mr)\left(\frac{1}{2}iC_3 + \frac{1}{2}C_4\right) = 0 \tag{10.12}$$

As a result,

$$C_4 = -iC_3 \tag{10.13}$$

Equation (10.11) becomes

$$f(r)g(z) = -iC_3 J_0(mr)e^{-mz} \tag{10.14}$$

after using Eq. (10.13). Another form of Eq. (10.14) can be applied:

$$f(r)g(z) = A(m)J_0(mr)e^{-mz} \tag{10.15}$$

The **principle of superposition** gives

$$V(r,z) = h(r) + q(z) + \int_0^\infty (A(m)J_0(mr)e^{-mz})dm \tag{10.16}$$

Equation (10.16) takes the form

$$V(r,z) = C_1 + C_2 \ln(r) + B_1 z + B_2 + \int_0^\infty (A(m)J_0(mr)e^{-mz})dm \tag{10.17}$$

after solving Eqs. (10.5) and (10.6) for $h(r)$ and $q(z)$, respectively. As before, C_2 is set equal to zero to avoid singularity at $r = 0$; $B_1 = 0$ so that the solution is finite at large z. The constant B_2 can be removed to yield

$$V(r,z) = C_1 + \int_0^{\infty} (A(m)J_0(mr)e^{-mz})dm \tag{10.18}$$

Given that the electrostatic potential must be zero far away from the disc, we have $C_1 = 0$ resulting in

$$V(r,z) = \int_0^{\infty} (A(m)J_0(mr)e^{-mz})dm \tag{10.19}$$

To determine the function $A(m)$, it is important to recall that the electric charge density σ is non-vanishing at $z = 0$ for $r < a$. In the Gaussian units, the electric charge density is defined by

$$\sigma(r) = \frac{1}{4}\frac{\left.\left(\frac{\partial}{\partial z}V(r,z)\right)\right|_{z=0}}{\pi} \tag{10.20}$$

The substitution of $V(r, z)$ in Eq. (10.20) produces

$$\sigma(r) = \frac{1}{4}\left(\frac{1}{\pi}\int_0^{\infty} (A(m)\, mJ_0(mr))dm\right) \tag{10.21}$$

The integral in Eq. (10.21) must be equal to zero for $r > a$. To meet this constraint, the **Weber's integral** is introduced:

$$\int_0^{\infty} (J_0(mr)\sin(ma))dm = \begin{cases} \dfrac{1}{\sqrt{a^2 - r^2}} & r < a \\ 0 & a < r \end{cases} \tag{10.22}$$

which yields

$$\int_0^{\infty} (A(m)mJ_0(mr))dm = F\int_0^{\infty} (J_0(mr)\sin(ma))dm \tag{10.23}$$

where F is an arbitrary constant. Based on Eq. (10.23), the following equality holds:

$$A(m)mJ_0(mr) = FJ_0(mr)\sin(ma) \tag{10.24}$$

Therefore,

$$A(m) = \frac{F\sin(ma)}{m} \tag{10.25}$$

Replacing $A(m)$ in Eq. (10.19) gives

$$V(r,z) = \int_0^{\infty} \left(\frac{F\sin(ma)\,J_0(mr)e^{-mz}}{m}\right)dm \tag{10.26}$$

Similarly,

$$\sigma(r) = \frac{1}{4}\left(\frac{1}{\pi}\int_0^{\infty} (F\sin(ma)\,J_0(mr))dm\right) \tag{10.27}$$

when the expression for $A(m)$ is inserted into Eq. (10.21). Then, using Eq. (10.22), it is clear that

$$\sigma(r) = \frac{F}{4\pi\sqrt{a^2 - r^2}} \tag{10.28}$$

Now, the total charge on the disc, Q is defined by

$$Q = 4\pi \int_0^a r\sigma(r)dr \tag{10.29}$$

Substituting Eq. (10.28) into Eq. (10.29) gives

$$Q = Fa \tag{10.30}$$

or

$$F = \frac{Q}{a} \tag{10.31}$$

Finally, Eq. (10.26) is obtained as

$$V(r,z) = \frac{Q}{a}\int_0^\infty \left(\frac{\sin(ma)\,J_0(mr)}{me^{mz}}\right)dm \tag{10.32}$$

10.2 STEADY-STATE DIFFUSION-LIMITED LIGAND BINDING

Consider the steady-state diffusion of a ligand in a cylinder (Fig. 10.1):

$$\frac{1}{r}\frac{\partial}{\partial r}\left(r\frac{\partial C}{\partial r}\right) + \frac{\partial^2 C}{\partial z^2} = 0, \quad z > 0 \tag{10.33}$$

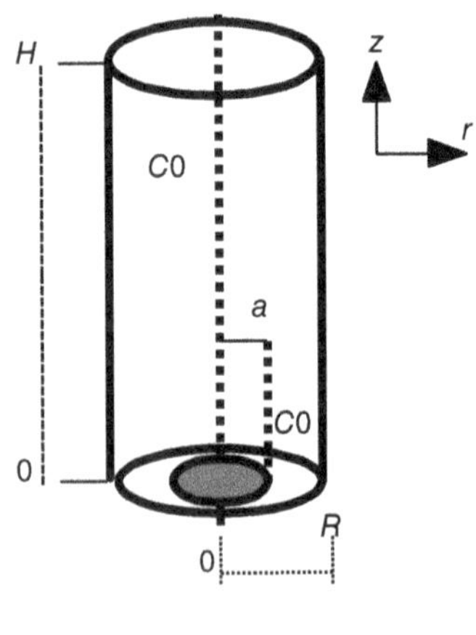

Figure 10.1 Diffusion of the ligand in the cylinder. The target disc, uniformly covered with receptors, is situated at the bottom of the cylinder.

where C is the concentration of the ligand. The radial and axial directions are represented by r and z, respectively. A target disc, uniformly covered with receptors, is located at the bottom of the cylinder. The boundary conditions are

$$C(r \leq a, 0) = 0 \tag{10.34}$$

$$\left.\frac{\partial C\,(r > a, z)}{\partial z}\right|_{z=0} = 0 \tag{10.35}$$

$$C(r \geq 0, z \rightarrow \infty) = C_0 \tag{10.36}$$

$$C(r \rightarrow \infty, z \geq 0) = C_0 \tag{10.37}$$

$$C(r, z)|_{r\rightarrow 0} = \text{ finite value} \tag{10.38}$$

Equation (10.34) assumes complete reaction of the ligand at the cell's surface, i.e., perfect sink target (Wang et al., 1992); a represents the radius of the disc. The bulk ligand concentration is held constant far from the disc (Zwanzig & Szabo, 1991), as described by Eqs. (10.36) and (10.37). Equation (10.35) assumes that the reaction only takes place on the disc. This problem is referred to as the **Weber's disc** (Crank, 1975). Processes involving ligand diffusion and binding to receptors have important applications in biology (Berezhkovskii et al., 2004; Traytak, 2013). Mathematical representations and analyses of the systems are important and may provide researchers with additional insights.

Using a procedure similar to the one presented in the Section 10.1, a solution of the form

$$C(r, z) = C_1 + C_2 \ln(r) + B_1 z + B_2 + \int_0^{\infty} (A(m)J_0(mr)\mathrm{e}^{-mz})dm \tag{10.39}$$

is assumed. The constant C_2 is set equal to zero to satisfy Eq. (10.38). Condition (10.36) is satisfied by making $B_1 = 0$. In addition, B_2 is assigned a value of zero, leading to

$$C(r, z) = C_1 + \int_0^{\infty} (A(m)J_0(mr)\mathrm{e}^{-mz})dm \tag{10.40}$$

Application of the boundary condition (10.37) results in

$$C(r, z) = C_0 + \int_0^{\infty} (A(m)J_0(mr)\mathrm{e}^{-mz})dm \tag{10.41}$$

Now, using Eq. (10.35) for $r > a$, we obtain

$$\int_0^{\infty} (-A(m)mJ_0(mr))dm = 0 \tag{10.42}$$

The Weber's integral, defined by Eq. (10.22), requires that

$$\int_0^{\infty} (-A(m)mJ_0(mr))dm = F \int_0^{\infty} (J_0(mr) \sin(ma))dm \tag{10.43}$$

where F is an arbitrary constant. The expression for $A(m)$ is deduced from Eq. (10.43):

$$A(m) = -\frac{F \sin(ma)}{m} \tag{10.44}$$

Therefore, $C(r, z)$ is

$$C(r, z) = C_0 - \int_0^{\infty} \left(\frac{F \sin(ma) J_0(mr) e^{-mz}}{m} \right) dm \tag{10.45}$$

Furthermore, application of the boundary condition (10.34) gives

$$C_0 - \int_0^{\infty} \left(\frac{F \sin(ma) J_0(mr)}{m} \right) dm = 0 \tag{10.46}$$

or

$$C_0 - \frac{\pi F}{2} = 0 \tag{10.47}$$

after computing the integral in Maple. The value of F is then obtained as

$$F = \frac{2C_0}{\pi} \tag{10.48}$$

which leads to

$$C(r, z) = C_0 - \frac{2C_0}{\pi} \int_0^{\infty} \left(\frac{\sin(ma) J_0(mr) e^{-mz}}{m} \right) dm \tag{10.49}$$

The concentration gradient at the surface $z = 0$ is given by

$$\left(\frac{\partial}{\partial z} C(r, z) \right) \bigg|_{z=0} = \frac{2C_0}{\pi \sqrt{a^2 - r^2}}, \quad r < a \tag{10.50}$$

which is the same result in (Crank, 1975). Figures 10.2 and 10.3 show the concentration and concentration gradient, respectively. The following parameters were used in the simulations: $C_0 = 1$ and $a = 0.1$.

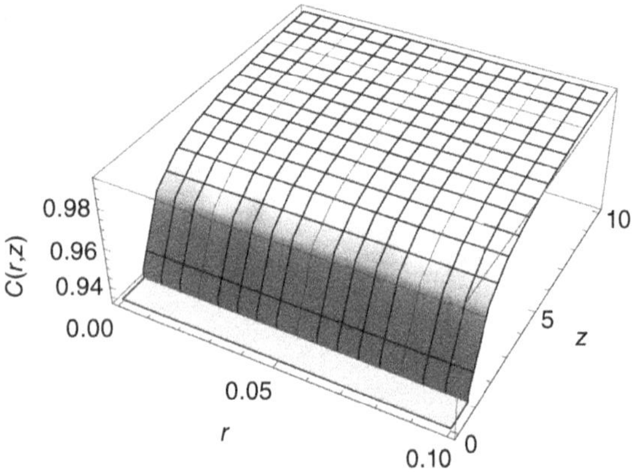

Figure 10.2 Steady-state ligand concentration using the following parameters: $C_0 = 1$ and $a = 0.1$.

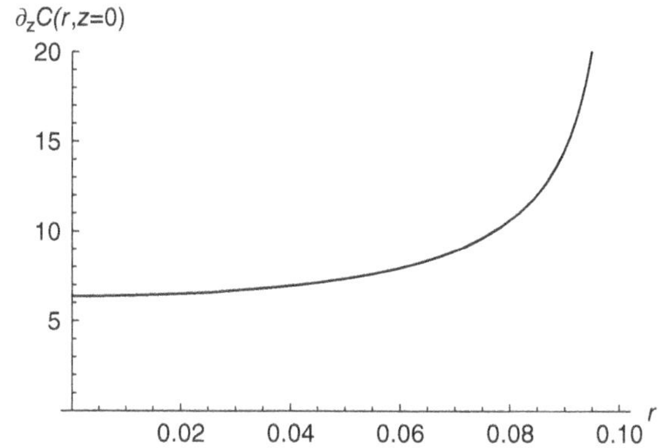

Figure 10.3 Steady-state surface concentration gradient using the following parameters: $C_0 = 1$ and $a = 0.1$.

10.3 TRANSIENT DIFFUSION-CONTROLLED LIGAND BINDING IN 2-D CYLINDRICAL COORDINATES

The transient version of the ligand binding problem in Section 10.2 is

$$\frac{\partial c}{\partial t} = \eta\left(\frac{1}{r}\frac{\partial}{\partial r}\left(r\frac{\partial c}{\partial r}\right) + \frac{\partial^2 c}{\partial z^2}\right), \quad z > 0 \tag{10.51}$$

$$c(r \leq a, 0, t) = 0 \tag{10.52}$$

$$\left.\frac{\partial c\,(r > a, z, t)}{\partial z}\right|_{z=0} = 0 \tag{10.53}$$

$$c(r \geq 0, z \to \infty, t) = c_0 \tag{10.54}$$

$$c(r \to \infty, z \geq 0, t) = c_0 \tag{10.55}$$

$$c(r, z, t)|_{r\to 0} = \text{ finite value} \tag{10.56}$$

$$c(r, z, 0) = c_0 \tag{10.57}$$

To solve the system formed by Eqs. (10.51) to (10.57), Laplace transforms and the procedures outlined in Sections 10.1 and 10.2 were applied. The Laplace transform of Eq. (10.51) is

$$sC(r,z) - C_0 = \frac{\eta\frac{\partial}{\partial r}C(r,z)}{r} + \eta\frac{\partial^2}{\partial r^2}C(r,z) + \eta\frac{\partial^2}{\partial z^2}C(r,z) \tag{10.58}$$

where

$$C(r,z) = \int_0^\infty c(r,z,t)e^{-st}dt \tag{10.59}$$

The solution is of the form

$$C(r,z) = \frac{C_0}{s} + h(r) + q(z) + f(r)g(z) \tag{10.60}$$

where

$$\frac{\mathrm{d}^2}{\mathrm{d}r^2} f(r) = c_1 f(r) - \frac{\frac{\mathrm{d}}{\mathrm{d}r} f(r)}{r} \tag{10.61}$$

$$\frac{\mathrm{d}^2}{\mathrm{d}z^2} g(z) = -c_1 g(z) + \frac{s g(z)}{\eta} \tag{10.62}$$

$$s h(r) = \frac{\frac{\mathrm{d}}{\mathrm{d}r} h(r)}{r} + \frac{\mathrm{d}^2}{\mathrm{d}r^2} h(r) \tag{10.63}$$

and

$$s q(z) = \eta \frac{\mathrm{d}^2}{\mathrm{d}z^2} q(z) \tag{10.64}$$

Solving Eqs. (10.61) and (10.62) yields

$$f(r)g(z) = \left(C_3 \sin\left(z\sqrt{\frac{c_1\eta - s}{\eta}} \right) + C_4 \cos\left(z\sqrt{\frac{c_1\eta - s}{\eta}} \right) \right) \times \left(C_1 J_0\left(r\sqrt{-c_1} \right) + C_2 Y_0\left(r\sqrt{-c_1} \right) \right) \tag{10.65}$$

To avoid singularities at $r=0$, we set $C_2 = 0$. The constant C_1 is also set equal to 1. Equation (10.65) is reduced to

$$f(r)g(z) = J_0(r\sqrt{-c_1}) \left(C_3 \sin\left(z\sqrt{\frac{c_1\eta - s}{\eta}} \right) + C_4 \cos\left(z\sqrt{\frac{c_1\eta - s}{\eta}} \right) \right) \tag{10.66}$$

As before, the following equation holds:

$$\sqrt{-c_1} = m \tag{10.67}$$

where m is an arbitrary positive real number. Equation (10.66) becomes

$$f(r)g(z) = J_0(mr) \left(C_3 \sin\left(z\sqrt{\frac{-m^2\eta - s}{\eta}} \right) + C_4 \cos\left(z\sqrt{\frac{-m^2\eta - s}{\eta}} \right) \right) \tag{10.68}$$

or

$$f(r)g(z) = J_0(mr) \times \left(iC_3 \left(\frac{1}{2} \mathrm{e}^{iz\sqrt{\frac{-m^2\eta - s}{\eta}}} - \frac{1}{2} \mathrm{e}^{-iz\sqrt{\frac{-m^2\eta - s}{\eta}}} \right) + C_4 \left(\frac{1}{2} \mathrm{e}^{iz\sqrt{\frac{-m^2\eta - s}{\eta}}} + \frac{1}{2} \mathrm{e}^{-iz\sqrt{\frac{-m^2\eta - s}{\eta}}} \right) \right) \tag{10.69}$$

To satisfy Eq. (10.54), Eq. (10.69) is written as

$$J_0(mr) \left(\frac{1}{2} iC_3 + \frac{1}{2} C_4 \right) = 0 \tag{10.70}$$

Consequently,

$$C_4 = -iC_3 \tag{10.71}$$

Equation (10.69) becomes

$$f(r)g(z) = A(m)J_0(mr)e^{-z\sqrt{\frac{m^2\eta+s}{\eta}}} \tag{10.72}$$

Now, solving Eq. (10.63), we obtain

$$h(r) = C_1 J_0\left(r\sqrt{-\frac{s}{\eta}}\right) + C_2 Y_0\left(r\sqrt{-\frac{s}{\eta}}\right) \tag{10.73}$$

To satisfy the condition (10.56), C_2 is set to zero, leading to

$$h(r) = C_1 J_0\left(r\sqrt{-\frac{s}{\eta}}\right) \tag{10.74}$$

The solution to Eq. (10.64) is

$$q(z) = B_1 e^{z\sqrt{\frac{s}{\eta}}} + B_2 e^{-z\sqrt{\frac{s}{\eta}}} \tag{10.75}$$

Equation (10.54) is satisfied by setting $B_1 = 0$, yielding

$$q(z) = B_2 e^{-z\sqrt{\frac{s}{\eta}}} \tag{10.76}$$

Application of the principle of superposition gives

$$C(r,z) = \frac{C_0}{s} + C_1 J_0\left(r\sqrt{-\frac{s}{\eta}}\right) + B_2 e^{-z\sqrt{\frac{s}{\eta}}} + \int_0^\infty \left(A(m)\,J_0(mr)e^{-z\sqrt{\frac{m^2\eta+s}{\eta}}}\right)dm \tag{10.77}$$

Using the boundary condition (10.54) in the Laplace domain, the following equation is obtained:

$$\frac{C_0}{s} + C_1 J_0\left(r\sqrt{-\frac{s}{\eta}}\right) = \frac{C_0}{s} \tag{10.78}$$

resulting in

$$C_1 = 0 \tag{10.79}$$

and

$$C(r,z) = \frac{C_0}{s} + B_2 e^{-z\sqrt{\frac{s}{\eta}}} + \int_0^\infty \left(A(m)\,J_0(mr)e^{-z\sqrt{\frac{m^2\eta+s}{\eta}}}\right)dm \tag{10.80}$$

Applying the boundary condition (10.54), we have

$$-B_2\sqrt{\frac{s}{\eta}} - \int_0^\infty \left(A(m)\,J_0(mr)\sqrt{\frac{m^2\eta+s}{\eta}}\right)dm = 0 \tag{10.81}$$

From Eq. (10.81), the following equation holds:

$$B_2 = 0 \tag{10.82}$$

and

$$-\int_0^\infty \left(A(m)\, J_0(mr) \sqrt{\frac{m^2\eta + s}{\eta}} \right) dm = 0 \tag{10.83}$$

From Eq. (10.83) and after using the Weber's integral, we get

$$-\int_0^\infty \left(A(m)\, J_0(mr) \sqrt{\frac{m^2\eta + s}{\eta}} \right) dm = F \int_0^\infty (\sin(ma) J_0(mr)) dm \tag{10.84}$$

Then,

$$A(m) = \frac{-\sin(ma) F \sqrt{\eta}}{\sqrt{m^2\eta + s}} \tag{10.85}$$

Replacing Eqs. (10.82) and (10.85) into Eq. (10.80) results in

$$C(r,z) = \frac{C_0}{s} - \int_0^\infty \left(\frac{\sin(ma)\, F \sqrt{\eta} J_0(mr) e^{-z\sqrt{\frac{m^2\eta+s}{\eta}}}}{\sqrt{m^2\eta + s}} \right) dm \tag{10.86}$$

The boundary condition (10.52) gives

$$\frac{C_0}{s} - \int_0^\infty \left(\frac{\sin(ma)\, F \sqrt{\eta} J_0(mr)}{\sqrt{m^2\eta + s}} \right) dm = 0 \tag{10.87}$$

which allows the computation of F:

$$F = \frac{C_0}{s\sqrt{\eta} \int_0^\infty \left(\frac{\sin(ka)\, J_0(kr)}{\sqrt{k^2\eta + s}} \right) dk} \tag{10.88}$$

after setting $k = m$. Note that the integral in Eq. (10.88) is independent of r given that we are considering the region $r < a$. Finally, the concentration $C(r, z)$ is

$$C(r,z) = \frac{C_0}{s} - \frac{C_0 \int_0^\infty \left(\frac{\sin(ma)\, J_0(mr) e^{-z\sqrt{\frac{m^2\eta+s}{\eta}}}}{\sqrt{m^2\eta + s}} \right) dm}{s \int_0^\infty \left(\frac{\sin(ka)\, J_0(kr)}{\sqrt{k^2\eta + s}} \right) dk} \tag{10.89}$$

The inverse Laplace transform of Eq. (10.89) is computed using the **Bromwich integral** and the **residue theorem**:

$$c(r,z,t) = C_0 - \frac{2C_0}{\pi}\int_0^\infty \left(\frac{\sin(ma)\,J_0(mr)e^{-mz}}{m}\right)dm - \frac{1}{2\pi i}\int \left[\frac{C_0\displaystyle\int_0^\infty \left(\frac{\sin(ma)\,J_0(mr)\mathrm{e}^{-z\sqrt{\frac{m^2\eta+s}{\eta}}}}{\sqrt{m^2\eta+s}}\right)dm}{s\displaystyle\int_0^\infty\left(\frac{\sin(ka)\,J_0(kr)}{\sqrt{k^2\eta+s}}\right)dk}\right]\mathrm{e}^{st}\,\mathrm{d}s,\quad s\neq 0 \tag{10.90}$$

The concentration gradient at the surface is given by

$$\left(\frac{\partial}{\partial z}\,C(r,z)\right)\bigg|_{z=0} = \frac{C_0\displaystyle\int_0^\infty\left(\frac{\sin(ma)\,J_0(mr)}{\sqrt{\eta}}\right)dm}{s\displaystyle\int_0^\infty\left(\frac{\sin(ka)\,J_0(kr)}{\sqrt{k^2\eta+s}}\right)dk} \tag{10.91}$$

in the Laplace domain or

$$\left(\frac{\partial}{\partial z}c(r,z,t)\right)\bigg|_{z=0} = \frac{2C_0}{\pi\sqrt{a^2-r^2}} + \frac{1}{2\pi i}\int\left[\frac{C_0}{s\sqrt{\eta}\sqrt{a^2-r^2}\displaystyle\int_0^\infty\left(\frac{\sin(ka)\,J_0(kr)}{\sqrt{k^2\eta+s}}\right)dk}\right]\mathrm{e}^{st}\mathrm{d}s,\quad s\neq 0 \tag{10.92}$$

in the time domain. Equation (10.91) is implemented in Mathematica as

$$J2a(r\_,\ s\_) := \frac{C0\left(\frac{1}{\sqrt{(a^2-r^2)\eta}}\right)}{s\text{Assuming}[r>0\&\&r<a,\ \text{NIntegrate}\left[\left(\frac{Sin[ka]BesselJ[0,kr]}{\sqrt{k^2\eta+s}}\right),\ \{k,0,1000\}\right]}; \tag{10.93}$$

For the following parameters $C_0 = 1$, $a = 0.1$ and $\eta = 1$, after using the Zakian method (Rice & Do, 1995) , we obtain $\left(\frac{\partial}{\partial z}c(0.05,z,100)\right)\Big|_{z=0} = 7.36$.

10.4 SUMMARY

A classical problem in Mathematical Physics and Electrostatics was used to introduce Weber's integral. The expression for the voltage at any point in space provided by a point charge on a disc was derived in cylindrical coordinates. We applied a similar method to analyze steady-state and transient diffusion of a ligand in a cylinder. A disc, covered with receptors, was placed at the center of the cylinder's bottom. Both the stationary and dynamic solutions were developed using a linear combination of additive and multiplicative separation of variables. The inverse Laplace transform was determined by the Bromwich integral and the residue theorem. It was necessary to apply numerical integration and the Zakian method to predict the transient behavior.

10.5 APPENDIX: MAPLE, MATHEMATICA AND MAXIMA CODE LISTINGS

10.5.1 Maple Code: Weberdiscsteady.mws (Section 10.2)

```
> restart:with(VectorCalculus):with(PDETools):
> eq:=expand(Laplacian(C(r,z),cylindrical[r,theta,z]))=0;
```

$$eq := \frac{\frac{\partial}{\partial r} C(r,z)}{r} + \frac{\partial^2}{\partial r^2} C(r,z) + \frac{\partial^2}{\partial z^2} C(r,z) = 0$$

We assume a solution of the form C(r,z) = h(r) + q(z) + f(r)*g(z)

```
> eq1:=pdsolve(eq,HINT=f(r)*g(z)):
> eq2:=factor(build(eq1)):
> eq3:=subs(_C2=0,eq2):
> eq4:=subs(_C1=1,eq3):
> eq5:=(-_c[1])^(1/2)=m:
> eq6:=isolate(eq5,_c[1]):
> eq7:=simplify(subs(eq6,eq4),power,symbolic):
> eq8:=convert(eq7,exp):
> eq9:=coeff(rhs(eq8),exp(m*z),1)=0:
> eq10:=isolate(eq9,_C4):
> eq11:=simplify(subs(eq10,eq8)):
> eq12:=subs(_C3=I*A(m),eq11):
> eq13:=C(r,z)=h(r)+q(z)+Int(rhs(eq12),m=0..infinity):
> eq14:=eval(subs(C(r,z)=h(r),eq)):
> eq14A:=dsolve(eq14):
> eq15:=eval(subs(C(r,z)=q(z),eq)):
> eq15A:=dsolve(eq15):
> eq15B:=subs(_C1=B[1],_C2=B[2],eq15A):
> eq16:=subs(eq14A,eq15B,eq13):
> eq17:=subs(_C2=0,eq16):
> eq18:=subs(B[1]=0,eq17):
> eq19:=subs(B[2]=0,eq18):
> eq20:=subs(_C1=C[0],eq19):
> eq22:=eval(subs(z=0,diff(rhs(eq20),z)))=0:
```

```
> eq23:=Int(BesselJ(0,m*r)*sin(m*a),m=0..infinity)=piecewise(r < a,
    simplify(int(BesselJ(0,m*r)*sin(m*a),m=0..infinity),power,
    symbolic), r>a,0):
> eq24:=lhs(eq22)=F*lhs(eq23):
> eq25:=-A(m)*m*BesselJ(0,m*r)=BesselJ(0,m*r)*sin(m*a)*F:
> eq26:=isolate(eq25,A(m)):
> eq27:=subs(eq26,eq20):
> eq28:=eval(subs(z=0,rhs(eq27)))=0:
> eq29:=eval(subs(Int=int,eq28)):
> eq30:=isolate(eq29,F):
> eq31:=expand(subs(eq30,eq27));
```

$$eq31 := C(r, z) = C_0 - \frac{2C_0\left(\int_0^{\infty} \frac{\text{BesselJ}(0, m\,r)\ \sin(m\ a)}{m\ e^{m\ z}}\, dm\right)}{\pi}$$

```
> eq32:=Eval(diff(C(r,z),z),z=0)=simplify(eval(subs(Int=int,
    eval(diff(rhs(eq31),z),z=0))),power,symbolic);
>
```

$$eq32 := \left.\frac{\partial}{\partial z}\, C(r, z)\right|_{z=0} = \frac{2\ C_0}{\pi\sqrt{a^2-r^2}}$$

```
>
```

10.5.2 Maple Code: Weberdisctransient.mws (Section 10.3)

```
> restart:with(VectorCalculus):with(PDETools):with(inttrans):
> eq:=diff(C(r,z,t),t)=eta*expand(Laplacian(C(r,z,t),
    cylindrical[r,theta,z]));
```

$$eq := \frac{\partial}{\partial t}\, C(r, z, t) = \eta\ \left(\frac{\frac{\partial}{\partial r}\, C(r,z,t)}{r} + \frac{\partial^2}{\partial r^2}\, C(r, z, t) + \frac{\partial^2}{\partial z^2}\, C(r, z, t)\right)$$

```
> eq0:=laplace(eq,t,s):
> eq0A:=C(r,z,0)=C[0]:
> eq0B:=subs(eq0A,eq0):
> eq0C:=subs(laplace(C(r,z,t),t,s)=C(r,z),eq0B):
> eq0D:=subs(C[0]=0,eq0C):
> eq1:=pdsolve(eq0D,HINT=f(r)*g(z)):
> eq2:=factor(build(eq1)):
> eq3:=subs(_C2=0,eq2):
> eq4:=subs(_C1=1,eq3):
> eq5:=(-_c[1])^(1/2)=m:
> eq6:=isolate(eq5,_c[1]):
> eq7:=simplify(subs(eq6,eq4),power,symbolic):
> eq8:=(convert(eq7,exp)) assuming m>0 and eta>0 and s>0:
> eq9:=coeff(rhs(eq8),exp(-I/eta^(1/2)*(-m^2*eta-s)^(1/2)*z),1)=0:
> eq10:=isolate(eq9,_C4):
> eq11:=simplify(subs(eq10,eq8)):
> eq11A:=sqrt(-m^2*eta-s)=I*sqrt(m^2*eta+s):
> eq11B:=subs(eq11A,eq11):
> eq12:=subs(_C3=I*A(m),eq11):
> eq13:=C(r,z)=C[0]/s+h(r)+q(z)+Int(rhs(eq12),m=0..infinity):
```

```
> eq14:=eval(subs(C(r,z)=h(r),eq0D)):
> eq14A:=dsolve(eq14):
> eq15:=eval(subs(C(r,z)=q(z),eq0D)):
> eq15A:=dsolve(eq15):
> eq15B:=subs(_C1=B[1],_C2=B[2],eq15A):
> eq16:=subs(eq14A,eq15B,eq13):
> eq17:=subs(_C2=0,eq16):
> eq18:=subs(B[1]=0,eq17):
> eq20:=C[0]/s+_C1*BesselJ(0,(-s/eta)^(1/2)*r)=C[0]/s:
> eq20A:=isolate(eq20,_C1):
> eq20B:=subs(eq20A,eq18):
> eq22:=eval(subs(z=0,diff(rhs(eq20B),z)))=0:
>
> eq22A:=B[2]=0:
> eq22B:=subs(eq22A,eq22):
> eq23:=Int(BesselJ(0,m*r)*sin(m*a),m=0..infinity)=piecewise(r < a,
    simplify(int(BesselJ(0,m*r)*sin(m*a),m=0..infinity),power,
    symbolic), r>a,0):
> eq24:=lhs(eq22B)=F*lhs(eq23):
> eq25:=BesselJ(0,m*r)*A(m)/eta^(1/2)*(-m^2*eta-s)^(1/2)*
    I=BesselJ(0,m*r)*sin(m*a)*F:
> eq26:=isolate(eq25,A(m)):
> eq27:=subs(eq26,eq22A,eq20B):
> eq28:=eval(subs(z=0,rhs(eq27)))=0:
> eq29:=eval(subs(Int=int,eq28)):
> eq30:=F=subs(m=k,solve(eq29,F)):
> eq31:=expand(subs(eq30,eq27)):
> eq31A:=C(r,z,t)=C[0]+expand(simplify(subs(s=0,(rhs(eq31)-C[0]/s)*s),
    power,symbolic))+Eval((1/(2*Pi*I))*Int(exp(s*t)*(rhs(eq31)-
    C[0]/s),s),s <>0);
```

$$eq31A := C(r,z,t) = C_0 - \frac{2\ C_0 \left(\int_0^\infty \frac{\text{BesselJ}(0, mr)\ \sin(ma)}{m\,e^{mz}}\, dm \right)}{\pi}$$

$$+Eval\left(-\frac{\frac{1}{2}\ I \left(\int \left(-\frac{e^{st} C_0 \left(\int_0^\infty \frac{\text{BesselJ}(0, mr)\ \sin(ma)\ e^{\frac{I\sqrt{-\eta m^2 - s}\ z}{\sqrt{\eta}}}}{\sqrt{-\eta m^2 - s}}\, dm \right)}{s \left[\int_0^\infty \frac{\text{BesselJ}(0,\ kr)\ \sin(ka)}{\sqrt{-\eta k^2 - s}}\, dk \right]} \right) ds \right)}{\pi},\ s \neq 0 \right)$$

```
> eq32:=Eval(diff(C(r,z),z),z=0)=simplify(eval(subs(Int=int,
     eval(diff(C[0]-2*C[0]/Pi*Int(BesselJ(0,m*r)*sin(m*a)/m/exp(m*z),
     m =0 .. infinity),z),z=0))),power,symbolic)+Eval(-1/2*I/Pi*
     Int(-exp(s*t)*C[0]/int(BesselJ(0,k*r)*sin(k*a)/(-k^2*eta-s)^(1/2),
     k=0 .. infinity)/s*simplify(int(BesselJ(0,m*r)*sin(m*a)/
     (-m^2*eta-s)^(1/2)*eval(subs(z=0,diff(exp(1/eta^(1/2)*
     (-m^2*eta-s)^(1/2)*z*I),z))),m =0 .. infinity),power,symbolic),
     s),s <> 0)
> ;
```

$$\left.\frac{\partial}{\partial z}C(r,z)\right|_{z=0}=\frac{2\,C_0}{\pi\sqrt{a^2-r^2}}+Eval\left(-\frac{\frac{1}{2}\,I\left(\int\left(-\frac{I\,e^{s\,t}C_0}{\left(\int_0^\infty\frac{\text{BesselJ}(0,\,k\,r)\,\sin(k\,a)}{\sqrt{-\eta k^2-s}}\,dk\right)s\sqrt{a^2-r^2}\,\sqrt{\eta}}\right)ds\right)}{\pi},\;s\neq 0\right)$$

10.5.3 Mathematica Code: Weberdisintegral.nb (Section 10.1)

eq=Laplacian[V[r,z],{r,θ,z},"Cylindrical"]/ /Expand

$V^{(0,2)}[r,z]+\frac{V^{(1,0)}[r,z]}{r}+V^{(2,0)}[r,z]$

Multiplicative separation:

eq2 = eq/.V → Function[{r, z}, f[r] * g[z]]

$\frac{g[z]f'[r]}{r}+g[z]f''[r]+f[r]g''[z]$

eq3 = Expand[eq2/f[r] * g[z])] = 0

$\frac{f'[r]}{rf[r]}+\frac{f''[r]}{f[r]}+\frac{g''[z]}{g[z]}=0$

eq3A = $\frac{g''[z]}{g[z]}=-c[1]$

$\frac{g''[z]}{g[z]}=-c[1]$

eq4 = $\frac{f'[r]}{rf[r]}+\frac{f''[r]}{f[r]}-c[1]=0$

$-c[1]\;+\;\frac{f'[r]}{rf[r]}+\frac{f''[r]}{f[r]}=0$

eq5 = DSolve[eq4, f, r]

{{f → Function [{r}, BesselJ [0, $ir\sqrt{c[1]}$]C[1] + BesselY[0, $-ir\sqrt{c[1]}$]C[2]]}}

eq5A = (DSolve[eq3A, g, z] / .C[1]- > A[1] / .C[2] → B[1]

{{g → Function [{z}, A[1]Cos[$z\sqrt{c[1]}$] + B[1]Sin[$z\sqrt{c[1]}$]]}}

eq6 = (BesselJ[0, $-ir\sqrt{c[1]}$]C[1] + BesselY[0, $-ir\sqrt{c[1]}$]C[2])
(A[1]Cos[$z\sqrt{c[1]}$] + B[1]Sin[$z\sqrt{c[1]}$])

(BesselJ[0, $ir\sqrt{c[1]}$]C[1] + BesselY[0, $-ir\sqrt{c[1]}$]C[2])(A[1]Cos[$z\sqrt{c[1]}$] + B[1]Sin[$z\sqrt{c[1]}$])

eq6A = c[1] = −m ^ 2

$c[1] = -m^2$

eq6C = eq6 / .c[1] → −m ^ 2

(BesselJ[0, $i\sqrt{-m^2 r}$]C[1] + BesselY[0, $-i\sqrt{-m^2 r}$]C[2])
(A[1]Cos[$\sqrt{-m^2 r}$] + B[1]Sin[$\sqrt{-m^2 z}$])

eq6D = Assuming[m > 0, Simplify[eq6C]]

(BesselJ[0, mr]C[1] + BesselY[0, mr]C[2])(A[1]Cosh[mz] + iB[1]Sinh[mz])

eq6E = ((((((eq6D / .A[1] → 1) / .B[1] → 0) / .C[1]− > C[m]) / .C[2] → 0))) /
Cosh[mz] → Exp[−m ∗ z]

e^{-mz}BesselJ[0, mr]C[m]

eq6F = Integrate[e^{-mz}BesselJ[0, mr]C[m], {m, 0, ∞}]

$\int_0^\infty e^{-mz}\text{BesselJ}[0, mr]C[m]\,dm$

eq6G = V[r, z] = eq6F

$V[r, z] = \int_0^\infty e^{-mz}\text{BesselJ}[0, mr]C[m]\,dm$

$\text{eq6H} = \sigma[r] = \left((-1/(4\pi))\,\partial_z\int_0^\infty e^{-mz}\text{BesselJ}[0, mr]C[m]\,dm\right) /. z \to 0$

$\sigma[r] = -\frac{\int_0^\infty -m\text{BesselJ}[0, mr]C[m]\,dm}{4\pi}$

eq6I = Assuming[r > a > 0,
Simplify[Integrate[BesselJ[0, mr]Sin[ma], {m, 0, ∞}]
0

eq6I = Assuming[r < a & &r > 0,
Simplify[Integrate[BesselJ[0, mr]Sin[ma], {m, 0, ∞}]

$\frac{1}{\sqrt{a^2-r^2}}$

eq7 = mBesselJ[0, mr]C[m] = FBesselJ[0, mr]Sin[ma]

eq8 = Solve[eq7, C[m]]

$\left\{\left\{C[m] \to \frac{F\text{Sin}[am]}{m}\right\}\right\}$

eq9 = eq6G / .C[m] → $\frac{F\text{Sin}[am]}{m}$

$V[r,z] = \int_0^\infty \frac{e^{-mz}F\text{BesselJ}[0, mr]\text{Sin}[am]}{m}\,dm$

$\text{eq10} = \sigma[r] = \text{Assuming}\left[r < a \,\&\&\, r > 0, \text{Simplify}\left[\frac{\int_0^\infty m\text{BesselJ}[0, mr]\frac{F\text{Sin}[am]}{m}dm}{4\pi}\right]\right]$

$\sigma[r] = \frac{F}{4\pi\sqrt{a^2-r^2}}$

$\text{eq11} = Q = 4 * \pi * \text{Assuming}\left[a > 0, \text{Simplify}\left[\text{Integrate}\left[\frac{F*r}{4\pi\sqrt{a^2-r^2}}, \{r, 0, a\}\right]\right]\right]$

$Q = aF$

eq12 = Solve[eq11, F]

$\left\{\left\{F \to \frac{Q}{a}\right\}\right\}$

eq13 = eq9 / .$F \to \frac{Q}{a}$

$V[r,z] = \int_0^\infty \frac{e^{-mz} Q\,\texttt{BesselJ}[0, mr]\text{Sin}[am]}{am}\, dm$

eq14 = eq10 / .$F \to \frac{Q}{a}$

$\sigma[r] = \frac{Q}{4a\pi\sqrt{a^2-r^2}}$

10.5.4 Mathematica Code: Weberdiscsteadymat.nb (Section 10.2)

eq = `Laplacian`$[C[r,z], \{r, \theta, z\}$, `"Cylindrical"]//Expand`

$C^{(0,2)}[r,z] + \frac{C^{(1,0)}[r,z]}{r} + C^{(2,0)}[r,z]$

`Multiplicative separation:`

eq2 = eq / .$C \to$ `Function`$[\{r,z\}, f[r] * g[z]]$

$\frac{g[z]f'[r]}{r} + g[z]f''[r] + f[r]g''[z]$

eq3 = `Expand`$[eq2 / (f[r]g[z])] = 0$

$\frac{f'[r]}{rf[r]} + \frac{f''[r]}{f[r]} + \frac{g''[z]}{g[z]} = 0$

eq3A = $\frac{g''[r]}{g[r]} = -c[1]$

$\frac{g''[r]}{g[r]} = -c[1]$

eq4 = $\frac{f'[r]}{rf[r]} + \frac{f''[r]}{f[r]} - c[1] = 0$

$-c[1] + \frac{f'[r]}{rf[r]} + \frac{f''[r]}{f[r]} = 0$

eq5 = `DSolve`$[eq4, f, r]$

$\{\{f \to$ `Function`$[\{r\}$, `BesselJ`$[0, ir\sqrt{c[1]}]C[1]$ + `BesselY`$[0, -ir\sqrt{c[1]}]C[2]]\}\}$

eq5A = (`DSolve`$[eq3A, g, z] / .C[1]- > A[1]) / .C[2] \to B[1]$

$\{\{g \to$ `Function`$[\{z\}, A[1]\text{Cos}[z\sqrt{c[1]}] + B[1]\text{Sin}[z\sqrt{c[1]}]]\}\}$

eq6 = (`BesselJ`$[0, -ir\sqrt{c[1]}]C[1]$ + `BesselY`$[0, -ir\sqrt{c[1]}]C[2])$
$(A[1]\text{Cos}[z\sqrt{c[1]}] + B[1]\text{Sin}[z\sqrt{c[1]}])$

(`BesselJ`$[0, ir\sqrt{c[1]}]C[1]$ + `BesselY`$[0, -ir\sqrt{c[1]}]C[2])(A[1]\text{Cos}[z\sqrt{c[1]}] + B[1]\text{Sin}[z\sqrt{c[1]}])$

eq6A = $c[1] = -m\hat{\ }2$

$c[1] = -m^2$

eq6C = eq6 / .$c[1] \to -m\hat{\ }2$

(`BesselJ`$[0, -i\sqrt{-m^2}r]C[1]$ + `BesselY`$[0, -i\sqrt{-m^2}r]C[2]) \times$
$(A[1]Cos[\sqrt{-m^2}z] + B[1]\text{Sin}[\sqrt{-m^2}z])$

eq6D = `Simplify`$[eq6C, m > 0]$

(`BesselJ`$[0, mr]C[1]$ + `BesselY`$[0, mr]C[2])(A[1]\text{Cosh}[mz] + iB[1]\text{Sinh}[mz])$

eq6E = $((((((eq6D / .A[1] \to 1) / .B[1] \to 0) / .C[1]- > C[m]) / .C[2] \to 0))) / .$
$\text{Cosh}[mz] \to \text{Exp}[-m * z]$

e^{-mz}`BesselJ`$[0, mr]C[m]$

eq6F = `Integrate`$[e^{-mz}$`BesselJ`$[0, mr]C[m], \{m, 0, \infty\}]$

$\int_0^\infty e^{-mz}\texttt{BesselJ}[0, mr]C[m]\, dm$

eq6G = $C[r,z] = c[0] + \text{eq6F}$

$C[r,z] = c[0] + \int_0^\infty e^{-mz}\texttt{BesselJ}[0, mr]C[m]\,dm$

eq6H = $0 = \left(\partial_z \int_0^\infty e^{-mz}\texttt{BesselJ}[0, mr]\,C[m]dm\right) /.z \to 0$

$0 = \int_0^\infty -m\texttt{BesselJ}[0, mr]C[m]\,dm$

eq6I = Assuming[$r > a$&&$a > 0$,
Simplify[Integrate[BesselJ[0, mr]Sin[ma], {m, 0, ∞}]]]

0

eq6J = Assuming[$r > a$&&$r > 0$,
Simplify[Integrate[BesselJ[0, mr]Sin[ma], {m, 0, ∞}]]]

$\frac{1}{\sqrt{a^2-r^2}}$

eq7 = $-m\texttt{BesselJ}[0, mr]C[m] = F\texttt{BesselJ}[0, mr]\text{Sin}[ma]$

$-m\texttt{BesselJ}[0, mr]C[m] = F\texttt{BesselJ}[0, mr]\text{Sin}[am]$

eq8 = Solve[eq7, $C[m]$]

$\left\{\left\{C[m] \to -\frac{F\text{Sin}[am]}{m}\right\}\right\}$

eq9 = eq6G /.$C[m] \to \frac{F\text{Sin}[am]}{m}$

$C[r,z] = c[0] + \int_0^\infty \frac{e^{-mz}F\texttt{BesselJ}[0, mr]\text{Sin}[am]}{m}\,dm$

eq10 = Assuming[$r < a$&&$r > 0$,

$\texttt{Simplify}\left[\left(c[0] + \int_0^\infty \frac{e^{-mz}F\texttt{BesselJ}[0, mr]\text{Sin}[am]}{m}\,dm\right) /.z \to 0\right]\Big] = 0$

$\frac{F\pi}{2} + c[0] = 0$

eq11 = Solve[eq10, F]

$\left\{\left\{F \to -\frac{2c[0]}{\pi}\right\}\right\}$

eq13 = eq9 /.$F \to -\frac{2c[0]}{\pi}$

$C[r,z] = c[0] + \int_0^\infty -\frac{2e^{-mz}\texttt{BesselJ}[0, mr]c[0]\text{Sin}[am]}{m\pi}\,dm$

eq14 = $J[r]$ = Assuming[$r < a$&&$r > 0$,

$\texttt{Simplify}\left[\left(\partial_z\left(c[0] + \int_0^\infty -\frac{2e^{-mz}\texttt{BesselJ}[0, mr]c[0]\text{Sin}[am]}{m\pi}\,dm\right)\right) /.\right.$
$z -> 0]]$

$\frac{2c[0]}{\pi\sqrt{a^2-r^2}}$

10.5.5 Mathematica Code: Weberdisctransientmat.nb (Section 10.3)

Eq0D = $sC[r,z] = \eta\frac{\partial_r C[r,z]}{r} + \eta\partial_{r,r}C[r,z] + \eta\partial_{z,z}C[r,z]$

$sC[r,z] = \eta C^{(0,2)}[r,z] + \frac{\eta C^{(1,0)}[r,z]}{r} + \eta C^{(2,0)}[r,z]$

$\texttt{DSolve}\left[\partial_{r,r}f[r] = \text{c1}f[r] - \frac{\partial_r f[r]}{r}, f[r], r\right]$

$\{\{f[r] \to \texttt{BesselJ}[0, i\sqrt{\text{c1}}r]C[1] + \texttt{BesselY}[0, -i\sqrt{\text{c1}}r]C[2]\}\}$

$\texttt{Simplify}\left[\texttt{DSolve}\left[\partial_{z,z}g[z] = -\text{c1}g[z] + \frac{sg[z]}{\eta}, g[z], z\right]\right] /.$
$\{C[1] \to C[3, C[2] \to C[4]\}$

$$\left\{\left\{g[z] \to e^{\frac{z\sqrt{s-\text{c1}\eta}}{\sqrt{\eta}}}C[3] + e^{-\frac{z\sqrt{s-\text{c1}\eta}}{\sqrt{\eta}}}C[4]\right\}\right\}$$

$$c[\text{r\_},\text{z\_}] = (\texttt{BesselJ}[0,\ i\sqrt{\text{c1}}r]C[1] + \texttt{BesselY}[0,\ -i\sqrt{\text{c1}}r]C[2])\left(e^{\frac{z\sqrt{s-\text{c1}\eta}}{\sqrt{\eta}}}C[3] + e^{-\frac{z\sqrt{s-\text{c1}\eta}}{\sqrt{\eta}}}C[4]\right)$$

$$(\texttt{BesselJ}[0,\ i\sqrt{\text{c1}}r]C[1] + \texttt{BesselY}[0,\ -i\sqrt{\text{c1}}r]C[2])\left(e^{\frac{z\sqrt{s-\text{c1}\eta}}{\sqrt{\eta}}}C[3] + e^{-\frac{z\sqrt{s-\text{c1}\eta}}{\sqrt{\eta}}}C[4]\right)$$

$$c[\text{r\_},\ \text{z\_}] = \%4\ /.\{C[1]\ \to\ 1,\ C[2]\ \to\ 0\}$$

$$\texttt{BesselJ}[0, i\sqrt{\text{c1}}r]\left(e^{\frac{z\sqrt{s-\text{c1}\eta}}{\sqrt{\eta}}}C[3] + e^{-\frac{z\sqrt{s-\text{c1}\eta}}{\sqrt{\eta}}}C[4]\right)$$

$$c[r,z]\ /\ .\text{c1} \to -m^2$$

$$\texttt{BesselJ}[0, i\sqrt{-m^2}r]\left(e^{\frac{z\sqrt{s+m^2\eta}}{\sqrt{\eta}}}C[3] + e^{-\frac{z\sqrt{s+m^2\eta}}{\sqrt{\eta}}}C[4]\right)$$

$$\%6\ /\ .C[3] \to 0$$

$$e^{-\frac{z\sqrt{s+m^2\eta}}{\sqrt{\eta}}}\texttt{BesselJ}[0,\ i\sqrt{-m^2}r]C[4]$$

$$c[r,z] = e^{-\frac{z\sqrt{s+m^2\eta}}{\sqrt{\eta}}}\texttt{BesselJ}[0,\ i\sqrt{-m^2}r]C[4]$$

$$(*\ \texttt{BesselJ}[0,\ -x] = \texttt{BesselJ}[0, x]\ *)$$

$$c[\text{r\_},\ \text{z\_}] = A[m]e^{-\frac{z\sqrt{s+m^2\eta}}{\sqrt{\eta}}}\texttt{BesselJ}[0,\ mr]$$

$$e^{-\frac{z\sqrt{s+m^2\eta}}{\sqrt{\eta}}}A[m]\texttt{BesselJ}[0,\ mr]$$

$$c[\text{r\_},\ \text{z\_}] = \frac{\text{C0}}{s} + h[r] + q[z] + \int_0^\infty A[m]e^{-\frac{z\sqrt{s+m^2\eta}}{\sqrt{\eta}}}\texttt{BesselJ}[0,\ mr]dm$$

$$\frac{\text{C0}}{s} + h[r] + \int_0^\infty e^{-\frac{z\sqrt{s+m^2\eta}}{\sqrt{\eta}}}A[m]\texttt{BesselJ}[0,\ mr]\,dm\ +\ q[z]$$

$$\texttt{DSolve}\left[sh[r] = \eta\frac{\partial_r h[r]}{r} + \eta\partial_{r,r}h[r] + \eta\partial_{z,z}h[r],\ h[r],\ r\right]$$

$$\left\{\left\{h[r] \to \texttt{BesselJ}\left[0,\ \frac{ir\sqrt{s}}{\sqrt{\eta}}\right]C[1] + \texttt{BesselY}\left[0,\ -\frac{ir\sqrt{s}}{\sqrt{\eta}}\right]C[2]\right\}\right\}$$

$$\texttt{DSolve}\left[sq[z] = \eta\frac{\partial_r q[z]}{r} + \eta\partial_{r,r}q[z] + \eta\partial_{z,z}q[z],\ q[z],\ z\right]\ /.\ \{C[1] \to B[1],\ C[2] \to B[2]\}$$

$$\left\{\left\{q[z] \to e^{\frac{\sqrt{s}z}{\sqrt{\eta}}}B[1] + e^{-\frac{\sqrt{s}z}{\sqrt{\eta}}}B[2]\right\}\right\}$$

$$c[\text{r\_},\ \text{z\_}] = \frac{\text{C0}}{s} + \left(\texttt{BesselJ}\left[0,\ \frac{ir\sqrt{s}}{\sqrt{\eta}}\right]C[1] + \texttt{BesselY}\left[0,\ -\frac{ir\sqrt{s}}{\sqrt{\eta}}\right]C[2]\right) + \left(e^{\frac{\sqrt{s}z}{\sqrt{\eta}}}B[1] + e^{-\frac{\sqrt{s}z}{\sqrt{\eta}}}B[2]\right) + \int_0^\infty A[m]e^{-\frac{z\sqrt{s+m^2\eta}}{\sqrt{\eta}}}\texttt{BesselJ}[0,\ mr]dm$$

$$\frac{\text{C0}}{s} + e^{\frac{\sqrt{s}z}{\sqrt{\eta}}}B[1] + e^{-\frac{\sqrt{s}z}{\sqrt{\eta}}}B[2] + \texttt{BesselJ}\left[0,\ \frac{ir\sqrt{s}}{\sqrt{\eta}}\right]C[1] + \texttt{BesselY}\left[0,\ -\frac{ir\sqrt{s}}{\sqrt{\eta}}\right]C[2] + \int_0^\infty e^{-\frac{z\sqrt{s+m^2\eta}}{\sqrt{\eta}}}A[m]\texttt{BesselJ}[0,\ mr]\,dm$$

$c[r,z]\ /.\{C[2] \to 0,\ B[1] \to 0\}$

$\frac{C0}{s} + e^{-\frac{\sqrt{s}z}{\sqrt{\eta}}} B[2] + \text{BesselJ}\left[0,\ \frac{ir\sqrt{s}}{\sqrt{\eta}}\right] C[1] + \int_0^\infty e^{-\frac{z\sqrt{s+m^2\eta}}{\sqrt{\eta}}} A[m]\text{BesselJ}[0,\ mr]\, dm$

$\left(*\ c[r,\ \infty] = \frac{C0}{s}\ *\right)$

$\text{Solve}\left[\frac{C0}{s} + \text{BesselJ}\left[0,\ \frac{ir\sqrt{s}}{\sqrt{\eta}}\right] C[1] = \frac{C0}{s},\ C[1]\right]$

$\{\{C[1] \to 0\}\}$

$c[r\_,\ z\_] = \frac{C0}{s} + e^{-\frac{\sqrt{s}z}{\sqrt{\eta}}} B[2] + \int_0^\infty e^{-\frac{z\sqrt{s+m^2\eta}}{\sqrt{\eta}}} A[m]\text{BesselJ}[0,\ mr] dm$

$\frac{C0}{s} + e^{-\frac{\sqrt{s}z}{\sqrt{\eta}}} B[2] + \int_0^\infty e^{-\frac{z\sqrt{s+m^2\eta}}{\sqrt{\eta}}} A[m]\text{BesselJ}[0,\ mr]\ dm$

$\partial_z c[r,z]\ /.\ z \to 0$

$-\frac{\sqrt{s}B[2]}{\sqrt{\eta}} + \int_0^\infty -\frac{\sqrt{s+m^2\eta}A[m]\text{BesselJ}[0,\ mr]}{\sqrt{\eta}}\, dm$

$(*\ B[2] = 0\ *)$

$\int_0^\infty -\frac{\sqrt{s+m^2\eta}A[m]\text{BesselJ}[0,\ mr]}{\sqrt{\eta}}\, dm = 0$

$\text{Assuming}\left[r > 0 \&\& r < a,\ \int_0^\infty \text{Sin}[ma]\,\text{BesselJ}[0,\ mr] dm\right]$

$\frac{1}{\sqrt{a^2-r^2}}$

$\text{Assuming}\left[a > 0 \&\& a < r,\ \int_0^\infty \text{Sin}[ma]\,\text{BesselJ}[0,\ mr] dm\right]$

0

$\int_0^\infty -\frac{\sqrt{s+m^2\eta}A[m]\text{BesselJ}[0,\ mr]}{\sqrt{\eta}}\, dm = F\int_0^\infty \text{Sin}[ma]\text{BesselJ}[0,\ mr] dm$

$\text{Solve}\left[-\frac{\sqrt{s+m^2\eta}A[m]\text{BesselJ}[0,\ mr]}{\sqrt{\eta}} = F\,\text{Sin}[ma]\text{BesselJ}[0,\ mr],\ A[m]\right]$

$\left\{\left\{A[m] \to -\frac{F\sqrt{\eta}\text{Sin}[am]}{\sqrt{s+m^2\eta}}\right\}\right\}$

$c[r\_,\ z\_] = \frac{C0}{s} + \int_0^\infty e^{-\frac{z\sqrt{s+m^2\eta}}{\sqrt{\eta}}} \left(-\frac{F\sqrt{\eta}\text{Sin}[am]}{\sqrt{s+m^2\eta}}\right) \text{BesselJ}[0,\ mr] dm;$

$\frac{C0}{s} + \int_0^\infty \left(-\frac{F\sqrt{\eta}\text{Sin}[am]}{\sqrt{s+m^2\eta}}\right) \text{BesselJ}[0,\ mr] dm = 0$

$F = \frac{C0}{s\sqrt{\eta}\int_0^\infty \left(\frac{\text{BesselJ}[0,\ kr]\ \text{Sin}[ak]}{\sqrt{s+k^2\eta}}\right) dk};$

$c[r\_,\ z\_] =$

$\frac{C0}{s} + \int_0^\infty e^{-\frac{z\sqrt{s+m^2\eta}}{\sqrt{\eta}}} \left(-\left(\frac{C0}{s\sqrt{\eta}\int_0^\infty \left(\frac{\text{BesselJ}[0,\ kr]\ \text{Sin}[ak]}{\sqrt{s+k^2\eta}}\right) dk}\right) \frac{\sqrt{\eta}\text{Sin}[am]}{\sqrt{s+m^2\eta}}\right)$

$\text{BesselJ}[0,\ mr] dm$

10.5.6 wxMaxima Code: Weberdisintegralxm.wxm (Section 10.1)

```
(%i1) load(vect);
(%i2) scalefactors([[r*cos(theta),r*sin(theta),z],r,theta,z]);
```

(%o2) *done*

```
(%i3) depends(V,[r,z]);
```

(%o3) $[V(r,z)]$

```
(%i4) assume_pos: r;
```

(%o4) r

```
(%i5) eq0: express(laplacian(V));
```

(%o5) $$\frac{\frac{d}{dz}\left(r\left(\frac{d}{dz}V\right)\right)+\frac{d}{d\theta}\frac{\frac{d}{d\theta}V}{r}+\frac{d}{dr}\left(r\left(\frac{d}{dr}V\right)\right)}{|r|}$$

```
Multiplicative separation: Maxima is not able to solve the equations
(%i27) eq8: V=f(r)*g(z)$
(%i28) eq9: (subst(eq8,eq0))$
(%i29) eq10: eq9,simp, diff;
```

(%o29) $$\frac{r\ f(r)\left(\frac{d^2}{d\ z^2}g(z)\right)+r\left(\frac{d^2}{d\ r^2}f(r)\right)g(z)+\left(\frac{d}{d\ r}f(r)\right)g(z)}{r}$$

```
(%i30) eq11: eq10/(f(r)*g(z)), simp, expand;
```

(%o30) $$\frac{\frac{d^2}{d\ z^2}g(z)}{g(z)}+\frac{\frac{d^2}{d\ r^2}f(r)}{f(r)}+\frac{\frac{d}{d\ r}f(r)}{r\ f(r)}$$

```
(%i35) eq12A: 'diff(g(z),z,2)/g(z)=m^2$
(%i36) eq12B: 'diff(f(r),r,2)/f(r)+'diff(f(r),r,1)/(r*f(r))+m^2$
(%i37) eq13: g(theta)=C(m)*exp(-m*z);
```

(%o37) $g(\theta)=C(m)\ \%e^{-m\ z}$

```
(%i39) eq14: f(r)=bessel_j(0,m*r);
```

(%o39) $f(r)=\mathtt{bessel\_j}(0,\ m\ r)$

```
(%i40) eq23: V(r,z)=integrate (rhs(eq13)*rhs(eq14), m, 0, inf);
```

(%o40) $$V(r,z)=\int_0^{\infty}\mathtt{bessel\_j}(0,m\,r)\,C(m)\ \%e^{-m\ z}\,dm$$

```
(%i41) eq23A: sigma(r)=(-1/(4*%pi))*diff(rhs(eq23),z);
```

(%o41) $$\sigma(r)=\frac{\int_0^{\infty}\mathtt{bessel\_j}(0,m\,r)\,m\,C(m)\ \%e^{-m\ z}\,dm}{4\ \pi}$$

```
(%i42) eq23B: subst(z=0,subst('diff(C(m),z,1)=0,eq23A));
```

(%o42) $$\sigma(r)=\frac{\int_0^{\infty}\mathtt{bessel\_j}(0,m\,r)\,m\ C(m)\ dm}{4\,\pi}$$

```
(%i43) eq23C: m*C(m)*bessel_j(0,m*r)=bessel_j(0,m*r)*sin(m*a)*F;
```

(%o43) $\mathtt{bessel\_j}(0,m\,r)\,m\,C(m)=\mathtt{bessel\_j}(0,m\,r)\,\sin(a\,m)\ F$

```
(%i44) eq24: solve(eq23C,C(m));
```

(%o44) $$\left[C(m)=\frac{\sin(a\,m)\,F}{m}\right]$$

```
(%i45) eq25: subst(eq24,eq23);
```

(%o45) $$V(r,z)=\int_0^{\infty}\frac{\mathtt{bessel\_j}(0,m\,r)\,\mathrm{Sin}(a\,m)\,\%e^{-m\,z}}{m}\,dm\,F$$

```
(%i46) eq26: subst(eq24,eq23B)$
(%i47) eq27: sigma(r)=(F/(4*%pi))*(1/sqrt(a^2-r^2));
```

(%o47) $$\sigma(r)=\frac{F}{4\,\pi\,\sqrt{a^2-r^2}}$$

```
(%i48) eq28: Q=4*%pi*integrate(r*rhs(eq27),r,0,a);
```

(%o48) $Q = a\,F$

```
(%i23) eq29: solve(eq28,F);
```

(%o23) $\left[F = \frac{Q}{a}\right]$

```
(%i49) eq30: subst(eq29,eq25);
```

(%o49) $V(r,\ z) = \frac{\int_0^{\infty} \frac{\text{bessel\_j}(0,\ m\ r)\ \text{Sin}(a\ m)\ \%e^{-m\ z}}{m}\, dm\ Q}{a}$

```
(%i50) eq31: subst(eq29,eq27);
```

(%o50) $\sigma(r) = \frac{Q}{4\ \pi\ a\ \sqrt{a^2-r^2}}$

10.5.7 wxMaxima Code: Weberdiscsteadyxm.wxm (Section 10.2)

```
(%i1) load(vect);
(%i2) scalefactors([[r*cos(theta),r*sin(theta),z],r,theta,z]);
```

(%o2) *done*

```
(%i3) depends(C,[r,z]);
```

(%o3) $[C(r,\ z)]$

```
(%i4) assume_pos: r;
```

(%o4) r

```
(%i5) eq0: express(laplacian(C));
```

(%o5) $\frac{\frac{d}{d\ z}\left(r\left(\frac{d}{dz}\ V\right)\right) + \frac{d}{d\theta}\ \frac{\frac{d}{d\theta}\ C}{r} + \frac{d}{d\ r}\ \left(r\left(\frac{d}{d\ r}\ C\right)\right)}{|r|}$

Multiplicative separation: Maxima is not able to solve the equations

```
(%i6) eq8: C=f(r)*g(z)$
(%i7) eq9: (subst(eq8,eq0))$
(%i8) eq10: eq9,simp, diff$
(%i9) eq11: eq10/(f(r)*g(z)), simp, expand;
```

(%o9) $\frac{\frac{d^2}{d\ z^2}\ g(z)}{g(z)} + \frac{\frac{d^2}{d\ r^2}\ f(r)}{f(r)} + \frac{\frac{d}{d\ r}\ f(r)}{r\ f(r)}$

```
(%i10) eq12A: 'diff(g(z),z,2)/g(z)=m^2$
(%i11) eq12B: 'diff(f(r),r,2)/f(r)+'diff(f(r),r,1)/(r*f(r))+m^2$
(%i12) eq13: g(theta)=C(m)*exp(-m*z);
```

(%o12) $g(\theta) = C(m)\ \%e^{-m\ z}$

```
(%i13) eq14: f(r)=bessel_j(0, m*r);
```

(%o13) $f(r) = \text{bessel\_j}(0,\ m\ r)$

```
(%i14) eq23: C(r,z)=c[0]+integrate (rhs(eq13)*rhs(eq14), m, 0, inf);
```

(%o14) $C(r,\ z) = \int_0^{\infty} \text{bessel\_j}(0,\ m\ r)\ C(m)\ \%e^{-m\ z}\ dm\ +\ c_0$

```
(%i15) eq23A: diff(rhs(eq23),z)$
(%i16) eq23B: subst(z=0,subst('diff(C(m),z,1)=0,eq23A))$
(%i19) eq23C: -m*C(m)*bessel_j(0,m*r)=bessel_j(0,m*r)*sin(m*a)*F;
```

(%o19) $-\text{bessel\_j}(0,\ m\ r)\ m\ C(m) = \text{bessel\_j}(0,\ m\ r)\ \sin\ (a\ m)\ F$

```
(%i20) eq24: solve(eq23C,C(m));
```

(%o20) $\left[C\ (m) = -\frac{\sin(a\ m)\ F}{m}\right]$

```
(%i21) eq25: subst(eq24,eq23)$
```

```
(%i22) eq26: subst(z=0,rhs(eq25))=0;
```

$$(\%o22)\quad c_0 - \int_0^{\infty} \frac{\text{bessel\_j}(0,\ m\ r)\ \text{Sin}(a\ m)}{m}\, dm\ F = 0$$

```
(%i23) eq27: integrate((sin(a*m)*J[0](m*r))/m,m,0,inf)=%pi/2;
```

$$(\%o23)\quad \int_0^{\infty} \frac{\sin(a\ m)\ J_0(m\ r)}{m} dm = \frac{\pi}{2}$$

```
(%i24) eq28: subst(eq27,eq26);
```

$$(\%o24)\quad c_0 - \int_0^{\infty} \frac{\text{bessel\_j}(0,\ m\ r)\ \sin(a\ m)}{m}\, dm\ F = 0$$

```
(%i25) eq29: solve(eq28,F);
```

$$(\%o25)\quad \left[F = \frac{c_0}{\int_0^{\infty} \frac{\text{bessel\_j}(0,\ m\ r)\ \sin(a\ m)}{m}\, dm} \right]$$

```
(%i26) eq30: subst(eq29,eq25);
```

$$(\%o26)\quad \text{C}(r,\ z) = c_0 - \frac{c_0 \int_0^{\infty} \frac{\text{bessel\_j}(0,\ m\ r)\ \sin(a\ m)\ \%e^{-m\ z}}{m}\, dm}{\int_0^{\infty} \frac{\text{bessel\_j}(0,\ m\ r)\ \sin(a\ m)}{m}\, dm}$$

```
(%i27) eq31: J(r)=subst(z=0,diff(rhs(eq30),z));
```

$$(\%o27)\quad \text{J}(r) = \frac{c_0 \int_0^{\infty} \text{bessel\_j}(0,\ m\ r)\ \sin(a\ m)\ dm}{\int_0^{\infty} \frac{\text{bessel\_j}(0,\ m\ r)\ \sin(a\ m)}{m}\, dm}$$

```
(%i28) eq32: J(r)=2*c[0]/(%pi*sqrt(a^2-r^2));
```

$$(\%o28)\quad \text{J}(r) = \frac{2\ c_0}{\pi \sqrt{a^2 - r^2}}$$

PROBLEMS

10.1. Plot the concentration and the surface concentration gradient (i.e., at $z = 0$) with $C_0 = 1$ and $a = 0.5$. Use the model for the steady-state diffusion-controlled ligand binding problem presented in Section 10.2.

10.2. What are the steady-state ligand concentration and surface concentration gradient when $r > a$? Use the model presented in Section 10.2.

10.3. Simplify the concentration:

$$C(r,\ 0) = C_0 - \frac{2C_0}{\pi} \int_0^{\infty} \left(\frac{\sin(ma)\, J_0(mr)}{m} \right) dm$$

Note: The final expression should not include the improper integral.

10.4. Plot the surface concentration gradient for the transient model presented in Section 10.3. Use the following parameters: $C_0 = 1$, $a = 0.5$ and $\eta = 1$. The simulation time is 20.

10.5. What is the ligand concentration $c(0.3,\ 1.0,\ 3.0)$ for the transient model presented in Section 10.3? Use the following parameters: $C_0 = 1$, $a = 0.5$ and $\eta = 1$.

REFERENCES

Berezhkovskii, A. M., Batsilas, L., & Shvartsman, S. Y. (2004). Ligand trapping in epithelial layers and cell cultures. *Biophys Chem*, 107(3), 221–227.

Crank, John. (1975). *The mathematics of diffusion* (2nd ed.). Oxford: Clarendon Press.

Rice, Richard G., & Do, Duong D. (1995). *Applied mathematics and modeling for chemical engineers*. New York: Wiley.

Stone, Michael, & Goldbart, Paul M. (2009). *Mathematics for physics: a guided tour for graduate students*. Cambridge, UK: Cambridge University Press.

Traytak, S. D. (2013). Ligand binding in a spherical region randomly crowded by receptors. *Phys Biol*, 10(4), 045009.

Wang, D., Gou, S. Y., & Axelrod, D. (1992). Reaction rate enhancement by surface diffusion of adsorbates. *Biophys Chem*, 43(2), 117–137.

Zwanzig, R., & Szabo, A. (1991). Time dependent rate of diffusion-influenced ligand binding to receptors on cell surfaces. *Biophys J*, 60(3), 671–678.

11

TWO-DIMENSIONAL ANALYSIS OF A CYLINDRICAL MATRIX DEVICE WITH A SMALL HOLE FOR DRUG RELEASE

11.1 MATHEMATICAL MODELING OF DRUG TRANSPORT THROUGH THE DEVICE

A model for a controlled-release device, first analyzed for prolonged delivery of a medicament, is solved in this chapter (Tojo, 1984; Tojo & Miyanami, 1983). The monolithic system is shown in Fig. 11.1. A polymer coating is applied to prevent mass transfer to the surrounding medium except for a small release area located at the center top of the system. Similar designs have been investigated by several authors (Ertan et al., 1997; Hsu et al., 1992; Simon & Ospina, 2012). The analytical approach may help improve the design of these devices.

The governing equation is

$$\frac{\partial}{\partial t}c(r, z, t) = \frac{\eta\left(\frac{\partial}{\partial r}c(r, z, t) + r\frac{\partial^2}{\partial r^2}c(r, z, t) + r\frac{\partial^2}{\partial z^2}c(r, z, t)\right)}{r} \tag{11.1}$$

where $c(r, z, t)$ is the concentration of the drug at the point (r, z) and η is its diffusion coefficient in the matrix. The initial concentration of the active pharmaceutical ingredient (API) is

$$c(r, z, 0) = c_{AS} \tag{11.2}$$

The boundary conditions at $r = R$ and $z = 0$ are

$$\left.\frac{\partial}{\partial r}c(r, z, t)\right|_{r=R} = 0 \tag{11.3}$$

Closed-form Solutions for Drug Transport through Controlled-Release Devices in Two and Three Dimensions, First Edition. Laurent Simon and Juan Ospina.

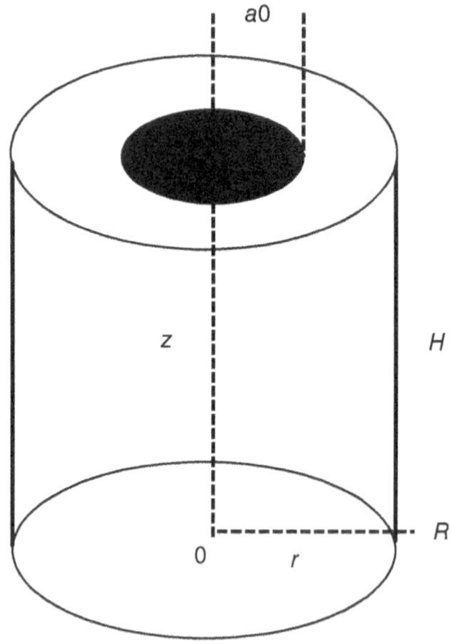

Figure 11.1 Cylindrical controlled-release device with an aperture of radius a_0.

and

$$\frac{\partial}{\partial z} c\left(r, z, t\right)\bigg|_{z=0} = 0 \tag{11.4}$$

The condition used by Tojo and Miyanami (1983), at $z = H$ is

$$\frac{\partial}{\partial z} c\left(r, z, t\right)\bigg|_{z=H} = \begin{cases} -\frac{k_m c(r,\, z,\, H)}{\eta} & r \le a_0 \\ 0 & a_0 < r \end{cases} \tag{11.5}$$

These boundary conditions, defined by Eqs. (11.3)–(11.5), mean that the drug is released only through the small circular area of radius a_0 drilled at $z = H$. The parameter k_m denotes a boundary-layer mass transfer coefficient. When k_m is high, there is a low mass transfer resistance as a result of factors, such as a vigorous mixing. A small k_m means that drug molecules exit the device at a slow rate.

This problem was solved analytically in (Simon & Ospina, 2012). Instead of applying Eq. (11.5), we consider two cases:

$$c(r,\ H,\ t) = 0 \tag{11.6}$$

and

$$\begin{cases} \frac{\partial}{\partial z} c\left(r, z, t\right)\Big|_{z=H} = 0 & a_0 < r \\ c(r, H, t) = 0 & r \le a_0 \end{cases} \tag{11.7}$$

Equation (11.6) represents the perfect sink condition at any location r while Eq. (11.7) applies this condition only when $r \le a_0$.

11.2 DRUG CONCENTRATION PROFILE INSIDE THE MATRIX

11.2.1 Concentration Profile Using Boundary Condition (11.6)

Given the boundary conditions, the concentration is not expected to be a function of r. Nonetheless, we describe a general method which is also applicable to the case when Eq. (11.7) is used. The Laplace transform of Eq. (11.1) with the initial (11.2) gives

$$sC(r,z) - c_{AS} = \eta \frac{1}{r}\frac{\partial C}{\partial r}(r,z) + \eta \frac{\partial^2}{\partial r^2} C(r,z) + \frac{\partial^2}{\partial z^2} C(r,z) \tag{11.8}$$

where

$$C(r,\, z) = \int_0^{\infty} c(r,\, z,\, t) e^{-st} dt \tag{11.9}$$

After writing $C(r, z)$ as $C(r, z) = C_{h1}(r, z) + C_{h2}(z) + c_{AS}/s$, the solution $C_{h1}(r, z)$, obtained from Eq. (11.8), is

$$\begin{aligned} C_{h1}(r,\, z) &= [C_1 J_0(r\sqrt{-c_1}) + C_2 Y_0(r\sqrt{-c_1})] \\ &\times \left[C_3 \sin\left(z\sqrt{\frac{c_1\eta - s}{\eta}}\right) + C_4 \cos\left(z\sqrt{\frac{c_1\eta - s}{\eta}}\right)\right] \end{aligned} \tag{11.10}$$

Because $Y_0(r)$ is singular around $r = 0$, the constant C_2 is set to zero for a finite solution at $r = 0$. Then, Eq. (11.10) becomes

$$C_{h1}(r,\, z) = J_0(r\sqrt{-c_1}) \left[C_3 \sin\left(z\sqrt{\frac{c_1\eta - s}{\eta}}\right) + C_4 \cos\left(z\sqrt{\frac{c_1\eta - s}{\eta}}\right)\right] \tag{11.11}$$

given that C_1 can be set equal to 1. Applying the Laplace transform of the boundary condition (11.3) to Eq. (11.11), gives

$$-\sqrt{-c_1} J_1(R\sqrt{-c_1}) \left[C_3 \sin\left(z\sqrt{\frac{c_1\eta - s}{\eta}}\right) + C_4 \cos\left(z\sqrt{\frac{c_1\eta - s}{\eta}}\right)\right] = 0 \tag{11.12}$$

Therefore,

$$J_1(R\sqrt{-c_1}) = 0 \tag{11.13}$$

which yields

$$c_1 = -\frac{\alpha_{1,n}^2}{R^2} \tag{11.14}$$

where $\alpha_{1,n}$ are the zeroes of $J_1(x)$. Equation (11.11) becomes

$$C_{h1}(r,\, z) = J_0\left(\frac{\alpha_{1,n} r}{R}\right) \left[C_3 i \sinh\left(\frac{z}{R}\sqrt{\frac{\alpha_{1,n}^2 \eta + sR^2}{\eta}}\right) + C_4 \cosh\left(\frac{z}{R}\sqrt{\frac{\alpha_{1,n}^2 \eta + sR^2}{\eta}}\right)\right] \tag{11.15}$$

Application of the **principle of superposition** results in

$$C_{h1}(r,z)=\sum_{n=1}^{\infty}J_0\left(\frac{\alpha_{1,n}r}{R}\right)\left[A_n i\sinh\left(\frac{z}{R}\sqrt{\frac{\alpha_{1,n}^2\eta+sR^2}{\eta}}\right)+B_n\cosh\left(\frac{z}{R}\sqrt{\frac{\alpha_{1,n}^2\eta+sR^2}{\eta}}\right)\right] \tag{11.16}$$

The solution $C_{h2}(z)$ takes the form

$$C_{h2}=C_1e^{z\sqrt{\frac{s}{\eta}}}+C_2e^{-z\sqrt{\frac{s}{\eta}}} \tag{11.17}$$

After applying the boundary condition (11.4), we obtain

$$C_1=C_2 \tag{11.18}$$

which leads to

$$C_{h2}=C_2e^{z\sqrt{\frac{s}{\eta}}}+C_1e^{-z\sqrt{\frac{s}{\eta}}} \tag{11.19}$$

Finally, the general solution of Eq. (11.8) is given by

$$C(r,z)=\frac{c_{AS}}{s}+C_1e^{z\sqrt{\frac{s}{\eta}}}+C_1e^{-z\sqrt{\frac{s}{\eta}}}$$

$$+\sum_{n=1}^{\infty}J_0\left(\frac{\alpha_{1,n}r}{R}\right)\left[A_n i\sinh\left(\frac{z}{R}\sqrt{\frac{\alpha_{1,n}^2\eta+sR^2}{\eta}}\right)+B_n\cosh\left(\frac{z}{R}\sqrt{\frac{\alpha_{1,n}^2\eta+sR^2}{\eta}}\right)\right] \tag{11.20}$$

The boundary condition (11.6) implies

$$0=\frac{c_{AS}}{s}+C_1e^{H\sqrt{\frac{s}{\eta}}}+C_1e^{-H\sqrt{\frac{s}{\eta}}}$$

$$+\sum_{n=1}^{\infty}J_0\left(\frac{\alpha_{1,n}r}{R}\right)\left[A_n i\sinh\left(\frac{H}{R}\sqrt{\frac{\alpha_{1,n}^2\eta+sR^2}{\eta}}\right)+B_n\cosh\left(\frac{H}{R}\sqrt{\frac{\alpha_{1,n}^2\eta+sR^2}{\eta}}\right)\right] \tag{11.21}$$

Multiplying both sides of Eq. (11.21) by $rJ_0\left(\frac{\alpha_{1,n}r}{R}\right)$ and integrating from 0 to R, we obtain

$$0=\left(\frac{c_{AS}}{s}+C_1e^{H\sqrt{\frac{s}{\eta}}}+C_1e^{-H\sqrt{\frac{s}{\eta}}}\right)\int_0^R rJ_0\left(\frac{\alpha_{1,n}r}{R}\right)dr$$

$$+\sum_{n=1}^{\infty}\int_0^R rJ_0\left(\frac{\alpha_{1,n}r}{R}\right)^2dr\left[A_n i\sinh\left(\frac{H}{R}\sqrt{\frac{\alpha_{1,n}^2\eta+sR^2}{\eta}}\right)+B_n\cosh\left(\frac{H}{R}\sqrt{\frac{\alpha_{1,n}^2\eta+sR^2}{\eta}}\right)\right] \tag{11.22}$$

or

$$0=\left(\frac{c_{AS}}{s}+C_1e^{H\sqrt{\frac{s}{\eta}}}+C_1e^{-H\sqrt{\frac{s}{\eta}}}\right)\frac{R^2J_1(\alpha_{1,n})}{\alpha_{1,n}}$$

$$+\sum_{n=1}^{\infty}\frac{R^2(\alpha_{1,n}J_0(\alpha_{1,n})^2+\alpha_{1,n}J_1(\alpha_{1,n})^2)}{2\alpha_{1,n}}\left[\begin{array}{l}A_n i\sinh\left(\frac{H}{R}\sqrt{\frac{\alpha_{1,n}^2\eta+sR^2}{\eta}}\right)\\+B_n\cosh\left(\frac{H}{R}\sqrt{\frac{\alpha_{1,n}^2\eta+sR^2}{\eta}}\right)\end{array}\right] \tag{11.23}$$

after computing the integrals. Given that $J_1(\alpha_{1,n})=0$, Eq.(11.23) reduces to

$$\sum_{n=1}^{\infty}\frac{R^2(\alpha_{1,n}J_0(\alpha_{1,n})^2)}{2\alpha_{1,n}}\left[A_n i\sinh\left(\frac{H}{R}\sqrt{\frac{\alpha_{1,n}^2\eta+sR^2}{\eta}}\right)+B_n\cosh\left(\frac{H}{R}\sqrt{\frac{\alpha_{1,n}^2\eta+sR^2}{\eta}}\right)\right]=0 \tag{11.24}$$

such that

$$\left[A_n i\sinh\left(\frac{H}{R}\sqrt{\frac{\alpha_{1,n}^2\eta+sR^2}{\eta}}\right)+B_n\cosh\left(\frac{H}{R}\sqrt{\frac{\alpha_{1,n}^2\eta+sR^2}{\eta}}\right)\right]=0 \tag{11.25}$$

Furthermore, the boundary condition (11.4) applied to Eq. (11.20) gives

$$\sum_{n=1}^{\infty}J_0\left(\frac{\alpha_{1,n}r}{R}\right)\left[\frac{A_n i}{R}\sqrt{\frac{\alpha_{1,n}^2\eta+sR^2}{\eta}}\right]=0 \tag{11.26}$$

From Eq. (11.26), we conclude that

$$A_n=0 \tag{11.27}$$

and

$$B_n=0 \tag{11.28}$$

after using Eq. (11.25). Therefore, the Laplace transformed concentration is

$$C(r,z)=\frac{c_{AS}}{s}+C_1e^{z\sqrt{\frac{s}{\eta}}}+C_1e^{-z\sqrt{\frac{s}{\eta}}} \tag{11.29}$$

Upon applying the boundary condition (11.6), the following equation holds:

$$\frac{c_{AS}}{s}+C_1e^{H\sqrt{\frac{s}{\eta}}}+C_1e^{-H\sqrt{\frac{s}{\eta}}}=0 \tag{11.30}$$

We deduce from Eq. (11.30) that

$$C_1=-\frac{c_{AS}}{s\left(e^{H\sqrt{\frac{s}{\eta}}}+C_1e^{-H\sqrt{\frac{s}{\eta}}}\right)} \tag{11.31}$$

or

$$C_1=-\frac{c_{AS}}{\frac{1}{2}s\cosh\left(H\sqrt{\frac{s}{\eta}}\right)} \tag{11.32}$$

Equation (11.29) becomes

$$C(r, z) = \frac{c_{AS}}{s} - \frac{c_{AS}}{\frac{1}{2} s \cosh\left(H\sqrt{\frac{s}{\eta}}\right)} \left(e^{z\sqrt{\frac{s}{\eta}}} + e^{-z\sqrt{\frac{s}{\eta}}}\right) \tag{11.33}$$

or

$$C(r, z) = \frac{c_{AS}\left(\cosh\left(H\sqrt{\frac{s}{\eta}}\right) - \cosh\left(z\sqrt{\frac{s}{\eta}}\right)\right)}{s \cosh\left(H\sqrt{\frac{s}{\eta}}\right)} \tag{11.34}$$

The inverse Laplace transform of $C(r, z)$ is computed using the following **Bromwich integral**

$$C(r, z, t) = \frac{1}{2\pi i} \int \left[\frac{c_{AS}\left(\cosh\left(H\sqrt{\frac{s}{\eta}}\right) - \cosh\left(z\sqrt{\frac{s}{\eta}}\right)\right)}{s \cosh\left(H\sqrt{\frac{s}{\eta}}\right)}\right] e^{st} ds \tag{11.35}$$

and the Bromwich integral is computed using the **residue theorem** as

$$C(r, z, t) = \sum_{pole} Residue\left(\left[\frac{c_{AS}\left(\cosh\left(H\sqrt{\frac{s}{\eta}}\right) - \cosh\left(z\sqrt{\frac{s}{\eta}}\right)\right)}{s \cosh\left(H\sqrt{\frac{s}{\eta}}\right)}\right] e^{st}\right)\Bigg|_{s=pole} \tag{11.36}$$

The residue at the pole $s = 0$ is 0. The other poles are determined by the equation

$$\cosh\left(H\sqrt{\frac{s}{\eta}}\right) = 0 \tag{11.37}$$

which yields

$$s = -\frac{(2n+1)^2 \pi^2 \eta}{4H^2} \tag{11.38}$$

where n is an integer ranging from 0 to ∞. The residue at the pole takes the form

$$Res = \frac{4c_{AS} \cos\left(\frac{(2n+1)\pi z}{2H}\right) e^{-\frac{(2n+1)^2 \pi^2 \eta t}{2H^2}}}{\pi(-1)^n(2n+1)} \tag{11.39}$$

As a result, the concentration in the time domain is

$$c(r, z, t) = \sum_{n=0}^{\infty} \left(\frac{4c_{AS} \cos\left(\frac{(2n+1)\pi z}{2H}\right) e^{-\frac{(2n+1)^2 \pi^2 \eta t}{2H^2}}}{\pi(-1)^n(2n+1)}\right) \tag{11.40}$$

The profile for $c(r, z, t)$ is plotted in Fig. 11.2 with the following parameter values: $c_{AS} = 1.0$ $H = 1.0$ and $\eta = 1.0$ for the first 20 terms.

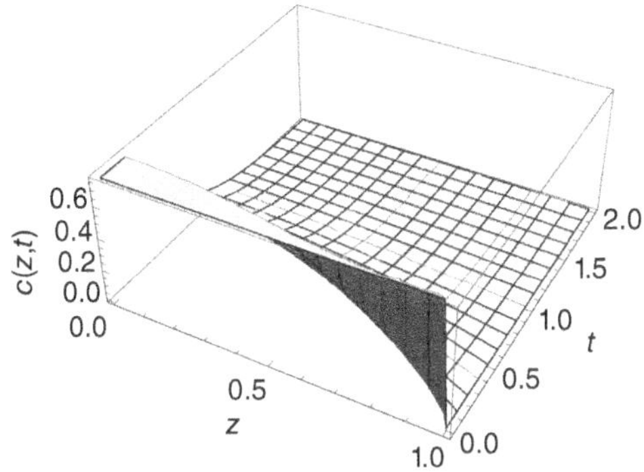

Figure 11.2 The profile of $c(r, z, t)$ with the boundary condition $c(r, H, t) = 0$ and the following parameter values: $c_{AS} = 1.0$ $H = 1.0$ and $\eta = 1.0$.

11.2.2 Concentration Profile Using Boundary Condition (11.7)

The boundary condition (11.7) is rewritten as

$$\frac{\partial}{\partial z} c(r, z, t)\bigg|_{z=H} \text{Heaviside}(r - a_0) = \beta c(r, z, t)(1 - \text{Heaviside}(r - a_0)) \tag{11.41}$$

with $\beta = -1$. Again, we look for a solution of the form (11.20), with A_n defined by Eq. (11.27):

$$C(r, z) = \frac{c_{AS}}{s} + C_1 e^{z\sqrt{\frac{s}{\eta}}} + C_1 e^{-z\sqrt{\frac{s}{\eta}}} + \sum_{n=1}^{\infty} J_0\left(\frac{\alpha_{1,n} r}{R}\right)\left[B_n \cosh\left(\frac{z}{R}\sqrt{\frac{\alpha_{1,n}^2 \eta + sR^2}{\eta}}\right)\right] \tag{11.42}$$

Eq. (11.41) becomes

$$\left(\begin{array}{c} C_1\sqrt{\frac{s}{\eta}} e^{H\sqrt{\frac{s}{\eta}}} - C_1\sqrt{\frac{s}{\eta}} e^{-H\sqrt{\frac{s}{\eta}}} \\ + \displaystyle\sum_{n=1}^{\infty} J_0\left(\frac{\alpha_{1,n} r}{R}\right)\left[B_n\left(\frac{1}{R}\sqrt{\frac{\alpha_{1,n}^2\eta + sR^2}{\eta}}\right)\sinh\left(\frac{H}{R}\sqrt{\frac{\alpha_{1,n}^2\eta + sR^2}{\eta}}\right)\right]\end{array}\right)\text{Heaviside}(r - a_0)$$
$$= \beta\left(\frac{c_{AS}}{s} + C_1 e^{H\sqrt{\frac{s}{\eta}}} + C_1 e^{-H\sqrt{\frac{s}{\eta}}} + \sum_{n=1}^{\infty} J_0\left(\frac{\alpha_{1,n} r}{R}\right)\left[B_n \cosh\left(\frac{H}{R}\sqrt{\frac{\alpha_{1,n}^2\eta + sR^2}{\eta}}\right)\right]\right)$$
$$\times (1 - \text{Heaviside}(r - a_0)) \tag{11.43}$$

Equation (11.43) generates a system of an infinite number of equations with an infinite number of constants C_1 and B_n. Some approximations are made to obtain explicit expressions for the concentration profile. A first-order estimation leads to

$$\left(\begin{array}{c} C_1\sqrt{\frac{s}{\eta}} e^{H\sqrt{\frac{s}{\eta}}} - C_1\sqrt{\frac{s}{\eta}} e^{-H\sqrt{\frac{s}{\eta}}} \\ + J_0\left(\frac{\alpha_{1,1} r}{R}\right)\left[B_1\left(\frac{1}{R}\sqrt{\frac{\alpha_{1,1}^2\eta + sR^2}{\eta}}\right)\sinh\left(\frac{H}{R}\sqrt{\frac{\alpha_{1,1}^2\eta + sR^2}{\eta}}\right)\right]\end{array}\right)\text{Heaviside}(r - a_0)$$

$$= \beta\left(\frac{c_{AS}}{s} + C_1 e^{H\sqrt{\frac{s}{\eta}}} + C_1 e^{-H\sqrt{\frac{s}{\eta}}} + J_0\left(\frac{\alpha_{1,1} r}{R}\right)\left[B_n \cosh\left(\frac{H}{R}\sqrt{\frac{\alpha_{1,1}^2 \eta + sR^2}{\eta}}\right)\right]\right)$$
$$\times\ (1 - \text{Heaviside}(r - a_0)) \tag{11.44}$$

and Eq. (11.42) reduces to

$$C(r, z) = \frac{c_{AS}}{s} + C_1 e^{z\sqrt{\frac{s}{\eta}}} + C_1 e^{-z\sqrt{\frac{s}{\eta}}} + J_0\left(\frac{\alpha_{1,1} r}{R}\right)\left[B_1 \cosh\left(\frac{z}{R}\sqrt{\frac{\alpha_{1,1}^2 \eta + sR^2}{\eta}}\right)\right] \tag{11.45}$$

After multiplying both sides of Eq. (11.44) by r and integrating from 0 to R, we get

$$-\frac{e^{-H\sqrt{\frac{s}{\eta}}}}{2\sqrt{\eta}\alpha_{1,1}}\left(\begin{array}{c} 2a_0 B_1 \sinh\left(\frac{H}{R}\sqrt{\frac{sR^2 + \alpha_{1,1}{}^2\eta}{\eta}}\right)\sqrt{sR^2 + \alpha_{1,1}{}^2\eta}\, J_1\left(\frac{a_0\alpha_{1,1}}{R}\right) e^{H\sqrt{\frac{s}{\eta}}} \\ + \alpha_{1,1} C_1 \sqrt{s} e^{2H\sqrt{\frac{s}{\eta}}} a_0{}^2 - \alpha_{1,1} C_1 \sqrt{s} a_0{}^2 - \alpha_{1,1} C_1 \sqrt{s} e^{2H\sqrt{\frac{s}{\eta}}} R^2 + \alpha_{1,1} C_1 \sqrt{s} R^2 \end{array}\right)$$
$$= \frac{e^{-H\sqrt{\frac{s}{\eta}}}}{2s\alpha_{1,1}}\left(\begin{array}{c} \beta a_0{}^2 c_{AS} e^{H\sqrt{\frac{s}{\eta}}}\alpha_{1,1} + \beta a_0{}^2 C_1 e^{2H\sqrt{\frac{s}{\eta}}} s\alpha_{1,1} + \beta a_0{}^2 C_1 s\alpha_{1,1} \\ + 2\beta a_0 R B_1 \cosh\left(\frac{H}{R}\sqrt{\frac{sR^2 + \alpha_{1,1}{}^2\eta}{\eta}}\right) J_1\left(\frac{a_0\alpha_{1,1}}{R}\right) s e^{H\sqrt{\frac{s}{\eta}}} \end{array}\right) \tag{11.46}$$

We also multiply both sides of Eq. (11.44) by $rJ_0\left(\frac{\alpha_{1,1} r}{R}\right)$ and integrating from 0 to R to give

$$-\frac{e^{-H\sqrt{\frac{s}{\eta}}}}{2\sqrt{\eta}\alpha_{1,1}R}\left(\begin{array}{c} 2a_0 R^2 C_1 \sqrt{s} J_1\left(\frac{a_0\alpha_{1,1}}{R}\right) e^{2H\sqrt{\frac{s}{\eta}}} - 2C_1\sqrt{s}R^2 a_0 J_1\left(\frac{a_0\alpha_{1,1}}{R}\right) \\ + B_1 \sinh\left(\frac{H}{R}\sqrt{\frac{sR^2 + \alpha_{1,1}{}^2\eta}{\eta}}\right)\sqrt{sR^2 + \alpha_{1,1}{}^2\eta}\,\alpha_{1,1} a_0{}^2 J_1\left(\frac{a_0\alpha_{1,1}}{R}\right)^2 e^{H\sqrt{\frac{s}{\eta}}} \\ + B_1 \sinh\left(\frac{H}{R}\sqrt{\frac{sR^2 + \alpha_{1,1}{}^2\eta}{\eta}}\right)\sqrt{sR^2 + \alpha_{1,1}{}^2\eta}\,\alpha_{1,1} a_0{}^2 J_0\left(\frac{a_0\alpha_{1,1}}{R}\right)^2 e^{H\sqrt{\frac{s}{\eta}}} \\ - B_1 \sinh\left(\frac{H}{R}\sqrt{\frac{sR^2 + \alpha_{1,1}{}^2\eta}{\eta}}\right)\sqrt{sR^2 + \alpha_{1,1}{}^2\eta}\,\alpha_{1,1} R^2 J_0(\alpha_{1,1})^2 e^{H\sqrt{\frac{s}{\eta}}} \end{array}\right)$$
$$= \frac{e^{-H\sqrt{\frac{s}{\eta}}}}{2s\alpha_{1,1}}\left(\begin{array}{c} 2\beta a_0 c_{AS} R J_1\left(\frac{a_0\alpha_{1,1}}{R}\right) e^{H\sqrt{\frac{s}{\eta}}} + 2\beta a_0 C_1 R J_1\left(\frac{a_0\alpha_{1,1}}{R}\right) s e^{2H\sqrt{\frac{s}{\eta}}} \\ + 2\beta a_0 C_1 R J_1\left(\frac{a_0\alpha_{1,1}}{R}\right) s \\ + \beta a_0{}^2 B_1 \cosh\left(\frac{H}{R}\sqrt{\frac{sR^2 + \alpha_{1,1}{}^2\eta}{\eta}}\right) J_0\left(\frac{a_0\alpha_{1,1}}{R}\right)^2 s\alpha_{1,1} e^{H\sqrt{\frac{s}{\eta}}} \\ + \beta a_0{}^2 B_1 \cosh\left(\frac{H}{R}\sqrt{\frac{sR^2 + \alpha_{1,1}{}^2\eta}{\eta}}\right) J_1\left(\frac{a_0\alpha_{1,1}}{R}\right)^2 s\alpha_{1,1} e^{H\sqrt{\frac{s}{\eta}}} \end{array}\right) \tag{11.47}$$

The solution of the system formed by Eq. (11.46) and (11.47) with respect to C_1 and B_1 gives very large expressions which are omitted here because of limited space. The values of C_1 and B_1 are substituted in the Laplace transform of the concentration profile given by Eq. (11.45). The inverse Laplace transform of Eq. (11.45) can be computed using the Bromwich integral and the residue theorem or the Zakian method.

11.3 NORMALIZED CUMULATIVE PERCENTAGE OF DRUG RELEASED

11.3.1 Percentage of Drug Released Using Boundary Condition (11.6)

The cumulative amount of API released into the environment at time t is the difference between the amount of drug initially dissolved in the device and the amount remaining:

$$M(t) = c_{AS}\pi R^2 H - 2\pi \int_0^H \int_0^R c(r, z, t) r dr dz \tag{11.48}$$

or

$$M(t) = c_{AS}\pi R^2 H - \pi R^2 \int_0^H c(r, z, t)\ dz \tag{11.49}$$

since the concentration is not a function of r. Equation (11.49) becomes

$$M(t) = \pi\ HR^2 c_{AS} - \sum_{n=0}^{\infty} \frac{8(-1)^{-n}\cos(\pi n) c_{AS} HR^2}{\pi(2n+1)} e^{-\frac{(2n+1)^2\pi^2\eta t}{4H^2}} \tag{11.50}$$

after using Eq. (11.40). Therefore, the percentage of drug released is

$$\frac{M(t)}{M(\infty)} = 1 - \sum_{n=0}^{\infty} \frac{8(-1)^{-n}\cos(\pi n)}{(2n+1)^2\pi^2} e^{-\frac{(2n+1)^2\pi^2\eta t}{4H^2}} \tag{11.51}$$

which is simplified to

$$\frac{M(t)}{M(\infty)} = 1 - \sum_{n=0}^{\infty} \frac{8e^{-\frac{(2n+1)^2\pi^2\eta t}{4H^2}}}{(2n+1)^2\pi^2} \tag{11.52}$$

Using the dimensionless time defined by

$$\tau = \frac{t\eta}{H^2} \tag{11.53}$$

Eq. (11.52) is written as

$$\frac{M(\tau)}{M(\infty)} = 1 - \sum_{n=0}^{\infty} \frac{8e^{-\frac{(2n+1)^2\pi^2\tau}{4}}}{(2n+1)^2\pi^2} \tag{11.54}$$

A plot of $M(\tau)/M(\infty)$ is shown in Fig. 11.3 for the first 20 terms.

11.3.2 Percentage of Drug Released Using Boundary Condition (11.7)

The Laplace transform of the total amount of drug released is given by

$$M(s) = \frac{\pi HR^2 c_{AS}}{s} - \frac{\pi R^2}{s^{3/2}}\left(Hc_{AS}\sqrt{s}e^{H\sqrt{\frac{s}{\eta}}} + C_2\sqrt{\eta}se^{2H\sqrt{\frac{s}{\eta}}} - C_1\sqrt{\eta s}\right)e^{-H\sqrt{\frac{s}{\eta}}} \tag{11.55}$$

and the percentage of drug released is

$$\frac{M(s)}{M(\infty)} = \frac{1}{s} - \frac{\pi R^2}{Hc_{AS}s^{3/2}}\left(Hc_{AS}\sqrt{s}e^{H\sqrt{\frac{s}{\eta}}} + C_1\sqrt{\eta}se^{2H\sqrt{\frac{s}{\eta}}} - C_1\sqrt{\eta s}\right)e^{-H\sqrt{\frac{s}{\eta}}} \tag{11.56}$$

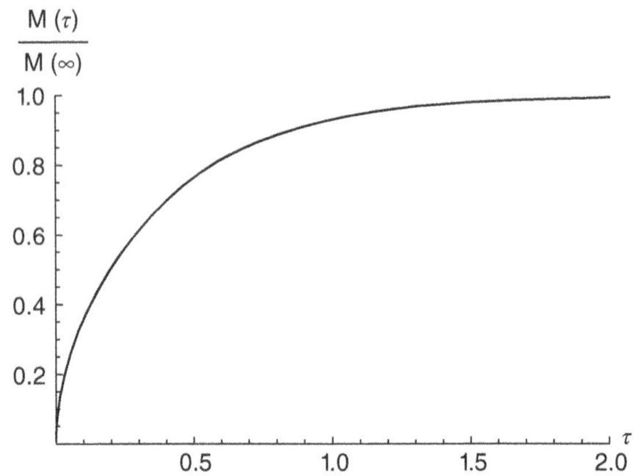

Figure 11.3 The percentage of drug released $M(\tau)/M(\infty)$ using a dimensionless time. The model with the boundary condition $c(r, H, t) = 0$ is applied.

After replacing C_1 in the above equation, a very large expression is obtained. The inverse Laplace transform of Eq. (11.56) can be computed using the Bromwich integral and residue theorem but, for practical applications, it is more efficient to use a numerical algorithm for the inversion.

11.4 SUMMARY

Two cases were presented for the release of a drug from a cylindrical device. In the first instance, the API was released through the bottom of the cylinder under perfect sink conditions. Although the model equation was written for two-dimensional transport, the resulting concentration profile depended only on the axial coordinate. Application of the Bromwich integral and the residue theorem yielded a series solution that could be easily implemented in Mathematica or Maple. The second example considered that the drug exited the matrix through a hole drilled at the bottom center of the device. Following a method similar to the one adopted in the first case, large expressions containing Bessel functions were developed in the Laplace domain. A first-order approximation required solving two equations for two parameters and taking the inverse Laplace transform. The cumulative percentages of drug released were derived for the two systems.

11.5 APPENDIX: MAPLE CODE LISTINGS

11.5.1 Maple Code: design2Dbound1.mws (Sections 11.2.1 and 11.3.1)

```
> restart:with(VectorCalculus):with(inttrans):with(PDETools):
> eq:=diff(C(r,z,t),t)=eta*Laplacian(C(r,z,t),cylindrical[r,theta,z]);
```

$$eq := \frac{\partial}{\partial t}C(r, z, t) = \frac{\eta\left(\frac{\partial}{\partial r}C(r, z, t) + r\left(\frac{\partial^2}{\partial r^2}C(r, z, t)\right) + r\left(\frac{\partial^2}{\partial z^2}C(r, z, t)\right)\right)}{r}$$

```
> eq1:=C(r,z,0)=c[AS]:
> eq2:=laplace(eq,t,s):
> eq3:=subs(eq1,eq2):
> eq4:=subs(laplace(C(r,z,t),t,s)=C(r,z),eq3):
> C(r,z)=int(C(r,z,t)*exp(-s*t),t=0..infinity):
> eq5:=subs(c[AS]=0,eq4):
> eq6:=pdsolve(eq5,HINT=f(r)*g(z)):
> eq7:=factor(build(eq6)):
> eq8:=subs(_C2=0,eq7):
> eq9:=subs(_C1=1,eq8):
> eq10:=subs(r=R,diff(rhs(eq9),r))=0:
> eq11:=BesselJ(1,(-_c[1])^(1/2)*R)=0:
> eq12:=(-_c[1])^(1/2)*R=alpha[1,n]:
>
> eq13:=isolate(eq12,_c[1]):
> eq14:=simplify(subs(eq13,eq9),power,symbolic):
> eq14A:=subs(_C3=A[n],_C4=B[n],C=C[n],eq14):
>
> eq14B:=C(r,z)=Sum(rhs(eq14A),n=1..infinity):
> eq15:=eval(subs(C(r,z)=h(z),eq5)):
> eq16:=dsolve(eq15):
> eq17:=subs(z=0,diff(rhs(eq16),z))=0:
>
> eq17A:=isolate(eq17,_C2):
> eq18:=subs(eq17A,eq16):
> eq19:=C(r,z)=c[AS]/s+rhs(eq18)+rhs(eq14B):
> eq20:=subs(z=H,rhs(eq19))=0:
>
> eq21:=Int(r*BesselJ(0,alpha[1,n]/R*r)^2,r=0..R)*(A[n]*sinh
(1/eta^(1/2)*(alpha[1,n]^2*eta+s*R^2)^(1/2)/R*H)*I+B[n]*cosh
(1/eta^(1/2)*(alpha[1,n]^2*eta+s*R^2)^(1/2)/R*H))+(c[AS]/s+_C1*exp
(1/eta^(1/2)*s^(1/2)*H)+_C1*exp(-1/eta^(1/2)*s^(1/2)*H))*Int
(r*BesselJ(0,alpha[1,n]/R*r),r=0..R)=0:
> eq22:=eval(subs(Int=int,eq21)):
> BesselJ(1,alpha[1,n])=0:
> eq23:=subs(BesselJ(1,alpha[1,n])=0,eq22):
>
> eq23A:=A[n]*sinh(1/eta^(1/2)*(alpha[1,n]^2*eta+s*R^2)^(1/2)/R*H)
*I+B[n]*cosh(1/eta^(1/2)*(alpha[1,n]^2*eta+s*R^2)^(1/2)/R*H)=0:
> eq23B:=eval(subs(z=0,diff(rhs(eq19),z)))=0:
> A[n]=0:
> eq24  := {A[n] = 0,  B[n] = 0}:
> eq25:=eval(subs(Sum=sum,subs(eq24,eq19))):
> eq26:=subs(z=H,rhs(eq25))=0:
> eq27:=isolate(eq26,_C1):
> eq28:=convert(eq27,cosh):
> eq29:=convert(simplify(subs(eq28,eq25)),cosh):
> C(r,z,t)=(1/(2*Pi*i))*Int(rhs(eq29)*exp(s*t),s):
> C(r,z,t)=Sum(Eval(Residue(rhs(eq29)*exp(s*t)),s=pole),pole);
```

```
>
```

$$C(r, z, t) = \sum_{pole} \left(Residue\left(-\frac{1}{2} \frac{c_{AS}\left(-2\cosh\left(\frac{\sqrt{s}H}{\sqrt{\eta}}\right) + 2\cosh\left(\frac{\sqrt{s}z}{\sqrt{\eta}}\right)\right) e^{st}}{s\cosh\left(\frac{\sqrt{s}H}{\sqrt{\eta}}\right)} \right)\Bigg|_{s=pole} \right)$$

```
> eq30:=residue(rhs(eq29),s=0):
> eq31:=cosh(1/eta^(1/2)*s^(1/2)*H)=0:
> eq32:=1/eta^(1/2)*s^(1/2)*H=I*(2*n+1)*Pi/2:
> eq33:=isolate(eq32,s):
> eq34:=numer(rhs(eq29))*exp(s*t)/diff(denom(rhs(eq29)),s):
> eq35:=simplify(subs(eq33,eq34),power,symbolic):
> eq36:= factor(eq35) assuming n::integer:
> eq36:
>
> eq37:=C(r,z,t)=eq30+Sum((eq36),n=0..infinity);
```

$$eq37 := C(r, z, t) = \sum_{n=0}^{\infty} \frac{4\ c_{AS}\cos\left(\frac{1}{2}\ \frac{(2n+1)\ \pi z}{H}\right)\ e^{-\frac{1}{4}\ \frac{(2n+1)^2\pi^2\eta\ t}{H^2}}}{\pi\ (-1)^n(2n+1)}$$

```
> eq38:=eval(subs(t=0,eq37)):
> eq39:=M(t)=Pi*H*R^2*c[AS]-Sum(factor(int(eq36,z=0..H)*Pi*R^2),n=0..
infinity):
> eq40:=M(t)/M(infinity)=1-Sum(factor(int(eq36,z=0..H)*Pi*R^2/(Pi*H*R^2
*c[AS])),n=0..infinity):
> eq41:=simplify(subs(cos(Pi*n)=(-1)^n,eq40),power,symbolic):
> t=tau*H^2/eta:
> isolate(t=tau*H^2/eta,tau):
> eq42:=M(tau)/M(infinity)=subs(t=tau*H^2/eta,rhs(eq41));
```

$$eq42 := \frac{M(\tau)}{M(\infty)} = 1 - \left(\sum_{n=0}^{\infty} \frac{8\ e^{-\frac{1}{4}\ (2n+1)^2\pi^2\tau}}{\pi^2(2n+1)^2} \right)$$

```
> eq43:=lhs(eq42)=eval(subs(Sum=sum,subs(infinity=20,rhs(eq42)))):
> plot(rhs(eq43),tau=0..2):
>
```

11.5.2 Maple Code: design2Dbound1.mws (Sections 11.2.2 and 11.3.2)

```
> restart:with(VectorCalculus):with(inttrans):with(PDETools):
> eq:=diff(C(r,z,t),t)=eta*Laplacian(C(r,z,t),cylindrical[r,theta,z]);
```

$$eq := \frac{\partial}{\partial t}C(r, z, t) = \frac{\eta\ \left(\frac{\partial}{\partial r}C(r, z, t) + r\left(\frac{\partial^2}{\partial r^2}C(r, z, t)\right) + r\left(\frac{\partial^2}{\partial z^2}C(r, z, t)\right)\right)}{r}$$

```
> eq1:=C(r,z,0)=c[AS]:
> eq2:=laplace(eq,t,s):
> eq3:=subs(eq1,eq2):
> eq4:=subs(laplace(C(r,z,t),t,s)=C(r,z),eq3):
```

```
>
 C(r,z)=int(C(r,z,t)*exp(-s*t),t=0..infinity):
>
> eq5:=subs(c[AS]=0,eq4):
> eq6:=pdsolve(eq5,HINT=f(r)*g(z)):
> eq7:=factor(build(eq6)):
> eq8:=subs(_C2=0,eq7):
> eq9:=subs(_C1=1,eq8):
> eq10:=subs(r=R,diff(rhs(eq9),r))=0:
> eq11:=BesselJ(1,(-_c[1])^(1/2)*R)=0:
> eq12:=(-_c[1])^(1/2)*R=alpha[1,n]:
>
> eq13:=isolate(eq12,_c[1]):
> eq14:=simplify(subs(eq13,eq9),power,symbolic):
> eq14A:=subs(_C3=A[n],_C4=B[n],C=C[n],eq14):
>
> eq14B:=C(r,z)=Sum(rhs(eq14A),n=1..infinity);
```

$$eq14B := C(r,z) = \sum_{n=1}^{\infty} \mathrm{BesselJ}\left[0, \frac{\alpha_{1,n} r}{R}\right] \left(IA_n \sinh\left(\frac{\sqrt{R^2 s + \eta\ \alpha_{1,n}^2}\, z}{\sqrt{\eta} R}\right) + B_n \cosh\left(\frac{\sqrt{R^2 s + \eta\ \alpha_{1,n}^2}\, z}{\sqrt{\eta} R}\right)\right)$$

```
> eq15:=eval(subs(C(r,z)=h(z),eq5)):
> eq16:=dsolve(eq15):
> eq17:=subs(z=0,diff(rhs(eq16),z))=0:
>
> eq17A:=isolate(eq17,_C2):
> eq18:=subs(eq17A,eq16):
> eq19:=C(r,z)=c[AS]/s+rhs(eq18)+rhs(eq14B):
> auxi:=Eval(diff(C[A](r,z,t),z),z=H)*Heaviside(r-a[0])=beta*Eval(C[A]
(r,z,t),z=H)*(1-Heaviside(r-a[0])):
>
> eq19A:=eval(subs(z=0,diff(rhs(eq19),z)))=0:
> eq19B:=subs(A[n]=0,eq19):
> eq19C:=subs(C[A](r,z,t)=C(r,z),auxi):
> eq20:=eval(subs(Eval=eval,subs(eq19B,eq19C))):
> eq20A:=eval(subs(Sum=sum,subs(infinity=1,eq20))):
> eq20B:=eval(subs(Sum=sum,subs(infinity=1,eq19B))):
> eq21:=subs(BesselJ(1,alpha[1,1])=0,int(r*lhs(eq20A),r=0..R)=int(r*
rhs(eq20A),r=0..R)) assuming R > a[0] and a[0]>0:
> r*BesselJ(0,alpha[1,1]/R*r):
> eq21A:=subs(BesselJ(1,alpha[1,1])=0,int(r*lhs(eq20A)*BesselJ(0,alpha
[1,1]/R*r),r=0..R)=int(r*rhs(eq20A)*BesselJ(0,alpha[1,1]/R*r),r=0..R))
assuming R > a[0] and a[0]>0:
> eq22 := solve({eq21, eq21A}, {_C1, B[1]}):
> eq20B;
```

$$C(r, z) = \frac{c_{AS}}{s} + \_C1\ e^{\frac{\sqrt{s}\,z}{\sqrt{\eta}}} + \_C1\ e^{-\frac{\sqrt{s}\,z}{\sqrt{\eta}}} + \text{BesselJ}\left(0, \frac{\alpha_{1,1}\,r}{R}\right) B_1 \cosh\left(\frac{\sqrt{R^2 s + \eta\,\alpha_{1,1}^2}\; z}{\sqrt{\eta}\,R}\right)$$

```
> eq39:=M(s)=Pi*H*R^2*c[AS]/s-2*Pi*factor(subs(BesselJ(1,alpha[1,1])
=0,int(int(rhs(eq20B)*r,z=0..H),r=0..R)));
>
>
```

$$eq39 := M(s) = \frac{\pi H R^2 c_{AS}}{s} - \frac{\pi R^2\left(H\, e^{\frac{\sqrt{s}H}{\sqrt{\eta}}}\sqrt{s}c_{AS} + \sqrt{\eta}\; e^{\frac{2\sqrt{s}H}{\sqrt{\eta}}}\ \_C1\ s - \sqrt{\eta}\_C1\, s\right) e^{-\frac{\sqrt{s}H}{\sqrt{\eta}}}}{s^{3/2}}$$

```
>
> eq40:=simplify(subs(eq22,eq39)):
>
> limit(s*rhs(eq40),s=0);
```

$H \pi R^2 c_{AS}$

```
>
```

PROBLEMS

11.1. Plot the profile for $c(r, z, t)/c_{AS}$ using the following parameter values: $H = 1.5$, $\eta = 1.0$ and a simulation time of 2.

$$\frac{c(r, z, t)}{c_{AS}} = \sum_{n=0}^{\infty}\left(\frac{4\cos\left(\frac{(2n+1)\pi z}{2H}\right) e^{-\frac{(2n+1)^2\pi^2\eta t}{4H^2}}}{\pi(-1)^n(2n+1)}\right) \quad (1)$$

11.2. Derive an expression for the flux defined by

$$J(t) = \frac{dM(t)}{dt} \quad (1)$$

for the model in Sections 11.2.1, i.e.,

$$c(r, z, t) = \sum_{n=0}^{\infty}\left(\frac{4c_{AS}\cos\left(\frac{(2n+1)\pi z}{2H}\right) e^{-\frac{(2n+1)^2\pi^2\eta t}{4H^2}}}{\pi(-1)^n(2n+1)}\right) \quad (2)$$

11.3. Show that the flux for the model defined in Section 11.2.2, i.e., with the following boundary condition:

$$\begin{cases} \frac{\partial}{\partial z}c(r, z, t)\Big|_{z=H} = 0 & a_0 < r \\ c(r, H, t) = 0 & r \le a_0 \end{cases} \quad (1)$$

approaches zero as the time t goes to infinity.
Hint: Use the final value theorem.

11.4. Plot the profile for $c(r, 1.0, t)$ using the following parameter values: $c_{AS} = 1.0, a_0 = 0.1$, $R = 0.3$, $H = 1.5$, $\eta = 1.0$ and a simulation time of 2. The model is defined in Section 11.2.2, i.e., with the following boundary condition:

$$\begin{cases} \frac{\partial}{\partial z}c(r, z, t)\Big|_{z=H} = 0 & a_0 < r \\ c(r, H, t) = 0 & r \leq a_0 \end{cases} \tag{1}$$

11.5. Solve the following problem:

$$\frac{\partial}{\partial t}c(r, z, t) = \frac{\eta\left(\frac{\partial}{\partial r}c(r, z, t) + r\frac{\partial^2}{\partial r^2}c(r, z, t) + r\frac{\partial^2}{\partial z^2}c(r, z, t)\right)}{r} \tag{1}$$

$$c(r, z, 0) = c_{AS} \tag{2}$$

$$\frac{\partial}{\partial r}c(r, z, t)\Big|_{r=R} = 0 \tag{3}$$

$$\frac{\partial}{\partial z}c(r, z, t)\Big|_{z=0} = 0 \tag{4}$$

and

$$\frac{\partial}{\partial z}c(r, z, t)\Big|_{z=H} = \begin{cases} -\frac{k_m c(r, z, H)}{\eta} & r \leq a_0 \\ 0 & a_0 < r \end{cases} \tag{5}$$

Note: The dimensionless form of this problem was solved in Simon & Ospina (2012).

REFERENCES

Ertan G, Karasulu E, Demirtas D, Arici M, Guneri T. Release characteristics of implantable cylindrical polyethylene matrices. *J Pharm Pharmacol* 1997;49(3):229–235.

Hsu JP, Ting C, Lin MJ. A theoretical analysis of a new drug delivery system: a cylindrical device with a vertical opening on its surface. *J Pharm Sci* 1992;81(9):866–870.

Simon L Ospina J. Two-dimensional solution and analysis of a cylindrical matrix device with a circular release area. *Chemical Engineering Communications* 2012;200(1):115–138.

Tojo K. Prolonged drug release from a cylindrical matrix device with a small release hole. *Chem Eng Commun* 1984;30(6):311–322.

Tojo K, Miyanami K. Controlled Release from a Cylindrical Matrix Device. *Bull Univ Osaka Prefecture. Series A, Eng Nat Sci* 1983;31(2):149–157.

12

THREE-DIMENSIONAL DRUG DIFFUSION THROUGH MEMBRANES FROM DISTRIBUTED SOURCES

12.1 GOVERNING EQUATIONS OF THE TRANSPORT MODEL

The diagram for the spatially-distributed source model was given in Fig. 5.1 (Ley et al., 2006). The governing transport equation is

$$\frac{\partial C}{\partial t} = D_x \frac{\partial^2 C}{\partial x^2} + D_y \frac{\partial^2 C}{\partial y^2} + D_z \frac{\partial^2 C}{\partial z^2}; \quad 0 \le x \le \frac{b}{2}, \quad 0 \le y \le \frac{b}{2}, \quad 0 \le z \le L \tag{12.1}$$

where C represents the drug concentration in the membrane. Just as in Chapter 5, the diffusion coefficients in the $x-$, $y-$ and $z-$ directions are labeled D_x, D_y and D_z, respectively. The diffusion coefficients satisfy the following constraint: $D_x = D_y \neq D_z$ (Ley et al., 2006). In addition, the flux at the membrane surface is zero, except at the source regions:

$$C(x, y, 0, t) = C_{eq}; \quad 0 \le x \le \frac{d_s}{2}, \quad 0 \le y \le \frac{d_s}{2} \tag{12.2}$$

$$\left.\frac{\partial C(x, y, z, t)}{\partial z}\right|_{z=0} = 0; \quad \frac{d_s}{2} < x \le \frac{b}{2}, \quad \frac{d_s}{2} < y \le \frac{b}{2} \tag{12.3}$$

$$\left.\frac{\partial C(x, y, z, t)}{\partial x}\right|_{x=0} = 0; \quad 0 \le y \le \frac{b}{2}, \quad 0 \le z \le L \tag{12.4}$$

Closed-form Solutions for Drug Transport through Controlled-Release Devices in Two and Three Dimensions, First Edition. Laurent Simon and Juan Ospina.

$$\left.\frac{\partial C(x,y,z,t)}{\partial x}\right|_{x=\frac{b}{2}} = 0; \quad 0 \le y \le \frac{b}{2}, \quad 0 \le z \le L \tag{12.5}$$

$$\left.\frac{\partial C(x,y,z,t)}{\partial y}\right|_{y=0} = 0; \quad 0 \le x \le \frac{b}{2}, \quad 0 \le z \le L \tag{12.6}$$

$$\left.\frac{\partial C(x,y,z,t)}{\partial y}\right|_{y=\frac{b}{2}} = 0; \quad 0 \le x \le \frac{b}{2}, \quad 0 \le z \le L \tag{12.7}$$

$$\left.\frac{\partial C(x,y,z,t)}{\partial z}\right|_{z=L} + wC(x,y,L,t) = 0; \quad 0 \le x \le \frac{b}{2}, \quad 0 \le y \le \frac{b}{2} \tag{12.8}$$

where C_{eq} is the concentration in the membrane in equilibrium with the source. The initial concentration is zero:

$$C(x, y, z, 0) = 0 \tag{12.9}$$

While the no-flux statement at $z = L$ was used in Ley et al. (2006), the Robin boundary condition is applied here. The following transformations of variables

$$\theta = \frac{C}{C_{eq}}, \quad \xi = \frac{x}{b/2}, \quad \eta = \frac{y}{b/2}, \quad \zeta = \frac{z}{L}, \quad \delta = \frac{d_s}{b}, \quad \beta = \frac{b}{2L}\sqrt{\frac{D_z}{D_x}} \tag{12.10}$$

lead to these dimensionless equations:

$$\frac{\partial \theta}{\partial \tau} = \frac{1}{\beta^2}\left(\frac{\partial^2 \theta}{\partial \xi^2} + \frac{\partial^2 \theta}{\partial \eta^2}\right) + \frac{\partial^2 \theta}{\partial \zeta^2}; \quad 0 \le \xi \le 1, 0 \le \eta \le 1, 0 \le \zeta \le 1 \tag{12.11}$$

$$\theta(\xi, \eta, 0, \tau) = 1; \quad 0 \le \xi \le \delta, \quad 0 \le \eta \le \delta \tag{12.12}$$

$$\left.\frac{\partial \theta(\xi,\eta,\zeta,\tau)}{\partial \zeta}\right|_{\zeta=0} = 0; \quad \delta < \xi \le 1, \quad \delta < \eta \le 1 \tag{12.13}$$

$$\left.\frac{\partial \theta(\xi,\eta,\zeta,\tau)}{\partial \xi}\right|_{\xi=0} = 0; \quad 0 \le \eta \le 1, \quad 0 \le \zeta \le 1 \tag{12.14}$$

$$\left.\frac{\partial \theta(\xi,\eta,\zeta,\tau)}{\partial \xi}\right|_{\xi=1} = 0; \quad 0 \le \eta \le 1, \quad 0 \le \zeta \le 1 \tag{12.15}$$

$$\left.\frac{\partial \theta(\xi,\eta,\zeta,\tau)}{\partial \eta}\right|_{\eta=0} = 0; \quad 0 \le \xi \le 1, \quad 0 \le \zeta \le 1 \tag{12.16}$$

$$\left.\frac{\partial \theta(\xi,\eta,\zeta,\tau)}{\partial \eta}\right|_{\eta=1} = 0; \quad 0 \le \xi \le 1, \quad 0 \le \zeta \le 1 \tag{12.17}$$

$$\left.\frac{\partial \theta(\xi,\eta,\zeta,\tau)}{\partial \zeta}\right|_{\zeta=1} + w\theta(\xi, \eta, 1, \tau) = 0; \quad 0 \le \xi \le 1, \quad 0 \le \eta \le 1 \tag{12.18}$$

and

$$\theta(\xi, \eta, \zeta, 0) = 0 \tag{12.19}$$

12.2 ANALYTICAL SOLUTION OF THE DIFFUSION MODEL IN 3-D CARTESIAN SYSTEMS

The Laplace transform of Eq. (12.11), after applying the initial condition, is given by

$$s\theta(\xi, \eta, \zeta) = \frac{\frac{\partial^2}{\partial \xi^2}\theta(\xi, \eta, \zeta)}{\beta^2} + \frac{\frac{\partial^2}{\partial \eta^2}\theta(\xi, \eta, \zeta)}{\beta^2} + \frac{\partial^2}{\partial \zeta^2}\theta(\xi, \eta, \zeta) \tag{12.20}$$

where

$$\theta(\xi, \eta, \zeta) = \int_0^\infty \theta(\xi, \eta, \zeta, \tau)e^{-s\tau}d\tau \tag{12.21}$$

A solution to Eq. (12.20) of the form $\theta(\xi, \eta, \zeta) = f(\xi)g(\eta)h(\zeta)$ is assumed, where

$$\frac{d^2f(\xi)}{d\xi^2} = c_1 f(\xi) \tag{12.22}$$

$$\frac{d^2g(\eta)}{d\eta^2} = c_2 g(\eta) \tag{12.23}$$

and

$$\frac{d^2h(\zeta)}{d\zeta^2} = -\frac{c_1 h(\zeta)}{\beta^2} - \frac{c_2 h(\zeta)}{\beta^2} + sh(\zeta) \tag{12.24}$$

The solution of Eqs. (12.22)–(12.24) yields the following expression for the concentration:

$$\theta(\xi, \eta, \zeta) = (C_3 e^{\eta\sqrt{c_2}} + C_4 e^{-\eta\sqrt{c_2}})(C_1 e^{\xi\sqrt{c_1}} + C_2 e^{-\xi\sqrt{c_1}}) \times \left(C_5 \sin\left(\frac{\zeta\sqrt{-s\beta^2 + c_1 + c_2}}{\beta}\right) + C_6 \sin\left(\frac{\zeta\sqrt{-s\beta^2 + c_1 + c_2}}{\beta}\right)\right) \tag{12.25}$$

After applying the boundary condition (12.14) to Eq. (12.25), we have

$$(C_3 e^{\eta\sqrt{c_2}} + C_4 e^{-\eta\sqrt{c_2}})(C_1\sqrt{c_1} - C_2\sqrt{c_1}) \times \left(C_5 \sin\left(\frac{\zeta\sqrt{-s\beta^2 + c_1 + c_2}}{\beta}\right) + C_6 \cos\left(\frac{\zeta\sqrt{-s\beta^2 + c_1 + c_2}}{\beta}\right)\right) = 0 \tag{12.26}$$

As a result,

$$C_1 = C_2 \tag{12.27}$$

which leads to

$$\theta(\xi, \eta, \zeta) = 2(C_3 e^{\eta\sqrt{c_2}} + C_4 e^{-\eta\sqrt{c_2}})C_1 \cosh(\xi\sqrt{c_1}) \times \left(C_5 \sin\left(\frac{\zeta\sqrt{-s\beta^2 + c_1 + c_2}}{\beta}\right) + C_6 \cos\left(\frac{\zeta\sqrt{-s\beta^2 + c_1 + c_2}}{\beta}\right)\right) \tag{12.28}$$

or

$$\theta(\xi,\eta,\zeta) = 2(C_3 e^{\eta\sqrt{c_2}} + C_4 e^{-\eta\sqrt{c_2}})\cosh(\xi\sqrt{c_1}) \times \left(C_5 \sin\left(\frac{\zeta\sqrt{-s\beta^2 + c_1 + c_2}}{\beta}\right) + C_6 \cos\left(\frac{\zeta\sqrt{-s\beta^2 + c_1 + c_2}}{\beta}\right)\right) \quad (12.29)$$

after making $C_1 = 1$.

Applying the boundary condition (12.15) to (12.29), we have

$$2(C_3 e^{\eta\sqrt{c_2}} + C_4 e^{-\eta\sqrt{c_2}})\sinh(\sqrt{c_1})\sqrt{c_1} \times \left(C_5 \sin\left(\frac{\zeta\sqrt{-s\beta^2 + c_1 + c_2}}{\beta}\right) + C_6 \cos\left(\frac{\zeta\sqrt{-s\beta^2 + c_1 + c_2}}{\beta}\right)\right) = 0 \quad (12.30)$$

which implies

$$\sinh(\sqrt{c_1}) = 0 \quad (12.31)$$

or

$$c_1 = -n^2\pi^2 \quad (12.32)$$

Equation (12.29) becomes

$$\theta(\xi,\eta,\zeta) = 2(C_3 e^{\eta\sqrt{c_2}} + C_4 e^{-\eta\sqrt{c_2}})\cos(n\pi\xi) \times \left(C_5 \sin\left(\frac{\zeta\sqrt{-n^2\pi^2 - s\beta^2 + c_2}}{\beta}\right) + C_6 \cos\left(\frac{\zeta\sqrt{-n^2\pi^2 - s\beta^2 + c_2}}{\beta}\right)\right) \quad (12.33)$$

The boundary condition (12.16) results in

$$2(C_3\sqrt{c_2} - C_4\sqrt{c_2})\cos(n\pi\xi) \times \left(C_5 \sin\left(\frac{\zeta\sqrt{-n^2\pi^2 - s\beta^2 + c_2}}{\beta}\right) + C_6 \cos\left(\frac{\zeta\sqrt{-n^2\pi^2 - s\beta^2 + c_2}}{\beta}\right)\right) = 0 \quad (12.34)$$

and therefore,

$$C_3 = C_4 \quad (12.35)$$

Equation (12.33) becomes

$$\theta(\xi,\eta,\zeta) = 2C_3\cosh(n\sqrt{c_2})\cos(n\pi\xi) \times \left(C_5 \sin\left(\frac{\zeta\sqrt{-n^2\pi^2 - s\beta^2 + c_2}}{\beta}\right) + C_6 \cos\left(\frac{\zeta\sqrt{-n^2\pi^2 - s\beta^2 + c_2}}{\beta}\right)\right) \quad (12.36)$$

or

$$\theta(\xi, \eta, \zeta) = 4\cosh(n\sqrt{c_2})\cos(n\pi\xi) \times \left(C_5 \sin\left(\frac{\zeta\sqrt{-n^2\pi^2 - s\beta^2 + c_2}}{\beta}\right) + C_6 \cos\left(\frac{\zeta\sqrt{-n^2\pi^2 - s\beta^2 + c_2}}{\beta}\right)\right) \tag{12.37}$$

after setting $C_3 = 1$.

The boundary condition (12.17) yields

$$4\sinh(\sqrt{c_2})\sqrt{c_2}\cos(n\pi\xi) \times \left(C_5 \sin\left(\frac{\zeta\sqrt{-n^2\pi^2 - s\beta^2 + c_2}}{\beta}\right) + C_6 \cos\left(\frac{\zeta\sqrt{-n^2\pi^2 - s\beta^2 + c_2}}{\beta}\right)\right) = 0 \tag{12.38}$$

We deduce from Eq. (12.38) that

$$\sinh(\sqrt{c_2}) = 0 \tag{12.39}$$

or

$$c_2 = -m^2\pi^2 \tag{12.40}$$

After substituting for c_2 in Eq. (12.37), we obtain

$$\theta(\xi, \eta, \zeta) = 4\cosh(m\pi\eta)\cos(n\pi\xi) \times \left(C_5 \sin\left(\frac{\zeta\sqrt{-m^2\pi^2 - n^2\pi^2 - s\beta^2}}{\beta}\right) + C_6 \cos\left(\frac{\zeta\sqrt{-n^2\pi^2 - m^2\pi^2 - s\beta^2}}{\beta}\right)\right) \tag{12.41}$$

The **principle of superposition**, applied to Eq. (12.41), gives the following expression:

$$\theta(\xi, \eta, \zeta) = \sum_{m=0}^{\infty}\left(\sum_{m=0}^{\infty} 4\cosh(m\pi\eta)\cos(n\pi\xi)\right) \times \left(A_{n,m} \sin\left(\frac{\zeta\sqrt{-m^2\pi^2 - n^2\pi^2 - s\beta^2}}{\beta}\right) + B_{n,m} \cos\left(\frac{\zeta\sqrt{-n^2\pi^2 - m^2\pi^2 - s\beta^2}}{\beta}\right)\right)\Bigg)\Bigg) \tag{12.42}$$

The boundary condition (12.18) results in

$$4w\cos(m\pi\eta)\cos(n\pi\xi) \times \left(A_{n,m} \sin\left(\frac{\sqrt{-m^2\pi^2 - n^2\pi^2 - s\beta^2}}{\beta}\right) + B_{n,m} \cos\left(\frac{\sqrt{-m^2\pi^2 - n^2\pi^2 - s\beta^2}}{\beta}\right)\right)$$
$$+ 4\cos(m\pi\eta)\cos(n\pi\xi)$$

$$\times\left(\begin{array}{c}\frac{A_{n,m}\sqrt{-m^2\pi^2-n^2\pi^2-s\beta^2}}{\beta}\cos\left(\frac{\sqrt{-m^2\pi^2-n^2\pi^2-s\beta^2}}{\beta}\right)-\\ \frac{B_{n,m}\sqrt{-m^2\pi^2-n^2\pi^2-s\beta^2}}{\beta}\sin\left(\frac{\sqrt{-m^2\pi^2-n^2\pi^2-s\beta^2}}{\beta}\right)\end{array}\right)=0 \tag{12.43}$$

which allows the computation of $B_{n,m}$:

$$B_{n,m}=\left(\begin{array}{c}w\beta A_{n,m}\sin\left(\frac{\sqrt{-m^2\pi^2-n^2\pi^2-s\beta^2}}{\beta}\right)\\ +A_{n,m}\cos\left(\frac{\sqrt{-m^2\pi^2-n^2\pi^2-s\beta^2}}{\beta}\right)\sqrt{-m^2\pi^2-n^2\pi^2-s\beta^2}\end{array}\right)$$
$$\div\left(\begin{array}{c}-w\beta\cos\left(\frac{\sqrt{-m^2\pi^2-n^2\pi^2-s\beta^2}}{\beta}\right)\\ +\sin\left(\frac{\sqrt{-m^2\pi^2-n^2\pi^2-s\beta^2}}{\beta}\right)\sqrt{-m^2\pi^2-n^2\pi^2-s\beta^2}\end{array}\right) \tag{12.44}$$

We replace $B_{n,m}$ in Eq. (12.42) to get

$$\theta(\xi,\eta,\zeta)=\sum_{m=0}^{\infty}\left(\begin{array}{c}\sum_{n=0}^{\infty}4\cos(m\pi\eta)\cos(n\pi\xi)\\ \times\left(\begin{array}{c}A_{n,m}\sin\left(\frac{\sqrt{-m^2\pi^2-n^2\pi^2-s\beta^2}(\zeta)}{\beta}\right)\\ +\left(\begin{array}{c}w\beta A_{n,m}\sin\left(\frac{\sqrt{-m^2\pi^2-n^2\pi^2-s\beta^2}}{\beta}\right)\\ +A_{n,m}\cos\left(\frac{\sqrt{-m^2\pi^2-n^2\pi^2-s\beta^2}}{\beta}\right)\sqrt{-m^2\pi^2-n^2\pi^2-s\beta^2}\end{array}\right)\\ \times\cos\left(\frac{\zeta\sqrt{-m^2\pi^2-n^2\pi^2-s\beta^2}}{\beta}\right)\\ \div\left(\begin{array}{c}-w\beta\cos\left(\frac{\sqrt{-m^2\pi^2-n^2\pi^2-s\beta^2}}{\beta}\right)\\ +\sin\left(\frac{\sqrt{-m^2\pi^2-n^2\pi^2-s\beta^2}}{\beta}\right)\sqrt{-m^2\pi^2-n^2\pi^2-s\beta^2}\end{array}\right)\end{array}\right)\end{array}\right) \tag{12.45}$$

Now, the remaining boundary conditions (12.12) and (12.13) are rewritten in the following compact form:

$$\alpha(\xi,\eta)\ \theta\ (\xi,\eta,0)\ +\ k\ (\xi,\eta)\frac{\partial}{\partial\zeta}\ \theta(\xi,\eta,\zeta)\Big|_{\zeta=0}=f(\xi,\eta) \tag{12.46}$$

where

$$\alpha(\xi,\eta)=(Heaviside(\xi)\ -\ Heaviside(\xi-\delta))(Heaviside(\eta)\ -\ Heaviside(\eta-\delta)) \tag{12.47}$$

$$k(\xi,\eta)=1-(Heaviside(\xi)-Heaviside(\xi-\delta))$$
$$\times(Heaviside(\eta)-Heaviside(\eta-\ \delta)) \tag{12.48}$$

and

$$f(\xi,\eta)=(Heaviside(\xi)-Heaviside(\xi-\delta))(Heaviside(\eta)-Heaviside(\eta-\delta)) \tag{12.49}$$

To determine $A_{n,m}$, the application of Eq. (12.46)–(12.49) may be computationally intensive depending on the number of terms used in the series approximation.

12.2.1 Zeroth-Order Approximation

Using $m = n = 0$, Eq. (12.45) is written as

$$\theta(\xi, \eta, \zeta) = 4A_{0,0}\sin\left(\frac{\sqrt{-s\beta^2}\,(\zeta)}{\beta}\right) + \frac{4\left(w\beta A_{0,0}\sin\left(\frac{\sqrt{-s\beta^2}}{\beta}\right) + A_{0,0}\cos\left(\frac{\sqrt{-s\beta^2}}{\beta}\right)\sqrt{-s\beta^2}\right)}{\cos\left(\frac{\sqrt{-s\beta^2}(\zeta)}{\beta}\right)\left(-w\beta\cos\left(\frac{\sqrt{-s\beta^2}}{\beta}\right) + \sin\left(\frac{\sqrt{-s\beta^2}}{\beta}\right)\sqrt{-s\beta^2}\right)} \tag{12.50}$$

The application of Eq. (12.46) to Eq. (12.50) leads to

$$\frac{4\alpha(\xi, \eta)\left(w\beta A_{0,0}\sin\left(\frac{\sqrt{-s\beta^2}}{\beta}\right) + A_{0,0}\cos\left(\frac{\sqrt{-s\beta^2}}{\beta}\right)\sqrt{-s\beta^2}\right)}{-w\beta\cos\left(\frac{\sqrt{-s\beta^2}}{\beta}\right) + \sin\left(\frac{\sqrt{-s\beta^2}}{\beta}\right)\sqrt{-s\beta^2}} + 4\frac{k(\xi, \eta)A_{0,0}\sqrt{-s\beta^2}}{\beta} = \frac{f(\xi, \eta)}{s} \tag{12.51}$$

Both sides of Eq. (12.51) are integrated from 0 to 1 with respect to ξ and η to give

$$\int_0^1\int_0^1\left(\frac{4\alpha(\xi, \eta)\left(w\beta A_{0,0}\sin\left(\frac{\sqrt{-s\beta^2}}{\beta}\right) + A_{0,0}\cos\left(\frac{\sqrt{-s\beta^2}}{\beta}\right)\sqrt{-s\beta^2}\right)}{-w\beta\cos\left(\frac{\sqrt{-s\beta^2}}{\beta}\right) + \sin\left(\frac{\sqrt{-s\beta^2}}{\beta}\right)\sqrt{-s\beta^2}} + 4\frac{k(\xi, \eta)A_{0,0}\sqrt{-s\beta^2}}{\beta}\right)d\xi d\eta = \int_0^1\int_0^1\frac{f(\xi, \eta)}{s}d\xi d\eta \tag{12.52}$$

After evaluating the integrations in Eq. (12.52) and solving for $A_{0,0}$, we have

$$A_{0,0} = \frac{i\delta^2(w\cosh(\sqrt{s}) + \sinh(\sqrt{s})\sqrt{s})}{4s\begin{pmatrix}\delta^2 w\sinh\left(\sqrt{s}\right) + \delta^2\cosh(\sqrt{s})\sqrt{s} + \sqrt{s}\delta^2 w\cosh(\sqrt{s}) \\ -\sqrt{s}w\cosh(\sqrt{s}) - \sinh(\sqrt{s})s + \delta^2\sinh(\sqrt{s})s\end{pmatrix}} \tag{12.53}$$

The concentration in the Laplace domain is given by Eq. (12.50) with $A_{0,0}$ defined by Eq. (12.53):

$$\theta(\xi, \eta, \zeta) = \frac{i\delta^2(w\cosh(\sqrt{s}) + \sinh(\sqrt{s})\sqrt{s})}{s\begin{pmatrix}\delta^2 w\sinh\left(\sqrt{s}\right) + \delta^2\cosh(\sqrt{s})\sqrt{s} + \sqrt{s}\delta^2 w\cosh(\sqrt{s}) - \\ \sqrt{s}w\cosh(\sqrt{s}) - \sinh(\sqrt{s})s + \delta^2\sinh(\sqrt{s})s\end{pmatrix}}$$

$$\times \frac{\begin{pmatrix} -\sin\left(\frac{\zeta\sqrt{-s\beta^2}}{\beta}\right) w\beta\cos\left(\frac{\sqrt{-s\beta^2}}{\beta}\right) + \sin\left(\frac{\zeta\sqrt{-s\beta^2}}{\beta}\right)\sin\left(\frac{\sqrt{-s\beta^2}}{\beta}\right)\sqrt{-s\beta^2} \\ +\cos\left(\frac{\zeta\sqrt{-s\beta^2}}{\beta}\right)\beta w\sin\left(\frac{\sqrt{-s\beta^2}}{\beta}\right) + \cos\left(\frac{\zeta\sqrt{-s\beta^2}}{\beta}\right)\sqrt{-s\beta^2}\cos\left(\frac{\sqrt{-s\beta^2}}{\beta}\right) \end{pmatrix}}{\left(-w\beta\cos\left(\frac{\sqrt{-s\beta^2}}{\beta}\right) + \sin\left(\frac{\sqrt{-s\beta^2}}{\beta}\right)\sqrt{-s\beta^2}\right)} \tag{12.54}$$

Based on Eq. (12.54), the zeroth-order approximation is not a function of ξ and η.

As an example, when the parameters, $\beta = 0.1$, $\delta = 0.1$ and $w = 0$, are used, Eq. (12.54) becomes

$$\theta(\xi, \eta, \zeta) = \frac{1.0i\sinh(\sqrt{s})(\sin(\zeta\sqrt{-1.0s})\sin(\sqrt{-1.0s}) + \cos(\zeta\sqrt{-1.0s})\cos(\sqrt{-1.0s}))}{\sqrt{s}\sin(\sqrt{-1.0s})(\cosh(\sqrt{s})\sqrt{s} - 99.0\sinh(\sqrt{s})s)} \tag{12.55}$$

The steady-state value is

$$\lim_{s\to 0}[s\theta(\xi, \eta, \zeta)] = 1 \tag{12.56}$$

The **residue theorem** (Loney, 2007) is used to invert Eq. (12.55). The numerator and denominator of Eq. (12.55) are

$$P(s) = 1.0i\sinh(\sqrt{s})(\sin(\zeta\sqrt{-1.0s})\sin(\sqrt{-1.0s}) + \cos(\zeta\sqrt{-1.0s})\cos(\sqrt{-1.0s})) \tag{12.57}$$

and

$$Q(s) = \sqrt{s}\sin(\sqrt{-1.0s})(\cosh(\sqrt{s})\sqrt{s} - 99.0\sinh(\sqrt{s})s) \tag{12.58}$$

respectively. As a result,

$$\cosh(\sqrt{s})\sqrt{s} - 99.0\sinh(\sqrt{s})s = 0 \tag{12.59}$$

which implies

$$s = -\alpha_n^2 \tag{12.60}$$

and

$$\cos(\alpha_n) = -99.0\alpha_n\sin(\alpha_n) \tag{12.61}$$

after substituting the expression for s in Eq. (12.59). The first five values of α_n are 3.14, 6.28, 9.42, 12.56 and 15.71. Inversion of the Laplace transform gives

$$\theta(\xi, \eta, \zeta, \tau) = \theta_{ss} + \sum_{n=1}^{\infty} \frac{P(s_n)}{\left.\frac{dQ(s)}{ds}\right|_{s=s_n}} e^{s_n\tau} \tag{12.62}$$

or

$$\theta(\xi, \eta, \zeta, \tau) = 1.0 - \sum_{n=1}^{\infty} \frac{(2.0\sin(\alpha_n\zeta)\sin(\alpha_n) + 2.0\cos(\alpha_n\zeta)\cos(\alpha_n))e^{-\alpha_n{}^2\tau}}{2.0\cos(\alpha_n) + 296.0\alpha_n\sin(\alpha_n) - 9801.0\alpha_n{}^3\sin(\alpha_n)} \tag{12.63}$$

A plot of $\theta(\xi, \eta, 0.2, \tau)$ is shown in Fig. 12.1.

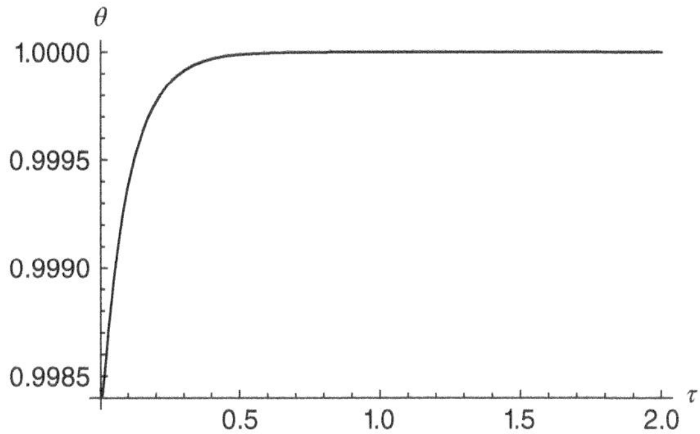

Figure 12.1 Concentration $\theta(\xi,\ \eta,\ 0.2,\ \tau)$ using a zeroth-order approximation and the following parameters: $\beta = 0.1$, $\delta = 0.1$ and $w = 0$.

12.2.2 First-Order Approximation

After using an upper limit of 1, Eq. (12.45) is written in terms of $A_{1,0}$, $A_{0,1}$, $A_{0,0}$ and $A_{1,1}$.
The application of Eq. (12.46) to the resulting expression yields

$$\alpha(\xi,\ \eta)\theta(\xi,\ \eta,\ 0)\ +\ k(\xi,\ \eta)\frac{\partial}{\partial\zeta}\theta(\xi,\ \eta,\ \zeta)\bigg|_{\zeta=0} = \frac{f(\xi,\ \eta)}{s} \tag{12.64}$$

Four equations are needed to solve for $A_{1,0}$, $A_{0,1}$, $A_{0,0}$ and $A_{1,1}$:

- Both sides of Eq. (12.64) are integrated from 0 to 1 with respect to ξ and η:

$$\int_0^1\int_0^1\left(\theta(\xi,\ \eta)\,\theta(\xi,\ \eta,\ 0) + k(\xi,\ \eta)\frac{\partial}{\partial\zeta}\theta(\xi,\ \eta,\ \zeta)\bigg|_{\zeta=0}\right)\ d\xi d\eta$$
$$= \int_0^1\int_0^1\frac{f(\xi,\ \eta)}{s}\ d\xi d\eta \tag{12.65}$$

- Both sides of Eq. (12.64) are multiplied by $\cos(\pi\eta)$ and integrated from 0 to 1 with respect to ξ and η:

$$\int_0^1\int_0^1\left(\alpha(\xi,\ \eta)\,\theta(\xi,\ \eta,\ 0) + k(\xi,\ \eta)\frac{\partial}{\partial\zeta}\theta(\xi,\ \eta,\ \zeta)\bigg|_{\zeta=0}\right)\ \cos(\pi\eta)\,d\xi d\eta$$
$$= \int_0^1\int_0^1\frac{f(\xi,\ \eta)\cos(\pi\eta)}{s}\ d\xi d\eta \tag{12.66}$$

- Both sides of Eq. (12.64) are multiplied by $\cos(\pi\xi)$ and integrated from 0 to 1 with respect to ξ and η:

$$\int_0^1\int_0^1\left(\alpha(\xi,\ \eta)\,\theta(\xi,\ \eta,\ 0) + k(\xi,\ \eta)\frac{\partial}{\partial\zeta}\theta(\xi,\ \eta,\ \zeta)\bigg|_{\zeta=0}\right)\ \cos(\pi\xi)\,d\xi d\eta$$
$$= \int_0^1\int_0^1\frac{f(\xi,\ \eta)\cos(\pi\xi)}{s}\ d\xi d\eta \tag{12.67}$$

- Both sides of Eq. (12.64) are multiplied by $\cos(\pi\xi)\cos(\pi\eta)$ and integrated from 0 to 1 with respect to ξ and η:

$$\int_0^1 \int_0^1 \left(\alpha(\xi, \eta)\,\theta(\xi, \eta, 0) + k(\xi, \eta) \frac{\partial}{\partial \zeta}\theta(\xi, \eta, \zeta)\Big|_{\zeta=0} \right) \cos(\pi\xi)\cos(\pi\eta)d\xi d\eta$$
$$= \int_0^1 \int_0^1 \frac{f(\xi, \eta)\cos(\pi\xi)\cos(\pi\eta)}{s} d\xi d\eta \tag{12.68}$$

The constants $A_{1,0}$, $A_{0,1}$, $A_{0,0}$ and $A_{1,1}$ are replaced in Eq. (12.45) to obtain the concentration in the Laplace domain. Finally, the solution in the time domain is developed using the numerical inversion techniques discussed in previous chapters.

12.3 AVERAGE DIMENSIONLESS CONCENTRATION AND FLUX

The average dimensionless concentration is defined by

$$< \theta(s) >= \int_0^1 \int_0^1 \int_0^1 \theta(\xi, \eta, \zeta) \mathrm{d}\xi \mathrm{d}\eta \mathrm{d}\zeta \tag{12.69}$$

The average dimensionless flux across a plane at ζ is defined by Ley et al. (2006)

$$< \psi(s) >= -\int_0^1 \int_0^1 \frac{\partial \theta(\xi, \eta, \zeta)}{\partial \zeta}\Big|_{\zeta} d\xi d\eta \tag{12.70}$$

Expressions for $< \theta(s) >$ and $< \psi(s) >$ are obtained after substituting for the concentration $\theta(\xi, \eta, \zeta)$. The inverse Laplace transforms are computed using an algorithm for numerical inversion or the Bromwich integral and the residue theorem. Explicit forms of Eqs. (12.69) and (12.70) can be determined for zeroth- and first- order approximations. These computations are left to the reader as an exercise.

12.4 SUMMARY

The diffusional transport of drug across a membrane was studied. Expressions were derived to help analyze the transient behavior of the system discussed in Chapter 5 and series solutions were developed in terms of Laplace variables. These representations contained coefficients that could be determined after solving a system of equations. Time-domain solutions were offered for zeroth- and first-order approximations. The chapter described numerical inversion methods that could be used in the first-order case to invert the Laplace-transformed concentration and discussed techniques for calculating the average concentration and flux.

12.5 APPENDIX: MAPLE AND MATHEMATICA CODE LISTINGS

12.5.1 Maple Code: 3Ddistribtransient.mws (Derivation of Eq. (12.45))

```
> restart:with(inttrans):with(PDETools):
> eq:=diff(theta(xi,eta,zeta,tau),tau)=(1/beta^2)*(diff(theta(xi,eta,
    zeta,tau),xi,xi)+diff(theta(xi,eta,zeta,tau),eta,eta))+diff(theta
    (xi,eta,zeta,tau),zeta,zeta);
```

$$eq := \frac{\partial}{\partial\tau}\theta(\xi,\ \eta,\ \zeta,\ \tau) = \frac{\frac{\partial^2}{\partial\xi^2}\theta(\xi,\ \eta,\ \zeta,\ \tau)+\frac{\partial^2}{\partial\eta^2}\theta(\xi,\ \eta,\ \zeta,\ \tau)}{\beta^2} + \frac{\partial^2}{\partial\zeta^2}\theta(\xi,\ \eta,\ \zeta,\ \tau)$$

```
> eq0:=laplace(eq,tau,s):
> eq0A:=theta(xi,eta,zeta,0)=0:
> eq0B:=subs(eq0A,eq0):
> eq0C:=subs(laplace(theta(xi,eta,zeta,tau),tau,s)=theta(xi,eta,zeta),
    eq0B):
> eq0D:=subs(Theta[0]=0,eq0C):
> eq1:=pdsolve(eq0D,HINT=f(xi)*g(eta)*h(zeta)):
> eq2:=factor(build(eq1)):
> eq2A:=_C1*exp(_c[1]^(1/2)*xi)^2+_C2=exp(_c[1]^(1/2)*xi)*(_C1*exp
    (_c[1]^(1/2)*xi)+_C2*exp(-_c[1]^(1/2)*xi)):
> eq2B:=exp(_c[2]^(1/2)*eta)^2*_C3+_C4=exp(_c[2]^(1/2)*eta)*(exp
    (_c[2]^(1/2)*eta)*_C3+_C4*exp(-_c[2]^(1/2)*eta)):
> eq3:=subs(eq2A,eq2B,eq2):
> eq4:=eval(subs(xi=0,diff(rhs(eq3),xi)))=0:
> eq5:=isolate(eq4,_C2):
> eq6:=subs(eq5,eq3):
> eq7:=_C1*exp(_c[1]^(1/2)*xi)+_C1*exp(-_c[1]^(1/2)*xi)=convert
    (factor(_C1*exp(_c[1]^(1/2)*xi)+_C1*exp(-_c[1]^(1/2)*xi)),cos):
> eq8:=subs(eq7,eq6):
> eq9:=subs(_C1=1,eq8):
> eq10:=subs(xi=1,diff(rhs(eq9),xi))=0:
> eq11:=sinh(_c[1]^(1/2))=0:
> eq12:=_c[1]^(1/2)=I*n*Pi:
> eq13:=isolate(eq12,_c[1]):
> eq14:=simplify(subs(eq13,eq9),power,symbolic):
> eq15:=eval(subs(eta=0,diff(rhs(eq14),eta)))=0:
> eq16:=isolate(eq15,_C4):
> eq17:=subs(eq16,eq14):
> eq18:=_C3*exp(_c[2]^(1/2)*eta)+_C3*exp(-_c[2]^(1/2)*eta)=convert
    (factor(_C3*exp(_c[2]^(1/2)*eta)+_C3*exp(-_c[2]^(1/2)*eta)),cos):
> eq19:=subs(eq18,eq17):
> eq20:=subs(_C3=1,eq19):
> eq21:=subs(eta=1,diff(rhs(eq20),eta))=0:
> eq22:=sinh(_c[2]^(1/2))=0:
> eq23:=_c[2]^(1/2)=I*m*Pi:
> eq24:=isolate(eq23,_c[2]):
```

```
> eq25:=simplify(subs(eq24,eq20),power,symbolic):
> eq26:=subs(theta=theta[n,m],_C5=A[n,m],_C6=B[n,m],eq25):
> eq27:=theta(xi,eta,zeta)=Sum(Sum(rhs(eq26),n=0..infinity),m=0..
    infinity):
> eq28:=w*theta(xi,eta,1)+Eval(diff(theta(xi,eta,zeta),zeta),zeta=1)=0:
> eq29:=w*subs(zeta=1,rhs(eq26))+subs(zeta=1,diff(rhs(eq26),zeta))=0:
> eq30:=isolate(eq29,B[n,m]):
> eq31:=subs(eq30,eq26):
> eq32:=eval(simplify(subs(n=0,m=0,eq31),power,symbolic)):
> eq33:=theta(xi,eta,zeta)=Sum(Sum(rhs(eq31),n=0..infinity),m=0..
    infinity);
```

$$eq33 := \theta(\xi,\ \eta,\ \zeta) = \sum_{m=0}^{\infty}\left(\sum_{m=0}^{\infty} 4\cos(m\pi\eta)\ \cos(n\pi\xi)\ \left(A_{n,m}\sin\left(\frac{\sqrt{-\pi^2 m^2-\pi^2 n^2-\beta^2 s}\ \zeta}{\beta}\right)\right.\right.$$
$$+\ \left(\left(\sin\left(\frac{\sqrt{-\pi^2 m^2-\pi^2 n^2-\beta^2 s}}{\beta}\right)\beta\, w\, A_{n,m}\right.\right.$$
$$\left.+\ A_{n,m}\cos\left(\frac{\sqrt{-\pi^2 m^2-\pi^2 n^2-\beta^2 s}}{\beta}\right)\ \sqrt{-\pi^2 m^2-\pi^2 n^2-\beta^2 s}\right)$$
$$\left.\cos\left(\frac{\sqrt{-\pi^2 m^2-\pi^2 n^2-\beta^2 s}\ \zeta}{\beta}\right)\right)\Bigg/\left(-\cos\left(\frac{\sqrt{-\pi^2 m^2-\pi^2 n^2-\beta^2 s}}{\beta}\right)\beta w\right.$$
$$\left.\left.\left.+\ \sin\left(\frac{\sqrt{-\pi^2 m^2-\pi^2 n^2-\beta^2 s}}{\beta}\right)\ \sqrt{-\pi^2 m^2-\pi^2 n^2-\beta^2 s}\right)\right)\right)$$

12.5.2 Maple Code: 3Ddistribzero.mws (Derivation of Eq. (12.63))

```
> restart:with(inttrans):with(PDETools):
> eq:=diff(theta(xi,eta,zeta,tau),tau)=(1/beta^2)*(diff(theta(xi,eta,
    zeta,tau),xi,xi)+diff(theta(xi,eta,zeta,tau),eta,eta))+diff(theta
    (xi,eta,zeta,tau),zeta,zeta):
> eq0:=laplace(eq,tau,s):
> eq0A:=theta(xi,eta,zeta,0)=0:
> eq0B:=subs(eq0A,eq0):
> eq0C:=subs(laplace(theta(xi,eta,zeta,tau),tau,s)=theta(xi,eta,zeta),
    eq0B):
> eq0D:=subs(Theta[0]=0,eq0C):
> eq1:=pdsolve(eq0D,HINT=f(xi)*g(eta)*h(zeta)):
> eq2:=factor(build(eq1)):
> eq2A:=_C1*exp(_c[1]^(1/2)*xi)^2+_C2=exp(_c[1]^(1/2)*xi)*(_C1*
    exp(_c[1]^(1/2)*xi)+_C2*exp(-_c[1]^(1/2)*xi)):
> eq2B:=exp(_c[2]^(1/2)*eta)^2*_C3+_C4=exp(_c[2]^(1/2)*eta)*(exp
    (_c[2]^(1/2)*eta)*_C3+_C4*exp(-_c[2]^(1/2)*eta)):
> eq3:=subs(eq2A,eq2B,eq2):
> eq4:=eval(subs(xi=0,diff(rhs(eq3),xi)))=0:
> eq5:=isolate(eq4,_C2):
> eq6:=subs(eq5,eq3):
```

```
> eq7:=_C1*exp(_c[1]^(1/2)*xi)+_C1*exp(-_c[1]^(1/2)*xi)=convert
    (factor(_C1*exp(_c[1]^(1/2)*xi)+_C1*exp(-_c[1]^(1/2)*xi)),cos):
> eq8:=subs(eq7,eq6):
> eq9:=subs(_C1=1,eq8):
> eq10:=subs(xi=1,diff(rhs(eq9),xi))=0:
> eq11:=sinh(_c[1]^(1/2))=0:
> eq12:=_c[1]^(1/2)=I*n*Pi:
> eq13:=isolate(eq12,_c[1]):
> eq14:=simplify(subs(eq13,eq9),power,symbolic):
> eq15:=eval(subs(eta=0,diff(rhs(eq14),eta)))=0:
> eq16:=isolate(eq15,_C4):
> eq17:=subs(eq16,eq14):
> eq18:=_C3*exp(_c[2]^(1/2)*eta)+_C3*exp(-_c[2]^(1/2)*eta)=convert
    (factor(_C3*exp(_c[2]^(1/2)*eta)+_C3*exp(-_c[2]^(1/2)*eta)),cos):
> eq19:=subs(eq18,eq17):
> eq20:=subs(_C3=1,eq19):
> eq21:=subs(eta=1,diff(rhs(eq20),eta))=0:
> eq22:=sinh(_c[2]^(1/2))=0:
> eq23:=_c[2]^(1/2)=I*m*Pi:
> eq24:=isolate(eq23,_c[2]):
> eq25:=simplify(subs(eq24,eq20),power,symbolic):
> eq26:=subs(theta=theta[n,m],_C5=A[n,m],_C6=B[n,m],eq25):
> eq27:=theta(xi,eta,zeta)=Sum(Sum(rhs(eq26),n=0..infinity),m=0..
    infinity):
> eq28:=w*theta(xi,eta,1)+Eval(diff(theta(xi,eta,zeta),zeta),zeta=1)=0:
> eq29:=w*subs(zeta=1,rhs(eq26))+subs(zeta=1,diff(rhs(eq26),zeta))=0:
> eq30:=isolate(eq29,B[n,m]):
> eq31:=subs(eq30,eq26):
> eq32:=eval(simplify(subs(n=0,m=0,eq31),power,symbolic)):
> eq33:=theta(xi,eta,zeta)=Sum(Sum(rhs(eq31),n=0..infinity),m=0..
    infinity):
```

Zeroth-order approximation:

```
> eq34:=subs(infinity=0,eq33);
```

$$eq34 := \theta(\xi, \eta, \zeta) = \sum_{m=0}^{0}\left(\sum_{n=0}^{0} 4\cos(m\pi\eta)\cos(n\pi\xi)\left(A_{n,m}\sin\left(\frac{\sqrt{-\pi^2 m^2-\pi^2 n^2-\beta^2 s}\ \zeta}{\beta}\right) + \left(\left(\sin\left(\frac{\sqrt{-\pi^2 m^2-\pi^2 n^2-\beta^2 s}}{\beta}\right)\beta w A_{n,m} + A_{n,m}\cos\left(\frac{\sqrt{-\pi^2 m^2-\pi^2 n^2-\beta^2 s}}{\beta}\right)\sqrt{-\pi^2 m^2-\pi^2 n^2-\beta^2 s}\right)\cos\left(\frac{\sqrt{-\pi^2 m^2-\pi^2 n^2-\beta^2 s}\ \zeta}{\beta}\right)\right) \Big/ \left(-\cos\left(\frac{\sqrt{-\pi^2 m^2-\pi^2 n^2-\beta^2 s}}{\beta}\right)\beta w + \sin\left(\frac{\sqrt{-\pi^2 m^2-\pi^2 n^2-\beta^2 s}}{\beta}\right)\sqrt{-\pi^2 m^2-\pi^2 n^2-\beta^2 s}\right)\right)\right)$$

```
> eq35:=subs(A[1,1]=0,eval(subs(Sum=sum,eq34))):
> eq36:=alpha(xi,eta)*theta(xi,eta,0)+k(xi,eta)*Eval(diff(theta(xi,eta,
    zeta),zeta),zeta=0)=f(xi,eta):
> eq37:=alpha(xi,eta)*eval(subs(zeta=0,rhs(eq35)))+k(xi,eta)*subs
    (zeta=0,diff(rhs(eq35),zeta))=f(xi,eta)/s:
```

```
> eq38:=Int(Int(lhs(eq37),xi=0..1),eta=0..1)=Int(Int(rhs(eq37),
    xi=0..1),eta=0..1):
>
>
> eq39:=alpha(xi,eta)=(Heaviside(xi)-Heaviside(xi-delta))*(Heaviside
    (eta)-Heaviside(eta-delta)):
> eq39A:=k(xi,eta)=1-rhs(eq39):
> eq39B:=f(xi,eta)=rhs(eq39):
> eq40 := eval(subs(Int = int, subs({eq39,eq39A,eq39B}, {eq38})))
assuming delta>0 and delta <1:
> eq41 := simplify(solve(eq40,{A[0,0]}), power, symbolic):
> eq42:=simplify(subs(eq41,eq35)):
>
>
> eq46:=simplify(factor(subs(beta =0.1,delta =0.1,w=0,eq42))):
> eq47:=1:
> eq48:=(denom(rhs(eq46))):
> eq49:=sin((-1.*s)^(1/2))=0:
> eq50:=((-1*s)^(1/2))=n*Pi:
> eq51:=isolate(eq50,s):
> eq52:=numer(rhs(eq46))*exp(s*tau)/diff(denom(rhs(eq46)),s):
> eq53:=eval(simplify(subs(eq51,eq52),power,symbolic)) assuming n::
    integer:
> eq54:=cosh(s^(1/2))*s^(1/2)-99.*sinh(s^(1/2))*s=0;
```

$$eq54 := \cosh(\sqrt{s})\ \sqrt{s} - 99.\ \sinh(\sqrt{s})\ s = 0$$

```
> eq55:=s^(1/2)=alpha[n]*I:
> eq57:=isolate(eq55,s):
> eq57A:=simplify(subs(eq57,eq54),power,symbolic):
> eq57B:=isolate(eq57A,cos(alpha[n])):
> eq58:=numer(rhs(eq46))*exp(s*tau)/diff(denom(rhs(eq46)),s):
> eq59:=(simplify(subs(eq57,eq58),power,symbolic)):
> eq60:=factor(subs(eq57B,eq59)):
> eq70:=theta(xi,eta,zeta,tau)=eq47+Sum((eq60),n=1..infinity);
```

$$eq70 := \theta(\xi,\ \eta,\ \zeta) = 1 + \sum_{n=1}^{\infty} \frac{0.0002040608101\ (\cos(1.\ \alpha_n)\ \ \cos(1.\ \alpha_n\zeta) + \sin(1.\ \alpha_n)\sin(1.\ \alpha_n\zeta))\ \ e^{-1.\ \ \alpha_n^2\tau}}{\alpha_n\ \ (\alpha_n + 0.09999489835)\ \ (\alpha_n - 0.09999489835)\ \ \sin(1.\ \alpha_n)}$$

```
>
```

12.5.3 Maple Code: 3Ddistribfirst.mws (Section 12.2.2)

```
> restart:with(inttrans):with(PDETools):
>
>
> eq:=diff(theta(xi,eta,zeta,tau),tau)=(1/beta^2)*(diff(theta(xi,eta,
    zeta,tau),xi,xi)+diff(theta(xi,eta,zeta,tau),eta,eta))+diff(theta
    (xi,eta,zeta,tau),zeta,zeta):
> eq0:=laplace(eq,tau,s):
> eq0A:=theta(xi,eta,zeta,0)=0:
```

```
> eq0B:=subs(eq0A,eq0):
> eq0C:=subs(laplace(theta(xi,eta,zeta,tau),tau,s)=theta(xi,eta,zeta),
    eq0B):
> eq0D:=subs(Theta[0]=0,eq0C):
> eq1:=pdsolve(eq0D,HINT=f(xi)*g(eta)*h(zeta)):
> eq2:=factor(build(eq1)):
> eq2A:=_C1*exp(_c[1]^(1/2)*xi)^2+_C2=exp(_c[1]
    ^(1/2)*xi)*(_C1*exp(_c[1]^(1/2)*xi)+_C2*exp(-_c[1]^(1/2)*xi)):
> eq2B:=exp(_c[2]^(1/2)*eta)^2*_C3+_C4=exp(_c[2]^(1/2)*eta)*(exp(_c[2]
    ^(1/2)*eta)*_C3+_C4*exp(-_c[2]^(1/2)*eta)):
> eq3:=subs(eq2A,eq2B,eq2):
> eq4:=eval(subs(xi=0,diff(rhs(eq3),xi)))=0:
> eq5:=isolate(eq4,_C2):
> eq6:=subs(eq5,eq3):
> eq7:=_C1*exp(_c[1]^(1/2)*xi)+_C1*exp(-_c[1]^(1/2)*xi)=convert(factor
    (_C1*exp(_c[1]^(1/2)*xi)+_C1*exp(-_c[1]^(1/2)*xi)),cos):
> eq8:=subs(eq7,eq6):
> eq9:=subs(_C1=1,eq8):
> eq10:=subs(xi=1,diff(rhs(eq9),xi))=0:
> eq11:=sinh(_c[1]^(1/2))=0:
> eq12:=_c[1]^(1/2)=I*n*Pi:
> eq13:=isolate(eq12,_c[1]):
> eq14:=simplify(subs(eq13,eq9),power,symbolic):
> eq15:=eval(subs(eta=0,diff(rhs(eq14),eta)))=0:
> eq16:=isolate(eq15,_C4):
> eq17:=subs(eq16,eq14):
> eq18:=_C3*exp(_c[2]^(1/2)*eta)+_C3*exp(-_c[2]^(1/2)*eta)=convert
    (factor(_C3*exp(_c[2]^(1/2)*eta)+_C3*exp(-_c[2]^(1/2)*eta)),cos):
> eq19:=subs(eq18,eq17):
> eq20:=subs(_C3=1,eq19):
> eq21:=subs(eta=1,diff(rhs(eq20),eta))=0:
> eq22:=sinh(_c[2]^(1/2))=0:
> eq23:=_c[2]^(1/2)=I*m*Pi:
> eq24:=isolate(eq23,_c[2]):
> eq25:=simplify(subs(eq24,eq20),power,symbolic):
> eq26:=subs(theta=theta[n,m],_C5=A[n,m],_C6=B[n,m],eq25):
> eq27:=theta(xi,eta,zeta)=Sum(Sum(rhs(eq26),n=0..infinity),m=0..
    infinity):
> eq28:=w*theta(xi,eta,1)+Eval(diff(theta(xi,eta,zeta),zeta),zeta=1)=0:
> eq29:=w*subs(zeta=1,rhs(eq26))+subs(zeta=1,diff(rhs(eq26),zeta))=0:
> eq30:=isolate(eq29,B[n,m]):
> eq31:=subs(eq30,eq26):
> eq32:=eval(simplify(subs(n=0,m=0,eq31),power,symbolic)):
> eq33:=theta(xi,eta,zeta)=Sum(Sum(rhs(eq31),n=0..infinity),m=0..
    infinity):
```

First-order approximation:

```
> eq34:=subs(infinity=1,eq33);
```

$$eq34 := \theta(\xi, \eta, \zeta) = \sum_{m=0}^{1}\left(\sum_{n=0}^{1} 4\cos(m\pi\eta)\cos(n\pi\xi)\left(A_{n,m}\sin\left(\frac{\sqrt{-\pi^2 m^2-\pi^2 n^2-\beta^2 s}\ \zeta}{\beta}\right) + \left(\left(\sin\left(\frac{\sqrt{-\pi^2 m^2-\pi^2 n^2-\beta^2 s}}{\beta}\right)\beta w A_{n,m} + A_{n,m}\cos\left(\frac{\sqrt{-\pi^2 m^2-\pi^2 n^2-\beta^2 s}}{\beta}\right)\sqrt{-\pi^2 m^2-\pi^2 n^2-\beta^2 s}\right)\cos\left(\frac{\sqrt{-\pi^2 m^2-\pi^2 n^2-\beta^2 s}\ \zeta}{\beta}\right)\right) \Big/ \left(-\cos\left(\frac{\sqrt{-\pi^2 m^2-\pi^2 n^2-\beta^2 s}}{\beta}\right)\beta w + \sin\left(\frac{\sqrt{-\pi^2 m^2-\pi^2 n^2-\beta^2 s}}{\beta}\right)\sqrt{-\pi^2 m^2-\pi^2 n^2-\beta^2 s}\right)\right)\right)$$

```
> eq35:=subs(A[1,1]=A[1,1],A[0,0]=A[0,0],eval(subs(Sum=sum,eq34)));
```

$$eq35 := \theta(\xi, \eta, \zeta) = 4\ A_{0,0}\sin\left(\frac{\sqrt{-\beta^2 s}\ \zeta}{\beta}\right) + \frac{4\left(\sin\left(\frac{\sqrt{-\beta^2 s}}{\beta}\right)\beta w A_{0,0} + A_{0,0}\cos\left(\frac{\sqrt{-\beta^2 s}}{\beta}\right)\sqrt{-\beta^2 s}\right)\cos\left(\frac{\sqrt{-\beta^2 s}\ \zeta}{\beta}\right)}{-\cos\left(\frac{\sqrt{-\beta^2 s}}{\beta}\right)\beta\ w + \sin\left(\frac{\sqrt{-\beta^2 s}}{\beta}\right)\sqrt{-\beta^2 s}}$$
$$+ 4\cos(\pi\xi)\left(A_{1,0}\sin\left(\frac{\sqrt{-\beta^2 s-\pi^2}\ \zeta}{\beta}\right) + \left(\left(\sin\left(\frac{\sqrt{-\beta^2 s-\pi^2}}{\beta}\right)\beta\ w\ A_{1,0} + A_{1,0}\cos\left(\frac{\sqrt{-\beta^2 s-\pi^2}}{\beta}\right)\sqrt{-\beta^2 s-\pi^2}\right)\cos\left(\frac{\sqrt{-\beta^2 s-\pi^2}\ \zeta}{\beta}\right)\right) \Big/ \left(-\cos\left(\frac{\sqrt{-\beta^2 s-\pi^2}}{\beta}\right)\beta\ w + \sin\left(\frac{\sqrt{-\beta^2 s-\pi^2}}{\beta}\right)\sqrt{-\beta^2 s-\pi^2}\right)\right)$$
$$+ 4\ \cos(\pi\eta)\left(A_{0,1}\sin\left(\frac{\sqrt{-\beta^2 s-\pi^2}\ \zeta}{\beta}\right) + \left(\left(\sin\left(\frac{\sqrt{-\beta^2 s-\pi^2}}{\beta}\right)\beta\ w\ A_{0,1} + A_{0,1}\cos\left(\frac{\sqrt{-\beta^2 s-\pi^2}}{\beta}\right)\sqrt{-\beta^2 s-\pi^2}\right)\cos\left(\frac{\sqrt{-\beta^2 s-\pi^2}\ \zeta}{\beta}\right)\right) \Big/ \left(-\cos\left(\frac{\sqrt{-\beta^2 s-\pi^2}}{\beta}\right)\beta\ w + \sin\left(\frac{\sqrt{-\beta^2 s-\pi^2}}{\beta}\right)\sqrt{-\beta^2 s-\pi^2}\right)\right)$$
$$+ 4\ \cos(\pi\eta)\ \cos(\pi\xi)\left(A_{1,1}\sin\left(\frac{\sqrt{-\beta^2 s-\pi^2}\ \zeta}{\beta}\right) + \left(\sin\left(\frac{\sqrt{-\beta^2 s-2\pi^2}}{\beta}\right)\beta w A_{1,1} + A_{1,1}\cos\left(\frac{\sqrt{-\beta^2 s-2\pi^2}}{\beta}\right)\sqrt{-\beta^2 s-2\pi^2}\right) - \cos\left(\frac{\sqrt{-\beta^2 s-2\pi^2}}{\beta}\right)\beta\ w + \sin\left(\frac{\sqrt{-\beta^2 s-2\pi^2}}{\beta}\right)\sqrt{-\beta^2 s-2\pi^2}\right)$$

```
> eq36:=alpha(xi,eta)*theta(xi,eta,0)+k(xi,eta)*Eval(diff(theta(xi,eta,
    zeta),zeta),zeta=0)=f(xi,eta):
> eq37:=eval(alpha(xi,eta)*eval(subs(zeta=0,rhs(eq35)))+k(xi,eta)*subs
    (zeta=0,diff(rhs(eq35),zeta))=f(xi,eta)/s):
> eq38:=Int(Int(lhs(eq37),xi=0..1),eta=0..1)=Int(Int(rhs(eq37),xi=0
    ..1),eta=0..1):
> eq38a:=Int(Int(lhs(eq37)*cos(Pi*xi),xi=0..1),eta=0..1)=Int(Int(rhs
    (eq37)*cos(Pi*xi),xi=0..1),eta=0..1):
> eq38A:=Int(Int(lhs(eq37)*cos(Pi*eta),xi=0..1),eta=0..1)=Int(Int(rhs
    (eq37)*cos(Pi*eta),xi=0..1),eta=0..1):
> eq38B:=Int(Int(lhs(eq37)*cos(Pi*eta)*cos(Pi*xi),xi=0..1),eta=0..1)
    =Int(Int(rhs(eq37)*cos(Pi*eta)*cos(Pi*xi),xi=0..1),eta=0..1):
>
```

```
> eq39:=alpha(xi,eta)=(Heaviside(xi)-Heaviside(xi-delta))*(Heaviside
      (eta)-Heaviside(eta-delta)):
> eq39A:=k(xi,eta)=1-rhs(eq39):
> eq39B:=f(xi,eta)=rhs(eq39):
> eq40 := eval(subs(Int = int, subs({eq39, eq39A, eq39B}, {eq38,
eq38a, eq38A, eq38B}))) assuming delta>0 and delta <1:
>
>
```

12.5.4 Mathematica Code: 3Ddistribzero1.nb

(* Section 12.2 *)

$\text{eq} = \partial_\tau \theta[\xi, \eta, \zeta, \tau] =$

$\frac{1}{\beta^2}(\partial_{\xi,\xi}\theta[\xi, \eta, \zeta, \tau] + \partial_{\eta,\eta}\theta[\xi, \eta, \zeta, \tau])+$

$\partial_{\zeta,\zeta}\theta[\xi, \eta, \zeta, \tau]$

$\theta^{(0,0,0,1)}[\xi, \eta, \zeta, \tau] = \theta^{(0,0,2,0)}[\xi, \eta, \zeta, \tau] + \frac{\theta^{(0,2,0,0)}[\xi,\eta,\zeta,\tau]+\theta^{(2,0,0,0)}[\xi,\eta,\zeta,\tau]}{\beta^2}$

$\text{eq1} = s * \theta[\xi, \eta, \zeta] - \frac{1}{\beta^2}(\partial_{\xi,\xi}\theta[\xi, \eta, \zeta]$
$\quad + \partial_{\eta,\eta}\theta[\xi, \eta, \zeta]) - \partial_{\zeta,\zeta}\theta[\xi, \eta, \zeta]$

$s\theta[\xi, \eta, \zeta] - \theta^{(0,0,2)}[\xi, \eta, \zeta] - \frac{\theta^{(0,2,0)}[\xi,\eta,\zeta]+\theta^{(2,0,0)}[\xi,\eta,\zeta]}{\beta^2}$

$\text{eq2} = \text{eq1} /. \theta \to \texttt{Function}[\{\xi, \eta, \zeta\}, f[\xi] * g[\eta] * h[\zeta]]$

$sf[\xi]g[\eta]h[\zeta] - \frac{g[\eta]h[\zeta]f''[\xi]+f[\xi]h[\zeta]g''[\eta]}{\beta^2} - f[\xi]g[\eta]h''[\zeta]$

$\text{eq3} = \texttt{Expand}[\text{eq2} / (f[\xi] * g[\eta] * h[\zeta])] = 0$

$S - \frac{f''[\xi]}{\beta^2 f[\xi]} - \frac{g''[\eta]}{\beta^2 g[\eta]} - \frac{h''[\zeta]}{h[\zeta]} = 0$

$\text{eq3A} = \frac{f''[\xi]}{f[\xi]} = c[1]$

$\frac{f''[\xi]}{f[\xi]} = c[1]$

$\text{eq4} = \frac{g''[\eta]}{g[\eta]} = c[2]$

$\frac{g''[\eta]}{g[\eta]} = c[2]$

$\text{eq4A} = s - \frac{c[1]}{\beta^2} - \frac{c[2]}{\beta^2} - \frac{h''[\zeta]}{h[\zeta]} = 0$

$s - \frac{c[1]}{\beta^2} - \frac{c[2]}{\beta^2} - \frac{h''[\zeta]}{h[\zeta]} = 0$

$\text{eq5} = \texttt{DSolve}[\text{eq3A}, f, \xi]$

$\{\{f \to \texttt{Function}[\{\xi\}, e^{\xi\sqrt{c[1]}}C[1] + e^{-\xi\sqrt{c[1]}}C[2]]\}\}$

$\text{eq5A} = (\texttt{DSolve}[\text{eq4}, g, \eta] /. C[1] -> A[1]) /. C[2] \to B[1]$

$\{\{g \to \texttt{Function}[\{\eta\}, e^{\eta\sqrt{c[2]}}A[1] + e^{-\eta\sqrt{c[2]}}B[1]]\}\}$

$\text{eq5B} = (\texttt{DSolve}[\text{eq4A}, h, \zeta] /. C[1] \to F[1]) /. C[2] \to G[1]$

$\left\{\left\{h \to \texttt{Function}\left[\{\zeta\}, e^{\frac{\zeta\sqrt{s\beta^2-c[1]-c[2]}}{\beta}}F[1] + e^{\frac{-\zeta\sqrt{s\beta^2-c[1]-c[2]}}{\beta}}G[1]\right]\right\}\right\}$

$\text{eq6} = (e^{\xi\sqrt{c[1]}}C[1] + e^{-\xi\sqrt{c[1]}}C[2])(e^{\eta\sqrt{c[2]}}A[1] + e^{-\eta\sqrt{c[2]}}B[1])$

$\left(F[1]\text{Cos}\left[\frac{\zeta\sqrt{-s\beta^2+c[1]+c[2]}}{\beta}\right] + G[1]\text{Sin}\left[\frac{\zeta\sqrt{-s\beta^2+c[1]+c[2]}}{\beta}\right]\right)$

$(e^{\eta\sqrt{c[2]}}A[1] + e^{-\eta\sqrt{c[2]}}B[1])(e^{\xi\sqrt{c[1]}}C[1] + e^{-\xi\sqrt{c[1]}}C[2])$

$\left(\text{Cos}\left[\frac{\zeta\sqrt{-s\beta^2+c[1]+c[2]}}{\beta}\right]F[1] + G[1]\text{Sin}\left[\frac{\zeta\sqrt{-s\beta^2+c[1]+c[2]}}{\beta}\right]\right)$

$\text{eq6A} = ((\partial_\xi \text{eq6}) /. \xi \to 0) = 0$

$$(e^{\eta\sqrt{c[2]}}A[1]+e^{-\eta\sqrt{c[2]}}B[1])(\sqrt{c[1]}C[1]-\sqrt{c[1]}C[2])\left(\text{Cos}\left[\frac{\zeta\sqrt{-s\beta^2+c[1]+c[2]}}{\beta}\right]F[1]+G[1]\text{Sin}\left[\frac{\zeta\sqrt{-s\beta^2+c[1]+c[2]}}{\beta}\right]\right)=0$$

eq6C = Solve[eq6A, $C[2]$]

$\{\{C[2]\rightarrow C[2]\}\}$

eq6D = eq6 / .$C[2]\rightarrow C[1]$

$$(e^{\eta\sqrt{c[2]}}A[1]+e^{-\eta\sqrt{c[2]}}B[1])(e^{-\xi\sqrt{c[1]}}C[1]+e^{\xi\sqrt{c[1]}}C[1])\left(\text{Cos}\left[\frac{\zeta\sqrt{-s\beta^2+c[1]+c[2]}}{\beta}\right]F[1]+G[1]\text{Sin}\left[\frac{\zeta\sqrt{-s\beta^2+c[1]+c[2]}}{\beta}\right]\right)$$

eq6E = (eq6D / .$(e^{-\xi\sqrt{c[1]}}C[1]+e^{\xi\sqrt{c[1]}}C[1])\rightarrow 2\ *\ C[1]\ *\ Cosh[\xi\sqrt{c[1]}]$) / . $C[1]\rightarrow 1$

$$2\ (e^{\eta\sqrt{c[2]}}A[1]+e^{-\eta\sqrt{c[2]}}B[1])\text{Cosh}[\xi\sqrt{c[1]}]\left(\text{Cos}\left[\frac{\zeta\sqrt{-s\beta^2+c[1]+c[2]}}{\beta}\right]F[1]+G[1]\text{Sin}\left[\frac{\zeta\sqrt{-s\beta^2+c[1]+c[2]}}{\beta}\right]\right)$$

eq6F = ((∂_ξeq6E) / .$\xi\rightarrow 1$) = 0

$$2\ (e^{\eta\sqrt{c[2]}}A[1]+e^{-\eta\sqrt{c[2]}}B[1])\sqrt{c[1]}\left(\text{Cos}\left[\frac{\zeta\sqrt{-s\beta^2+c[1]+c[2]}}{\beta}\right]F[1]+G[1]\text{Sin}\left[\frac{\zeta\sqrt{-s\beta^2+c[1]+c[2]}}{\beta}\right]\right)\text{Sinh}[\sqrt{c[1]}]=0$$

eq6G = $\text{Sinh}[\sqrt{c[1]}]=0$

$\text{Sinh}[\sqrt{c[1]}]=0$

eq6H = $\sqrt{c[1]}$ = Sqrt[−1]*n*Pi

$\sqrt{c[1]}=in\pi$

eq6I = Solve[eq6H, $c[1]$]

$\{\{c[1]\rightarrow -n^2\pi^2\}\}$

eq6J = eq6E / .$c[1]\rightarrow -n^2\pi^2$

$$2\ (e^{\eta\sqrt{c[2]}}A[1]+e^{-\eta\sqrt{c[2]}}B[1])\text{Cosh}[\sqrt{-n^2}\pi\xi]\left(\text{Cos}\left[\frac{\zeta\sqrt{-n^2\pi^2-s\beta^2+c[2]}}{\beta}\right]F[1]+G[1]\text{Sin}\left[\frac{\zeta\sqrt{-n^2\pi^2-s\beta^2+c[2]}}{\beta}\right]\right)$$

eq7 = ((∂_ηeq6J) / .$\eta\rightarrow 0$) = 0

$$2(A[1]\sqrt{c[2]}-B[1]\sqrt{c[2]})\text{Cosh}[\sqrt{-n^2}\pi\xi]\left(\text{Cos}\left[\frac{\zeta\sqrt{-n^2\pi^2-s\beta^2+c[2]}}{\beta}\right]F[1]+G[1]\text{Sin}\left[\frac{\zeta\sqrt{-n^2\pi^2-s\beta^2+c[2]}}{\beta}\right]\right)=0$$

eq8 = Solve[eq7, $B[1]$]

$\{\{B[1]\rightarrow A[1]\}\}$

eq9 = eq6J / .$B[1]\rightarrow A[1]$

$$2\ (e^{-\eta\sqrt{c[2]}}A[1]+e^{\eta\sqrt{c[2]}}A[1])\text{Cosh}[\sqrt{-n^2}\pi\xi]\left(\text{Cos}\left[\frac{\zeta\sqrt{-n^2\pi^2-s\beta^2+c[2]}}{\beta}\right]F[1]+G[1]\text{Sin}\left[\frac{\zeta\sqrt{-n^2\pi^2-s\beta^2+c[2]}}{\beta}\right]\right)$$

eq10 = (eq9 / .$(e^{-\eta\sqrt{c[2]}}A[1]+e^{\eta\sqrt{c[2]}}A[1])\rightarrow 2\ *\ A[1]\ *\ \text{Cosh}[\eta\sqrt{c[2]}]$) / . $A[1]\rightarrow 1$

$$4\text{Cosh}[\sqrt{-n^2}\pi\xi]\ \text{Cosh}[\eta\sqrt{c[2]}]\left(\text{Cos}\left[\frac{\zeta\sqrt{-n^2\pi^2-s\beta^2+c[2]}}{\beta}\right]F[1]+G[1]\text{Sin}\left[\frac{\zeta\sqrt{-n^2\pi^2-s\beta^2+c[2]}}{\beta}\right]\right)$$

eq11 = ((∂_ηeq10) / .$\eta\rightarrow 1$) = 0

$$4\sqrt{c[2]}\text{Cosh}[\sqrt{-n^2}\pi\xi]\ \left(\text{Cos}\left[\frac{\zeta\sqrt{-n^2\pi^2-s\beta^2+c[2]}}{\beta}\right]F[1]+G[1]\text{Sin}\left[\frac{\zeta\sqrt{-n^2\pi^2-s\beta^2+c[2]}}{\beta}\right]\right)$$

$\text{Sinh}[\sqrt{c[2]}]=0$

eq12 = $\text{Sinh}[\sqrt{c[2]}]=0$

eq12 = $\text{Sinh}[\sqrt{c[2]}]=0$

eq13 = $\sqrt{c[2]}$ = Sqrt[−1]*m*Pi

$\sqrt{c[2]}=im\pi$

eq14 = Solve[eq13, $c[2]$]

$\{\{c[2] \to -m^2\pi^2\}\}$

eq15 = Simplify[eq10 /. $c[2] \to -m^2\pi^2$, $n > 0 \&\& m > 0$]

$4\text{Cosh}[m\pi\eta]\text{Cos}[\eta\pi\xi]\left(\text{Cos}\left[\frac{\sqrt{-m^2\pi^2-n^2\pi^2-s\beta^2}\zeta}{\beta}\right]F[1]+G[1]\text{Sin}\left[\frac{\sqrt{-m^2\pi^2-n^2\pi^2-s\beta^2}\zeta}{\beta}\right]\right)$

eq16 = $((w^*\text{eq15}+\partial_\zeta\text{eq15})/.\zeta\to 1)=0$

$4\text{Cos}[m\pi\eta]\text{Cos}[n\pi\xi]$
$\left(\frac{\sqrt{-m^2\pi^2-n^2\pi^2-s\beta^2}\text{Cos}\left[\frac{\sqrt{-m^2\pi^2-n^2\pi^2-s\beta^2}}{\beta}\right]G[1]}{\beta}\right.$
$\left.\frac{\sqrt{-m^2\pi^2-n^2\pi^2-s\beta^2}F[1]\text{Sin}\left[\frac{\sqrt{-m^2\pi^2-n^2\pi^2-s\beta^2}}{\beta}\right]}{\beta}\right)$
$4w\text{Cos}[m\pi\eta]\text{Cos}[n\pi\xi]$
$\left(\text{Cos}\left[\frac{\sqrt{-m^2\pi^2-n^2\pi^2-s\beta^2}}{\beta}\right]F[1]+G[1]\text{Sin}\left[\frac{\sqrt{-m^2\pi^2-n^2\pi^2-s\beta^2}}{\beta}\right]\right)=0$

eq17 = Solve[eq16, F[1]]

$\left\{\left\{F[1]\to\left(-\sqrt{-m^2\pi^2-n^2\pi^2-s\beta^2}\text{Cos}\left[\frac{\sqrt{-m^2\pi^2-n^2\pi^2-s\beta^2}}{\beta}\right]G[1]-\right.\right.\right.$
$\left.w\beta G[1]\text{Sin}\left[\frac{\sqrt{-m^2\pi^2-n^2\pi^2-s\beta^2}}{\beta}\right]\right)/$
$\left.\left.\left(w\beta\text{Cos}\left[\frac{\sqrt{-m^2\pi^2-n^2\pi^2-s\beta^2}}{\beta}\right]-\sqrt{-m^2\pi^2-n^2\pi^2-s\beta^2}\text{Sin}\left[\frac{\sqrt{-m^2\pi^2-n^2\pi^2-s\beta^2}}{\beta}\right]\right)\right\}\right\}$

eq18 = (eq15 /. eq17) /. $G[1]\to A[n,m]$

$\left\{4\text{Cos}[m\pi\eta]\text{Cos}[n\pi\xi]\right.$
$\left(\text{Cos}\left[\frac{\sqrt{-m^2\pi^2-n^2\pi^2-s\beta^2}}{\beta}\right]\left(-\sqrt{-m^2\pi^2-n^2\pi^2-s\beta^2}A[n,m]\text{Cos}\left[\frac{\sqrt{-m^2\pi^2-n^2\pi^2-s\beta^2}}{\beta}\right]-\right.\right.$
$\left.\left.w\beta A[n,m]\text{Sin}\left[\frac{\sqrt{-m^2\pi^2-n^2\pi^2-s\beta^2}}{\beta}\right]\right)\right)/$
$\left(w\beta\text{Cos}\left[\frac{\sqrt{-m^2\pi^2-n^2\pi^2-s\beta^2}}{\beta}\right]-\sqrt{-m^2\pi^2-n^2\pi^2-s\beta^2}\text{Sin}\left[\frac{\sqrt{-m^2\pi^2-n^2\pi^2-s\beta^2}}{\beta}\right]\right)$
$\left.+A[n,m]\text{Sin}\left[\frac{\sqrt{-m^2\pi^2-n^2\pi^2-s\beta^2}\zeta}{\beta}\right]\right\}$

*eq*19 =

$\theta(\xi,\eta,\zeta)=$
$\sum_{n=0}^{\infty}\sum_{m=0}^{\infty}\{4\text{Cos}[m\pi\eta]\text{Cos}[n\pi\xi]$
$\left(\left(\text{Cos}\left[\frac{\sqrt{-m^2\pi^2-n^2\pi^2-s\beta^2}\zeta}{\beta}\right]\right.\right.$
$(-\sqrt{-m^2\pi^2-n^2\pi^2-s\beta^2}A[n,m]\text{Cos}\left[\frac{\sqrt{-m^2\pi^2-n^2\pi^2-s\beta^2}}{\beta}\right]-$
$\left.\left.w\beta A[n,m]\text{Sin}\left[\frac{\sqrt{-m^2\pi^2-n^2\pi^2-s\beta^2}\zeta}{\beta}\right]\right)\right)$
$\left(w\beta\text{Cos}\left[\frac{\sqrt{-m^2\pi^2-n^2\pi^2-s\beta^2}}{\beta}\right]-\right.$
$\left.\sqrt{-m^2\pi^2-n^2\pi^2-s\beta^2}\text{Sin}\left[\frac{\sqrt{-m^2\pi^2-n^2\pi^2-s\beta^2}}{\beta}\right]\right)$
$\left.\left.+A[n,m]\text{Sin}\left[\frac{\sqrt{-m^2\pi^2-n^2\pi^2-s\beta^2}}{\beta}\right]\right)\right\}$

```
(*End Section 12.2*)

(* Section 12.2.1*)
```

eq20 = (eq18 /. $n\to 0$) /. $m\to 0$

$$\left\{4\left(\frac{\mathrm{Cos}\left[\frac{\sqrt{-s\beta^2}\zeta}{\beta}\right]\left(-\sqrt{-s\beta^2}A[0,0]\mathrm{Cos}\left[\frac{\sqrt{-s\beta^2}}{\beta}\right]-w\beta A[0,0]\mathrm{Sin}\left[\frac{\sqrt{-s\beta^2}}{\beta}\right]\right)}{w\beta\mathrm{Cos}\left[\frac{\sqrt{-s\beta^2}}{\beta}\right]-\sqrt{-s\beta^2}\mathrm{Sin}\left[\frac{\sqrt{-s\beta^2}}{\beta}\right]}+A[0,0]\mathrm{Sin}\left[\frac{\sqrt{-s\beta^2}\zeta}{\beta}\right]\right)\right\}$$

eq21 = `alpha`$[\xi,\eta]$(eq20 /. $\zeta\to 0$) + $k[\xi,\eta]$((∂_ζeq20) /. $\zeta\to 0$) = $f[\xi,\eta]$ / s

$$\left\{\frac{4\sqrt{-s\beta^2}A[0,0]k[\xi,\eta]}{\beta}+\frac{4\mathtt{alpha}[\xi,\eta]\left(-\sqrt{-s\beta^2}A[0,0]\mathrm{Cos}\left[\frac{\sqrt{-s\beta^2}}{\beta}\right]-w\beta A[0,0]\mathrm{Sin}\left[\frac{\sqrt{-s\beta^2}}{\beta}\right]\right)}{w\beta\mathrm{Cos}\left[\frac{\sqrt{-s\beta^2}}{\beta}\right]-\sqrt{-s\beta^2}\mathrm{Sin}\left[\frac{\sqrt{-s\beta^2}}{\beta}\right]}\right\}=\frac{f[\xi,\eta]}{s}$$

eq22 =
`Integrate[Integrate[alpha`$[\xi,\eta]$ * (eq20 /. $\zeta\to 0$)+
$k[\xi,\eta]$ * ((∂_ζeq20) /. $\zeta\to 0$), $\{\xi, 0, 1\}$], $\{\eta, 0, 1\}$][[1]] =
`Integrate[Integrate[`$f[\xi,\eta]$ / s, $\{\xi, 0, 1\}$], $\{\eta, 0, 1\}$]

$$\int_0^1\left(\int_0^1\left(\frac{4\sqrt{-s\beta^2}A[0,0]\,k[\xi,\eta]}{\beta}+\frac{4\mathtt{alpha}[\xi,\eta]\left(-\sqrt{-s\beta^2}A[0,0]\mathrm{Cos}\left[\frac{\sqrt{-s\beta^2}}{\beta}\right]-w\beta A[0,0]\mathrm{Sin}\left[\frac{\sqrt{-s\beta^2}}{\beta}\right]\right)}{w\beta\mathrm{Cos}\left[\frac{\sqrt{-s\beta^2}}{\beta}\right]-\sqrt{-s\beta^2}\mathrm{Sin}\left[\frac{\sqrt{-s\beta^2}}{\beta}\right]}\right)d\xi\right)d\eta$$
$$=\int_0^1\left(\int_0^1\frac{f[\xi,\eta]}{s}d\xi\right)\,d\eta$$

eq23 = ((eq22)/.$f[\xi,\eta]$-> `alpha`$[\xi,\eta]$) /.$k[\xi,\eta]\to 1-$ `alpha`$[\xi,\eta]$

$$\int_0^1\left(\int_0^1\left(\frac{4\sqrt{-s\beta^2}A[0,0]\,(1-\mathtt{alpha}[\xi,\eta])}{\beta}+\frac{4\mathtt{alpha}[\xi,\eta]\left(-\sqrt{-s\beta^2}A[0,0]\mathrm{Cos}\left[\frac{\sqrt{-s\beta^2}}{\beta}\right]-w\beta A[0,0]\mathrm{Sin}\left[\frac{\sqrt{-s\beta^2}}{\beta}\right]\right)}{w\beta\mathrm{Cos}\left[\frac{\sqrt{-s\beta^2}}{\beta}\right]-\sqrt{-s\beta^2}\mathrm{Sin}\left[\frac{\sqrt{-s\beta^2}}{\beta}\right]}\right)d\xi\right)d\eta$$
$$=\int_0^1\left(\int_0^1\frac{\mathtt{alpha}\,[\xi,\eta]}{s}d\xi\right)\,d\eta$$

eq24 =
(eq23)/.
`alpha`$[\xi,\eta]\to$ (`UnitStep`$[\xi]$ − `UnitStep`$[\xi-\delta]$)*
(`UnitStep`$[\eta]$ − `UnitStep`$[\eta-\delta]$)

$$4\sqrt{-s\beta^2}\;A[0,0]\left(-\left(\left(\left(\sqrt{s}\mathrm{Cosh}\left[\sqrt{s}\right]+\;w\mathrm{Sinh}[\sqrt{s}]\right)(1+(-1+\delta^2)\mathtt{UnitStep}[1-\delta]-\delta^2\mathtt{UnitStep}[-\delta])\right)/(\sqrt{s}w\beta\mathrm{Cosh}[\sqrt{s}]+s\beta\mathrm{Sinh}[\sqrt{s}])\right)+\frac{-(-1+\delta^2)\mathtt{UnitStep}[1-\delta]+\delta^2\mathtt{UnitStep}[-\delta]}{\beta}\right)=\frac{(-1+\mathtt{UnitStep}[1-\delta](1-\delta+\delta\mathtt{UnitStep}[-\delta]))^2}{s}$$

eq25 = `Simplify`[eq24, $\delta>0$&& $\delta<1$ &&$s>0$ && $\beta>0$]

$$-\frac{4iA[0,0](\sqrt{s}(\delta^2+w(-1+\delta^2))\mathrm{Cosh}[\sqrt{s}]+(w\delta^2+s(-1+\delta^2))\mathrm{Sinh}[\sqrt{s}])}{w\mathrm{Cosh}[\sqrt{s}]+\;\sqrt{s}\mathrm{Sinh}[\sqrt{s}]}=\frac{\delta^2}{s}$$

eq26 = `Solve`[eq25, $A[0,0]$]

$$\left\{\left\{A\,[0,0]\to(i\delta^2(w\mathrm{Cosh}[\sqrt{s}]+\;\sqrt{s}\mathrm{Sinh}[\sqrt{s}]))/\left(4s\left(-\sqrt{s}w\mathrm{Cosh}\left[\sqrt{s}\right]+\sqrt{s}\delta^2\mathrm{Cosh}[\sqrt{s}]+\sqrt{s}w\delta^2\mathrm{Cosh}[\sqrt{s}]-s\mathrm{Sinh}[\sqrt{s}]+\;s\delta^2\mathrm{Sinh}[\sqrt{s}]+w\delta^2\mathrm{Sinh}[\sqrt{s}]\right)\right)\right\}\right\}$$

eq27 = `Simplify`[eq20/.eq26, $\delta>0$ && $\delta<1$ &&$s>$ 0 && $\beta>0$]

$$\left\{\left\{\frac{\delta^2\left(\sqrt{s}\mathrm{Cosh}\left[\sqrt{s}(-1+\zeta)\right]-w\mathrm{Sinh}[\sqrt{s}(-1+\zeta)]\right)}{s(\sqrt{s}(\delta^2+w(-1+\delta^2))\mathrm{Cosh}[\sqrt{s}]+(w\delta^2+s(-1+\delta^2))\mathrm{Sinh}[\sqrt{s}])}\right\}\right\}$$

```
eq28 = Simplify[(((eq27 /.w → 0) /.β → 0.1)/.δ → 0.1)]
```

$$\left\{\left\{-\frac{0.010101\mathrm{Cosh}\left[\sqrt{s}(-1+\ \zeta)\right]}{-0.010101s\mathrm{Cosh}[\sqrt{s}]+1.s^{3/2}\mathrm{Sinh}[\sqrt{s}]}\right\}\right\}$$

```
eq29 = ((s* Series[eq28[[1]], {s, 0, 1}])/.s → 0)[[1]]
```

1.

```
eq31 = Numerator[eq28]*Exp[s*τ]∂_s Denominator[eq28]
```

$$\left\{\left\{-\frac{0.010101e^{s\tau}\mathrm{Cosh}\left[\sqrt{s}(-1+\zeta)\right]}{-0.010101\mathrm{Cosh}[\sqrt{s}]+\ 0.5s\mathrm{Cosh}[\sqrt{s}]+1.49495\sqrt{s}\mathrm{Sinh}[\sqrt{s}]}\right\}\right\}$$

```
eq33 = Expand[Denominator[eq28]/s = 0
```

$$\{\{-0.010101\mathrm{Cosh}[\sqrt{s}] + 1.\sqrt{s}\mathrm{Sinh}[\sqrt{s}]\}\} = 0$$

```
eq34 = s^(1/2) = alpha[n] * Sqrt[-1]
```

$\sqrt{s} = i$ alpha[n]

```
eq35 = Solve[eq34, s]
```

{{s → −alpha[n]^2}}

```
eq35A = Simplify[eq33/.eq35, alpha[n] > 0]
```

{{{−0.010101Cos[alpha[n]] − 1 .alpha[n]Sin[alpha[n]]}} = 0}

```
eq35B = Solve[eq35A, Cos[alpha[n]]]
```

{{Cos[alpha[n]] → −99.alpha[n]Sin[alpha[n]]}}

```
eq36 =
FullSimplify[
(Factor[(Simplify[eq31/.eq35, alpha[n] > 0])/.eq35B]/.alpha[n] → α_n)]
```

$$\left\{\left\{\left\{\left\{-\frac{0.000204061e^{-\tau\alpha_n^2}\mathrm{Cos}[(-1+\zeta)\alpha_n]\mathrm{Csc}[\alpha_n]}{-0.00999898\alpha_n+1.\alpha_n^3}\right\}\right\}\right\}\right\}$$

```
eq37 =
θ[ξ, η, ζ, τ] =
```

$$\mathrm{eq29} + \sum_{n=0}^{\infty} -\frac{0.000204061e^{-\tau\alpha_n^2}\mathrm{Cos}[(-1+\zeta)\alpha_n]\mathrm{Csc}[\alpha_n]}{-\text{"0.00999898"}\,\alpha_n + \text{"1."}\,\alpha_n^3}$$

$$\theta[\xi,\ \eta,\ \zeta,\ \tau] = 1.\ + \sum_{n=0}^{\infty} -\frac{0.000204061e^{-\tau\alpha_n^2}\mathrm{Cos}[(-1+\zeta)\alpha_n]\mathrm{Csc}[\alpha_n]}{-0.00999898\alpha_n + 1.\alpha_n^3}$$

```
(* End of Section 12.2.1*)

(* Section 12.2.2 *)
eq38 = ((eq18/.n → 0)/.m → 0) + ((eq18/.n → 1)/.m → 0)+
((eq18/.n → 0)/.m → 1) + ((eq18/.n → 1)/.m → 1)
```

$$\left\{4\left(\frac{\mathrm{Cos}\left[\frac{\sqrt{-s\beta^2}\zeta}{\beta}\right]\left(-\sqrt{-s\beta^2}A[0,0]\mathrm{Cos}\left[\frac{\sqrt{-s\beta^2}}{\beta}\right]-w\beta A[0,0]\mathrm{Sin}\left[\frac{\sqrt{-s\beta^2}}{\beta}\right]\right)}{w\beta\mathrm{Cos}\left[\frac{\sqrt{-s\beta^2}}{\beta}\right]-\sqrt{-s\beta^2}\mathrm{Sin}\left[\frac{\sqrt{-s\beta^2}}{\beta}\right]}+A[0,0]\mathrm{Sin}\left[\frac{\sqrt{-s\beta^2}\zeta}{\beta}\right]\right)+\right.$$

$$4\mathrm{Cos}[\pi\eta]\mathrm{Cos}[\pi\xi]\left(\left(\mathrm{Cos}\left[\frac{\sqrt{-2\pi^2-s\beta^2}\zeta}{\beta}\right]\left(-\sqrt{-2\pi^2-s\beta^2}A[1,1]\mathrm{Cos}\left[\frac{\sqrt{-2\pi^2-s\beta^2}}{\beta}\right]-w\beta A[1,1]\mathrm{Sin}\left[\frac{\sqrt{-2\pi^2-s\beta^2}}{\beta}\right]\right)\right)\Big/\left(w\beta\mathrm{Cos}\left[\frac{\sqrt{-2\pi^2-s\beta^2}}{\beta}\right]-\sqrt{-2\pi^2-s\beta^2}\mathrm{Sin}\left[\frac{\sqrt{-2\pi^2-s\beta^2}}{\beta}\right]\right)+A[1,1]\mathrm{Sin}\left[\frac{\sqrt{-2\pi^2-s\beta^2}\zeta}{\beta}\right]\right)+$$

$$4\mathrm{Cos}[\pi\eta]\left(\left(\mathrm{Cos}\left[\frac{\sqrt{-\pi^2-s\beta^2}\zeta}{\beta}\right]\left(-\sqrt{-\pi^2-s\beta^2}A[0,1]\mathrm{Cos}\left[\frac{\sqrt{-\pi^2-s\beta^2}}{\beta}\right]-w\beta A[0,1]\mathrm{Sin}\left[\frac{\sqrt{-\pi^2-s\beta^2}}{\beta}\right]\right)\right)\Big/\left(w\beta\mathrm{Cos}\left[\frac{\sqrt{-\pi^2-s\beta^2}}{\beta}\right]-\sqrt{-\pi^2-s\beta^2}\mathrm{Sin}\left[\frac{\sqrt{-\pi^2-s\beta^2}}{\beta}\right]\right)+A[0,1]\mathrm{Sin}\left[\frac{\sqrt{-\pi^2-s\beta^2}\zeta}{\beta}\right]\right)+$$

$$\left.4\mathrm{Cos}[\pi\xi]\left(\left(\mathrm{Cos}\left[\frac{\sqrt{-\pi^2-s\beta^2}\zeta}{\beta}\right]\left(-\sqrt{-\pi^2-s\beta^2}A[1,0]\mathrm{Cos}\left[\frac{\sqrt{-\pi^2-s\beta^2}}{\beta}\right]-w\beta A[1,0]\mathrm{Sin}\left[\frac{\sqrt{-\pi^2-s\beta^2}}{\beta}\right]\right)\right)\Big/\left(w\beta\mathrm{Cos}\left[\frac{\sqrt{-\pi^2-s\beta^2}}{\beta}\right]-\sqrt{-\pi^2-s\beta^2}\mathrm{Sin}\left[\frac{\sqrt{-\pi^2-s\beta^2}}{\beta}\right]\right)+A[1,0]\mathrm{Sin}\left[\frac{\sqrt{-\pi^2-s\beta^2}\zeta}{\beta}\right]\right)\right\}$$

```
eq39 = alpha[ξ, η](eq38/.ζ → 0) + k[ξ, η]((∂ζeq38)/.ζ → 0) == f[ξ, η]/s;
eq40 =
Integrate[Integrate[alpha[ξ, η]*(eq38/.ζ → 0)+
k[ξ, η]*((∂ζeq38)/.ζ → 0), {ξ, 0, 1}], {η, 0, 1}] [[1]] ==
Integrate[Integrate[f[ξ, η]/s, {ξ, 0, 1}], {η, 0, 1}];
eq40A =
Simplify[
(((eq40)/.f[ξ, η]-> alpha[ξ, η])/.k[ξ, η] → 1 - alpha[ξ, η])/.
alpha[ξ, η] → (UnitStep[ξ] - UnitStep[ξ - δ])*
(UnitStep[η] - UnitStep[η - δ]), δ > 0&&δ < 1&&s > 0&&β > 0];
eq41 =
Integrate[
Integrate[(alpha[ξ, η](eq38/.ζ → 0) + k[ξ, η]((∂ζeq38)/.ζ → 0))
Cos[Pi* ξ], {ξ, 0, 1}], {η, 0, 1}][[1]] ==
Integrate[Integrate[f[ξ, η]* Cos[Pi* ξ]/s, {ξ, 0, 1}], {η, 0, 1}];
eq41A =
Simplify[
(((eq41)/.f[ξ, η]-> alpha[ξ, η])/.k[ξ, η] → 1 - alpha[ξ, η])/.
alpha[ξ, η] → (UnitStep[ξ] - UnitStep[ξ - δ])
(UnitStep[η] - UnitStep[η - δ]), δ > 0&&δ < 1&&s > 0&&β > 0];
eq42 =
Integrate[
Integrate [(alpha[ξ, η](eq38/.ζ → 0) + k[ξ, η]((∂ζeq38)/.ζ → 0))
Cos[Pi*η], {ξ, 0, 1}], {η, 0, 1}][[1]] ==
Integrate[Integrate[f[ξ, η]* Cos[Pi*η]/s, {ξ, 0, 1}], {η, 0, 1}];
```

```
eq42A =
Simplify[
(((eq42)/.f[ξ, η]- > alpha[ξ, η])/.k[ξ, η] → 1 - alpha[ξ, η])/.
alpha[ξ, η] → (UnitStep[ξ] - UnitStep[ξ - δ])
(UnitStep[η] - UnitStep[η - δ]), δ > 0&&δ < 1&&s > 0&&β > 0];
eq43 =
Integrate[
Integrate [(alpha[ξ, η](eq38/.ζ → 0) + k[ξ, η]((∂_ζeq38)/.ζ → 0))
Cos[Pi*η]Cos[Pi*ξ], {ξ, 0, 1}], {η, 0, 1}][[1]] =
Integrate[Integrate[f[ξ, η]* Cos[Pi*η]*Cos[Pi*ξ]/s, {ξ, 0, 1}],
{η, 0, 1}];
eq43A =
Simplify[
(((eq43)/.f[ξ, η]- > alpha[ξ, η])/.k[ξ, η] → 1 - alpha[ξ, η])/.
alpha[ξ, η] → (UnitStep[ξ] - UnitStep[ξ - δ])
(UnitStep[η] - UnitStep[η - δ]), δ > 0&&δ < 1&&s > 0&&β > 0];
```

PROBLEMS

12.1. The zeroth-order approximated concentration, when $\beta = 0.1$, $\delta = 0.1$ and $w = 0$, is

$$\theta(\xi, \eta, \zeta, \tau) = 1.0 - \sum_{n=1}^{\infty} \frac{(2.0\ \mathrm{Sin}\ (\alpha_n\zeta)\mathrm{Sin}\ (\alpha_n) + 2.0\cos(\alpha_n\zeta)\cos(\alpha_n))e^{-\alpha_n{}^2\tau}}{2.0\ \cos(\alpha_n) + 296.0\alpha_n\ \mathrm{Sin}\ (\alpha_n) - 9801.0\alpha_n{}^3\ \mathrm{Sin}(\alpha_n)} \quad (1)$$

Derive an expression for the average concentration.

12.2. Derive the zeroth-order approximated concentration for the following parameters: $\beta = 0.1$, $\delta = 0.1$ and $w = 0.2$.

12.3. Derive the zeroth-order average concentration for the following parameters: $\beta = 0.1$, $\delta = 0.1$ and $w = 0.2$.

12.4. Plot $\theta(\xi = 0.1,\ \eta = 0.2,\ \zeta = 0.3,\ \tau)$ using the first-order approximated solution and the following parameters: $\beta = 0.1$, $\delta = 0.1$ and $w = 0$.

12.5. Plot $\theta(\xi = 0.5,\ \eta = 0.5,\ \zeta = 0.5,\ \tau)$ using the first-order approximated solution and the following parameters: $\beta = 0.1$, $\delta = 0.1$ and $w = 0$.

REFERENCES

Ley EE, Goodyer CE, Bunge AL. Mathematical models of diffusion through membranes from spatially distributed sources. *J Membr Sci* 2006;283(1–2):399–410.

Loney NW. Applied mathematical methods for chemical engineers. 2nd ed. Boca Raton, FL: CRC/Taylor & Francis; 2007.

13

EFFECTIVE TIME CONSTANT FOR TWO- AND THREE-DIMENSIONAL CONTROLLED-RELEASED DRUG-DELIVERY MODELS

13.1 EFFECTIVE TIME CONSTANT IN CONTROLLED-RELEASE DRUG-DELIVERY SYSTEMS

One-dimensional diffusive models are often described by discrete eigenvalues (λ_n) and eigenfunctions (f_n) (Collins, 1980; Fernandes et al., 2005):

$$\phi(x, t) = \sum_{n=1}^{\infty} f_n(x) e^{-\lambda_n t} \tag{13.1}$$

where $\lambda_n = 1/t_n$. The constant t_n represents a characteristic time constant ($t_n > t_m$ for $n < m$). In general, the first few eigenvalues can adequately capture the system dynamics. If $2t_2 << t_1$, the effective time constant is approximately t_1 (Collins, 1980). The concept of a time constant is important to drug manufacturers who want to design controlled-release devices that are able to deliver drugs at a desired rate. Collins defined an effective relaxation time by

$$t_{eff} = \int_0^{\infty} t\Omega(t)\, dt \tag{13.2}$$

where the probability density function $\Omega(t)$ is

$$\Omega(t) = \frac{(\psi_{ss} - \psi(t))}{\displaystyle\int_0^{\infty} (\psi_{ss} - \psi(t))\, dt} \tag{13.3}$$

Closed-form Solutions for Drug Transport through Controlled-Release Devices in Two and Three Dimensions, First Edition. Laurent Simon and Juan Ospina.

Equation (13.2) is written in terms of Laplace variables as:

$$t_{eff} = \lim_{s\to 0}\left(\frac{\psi_{ss}}{s^2} + \frac{d\overline{\psi}(s)}{ds}\right)\left[\lim_{s\to 0}\left(\frac{\psi_{ss}}{s} - \overline{\psi}(s)\right)\right]^{-1} \tag{13.4}$$

where ψ_{ss} and $\overline{\psi}$ stand for the steady-state value and the Laplace transform of ψ, respectively. Note that the variable ψ may represent a transdermal flux or the percentage a drug released from a device. This method is applicable to systems approaching their steady states in such a way that $\psi_{ss} - \psi(t)$ does not change signs. In many applications, Eq. (13.4) is preferred over Eq. (13.2) because it is not necessary to invert Laplace transformed expressions to derive t_{eff}. Several examples on the use of t_{eff}, in assessing the performance of controlled-release devices, are available in the literature (Simon, 2009, 2012; Simon & Ospina, 2012; Wei et al., 2012). This performance index also helped researchers study dynamics of diffusive and stress-induced transport in polymers (Ferreira et al., 2011).

13.2 INTRAVITREAL DRUG DELIVERY USING A 2-D CYLINDRICAL MODEL

13.2.1 Governing Equations

A model of intravitreal drug delivery was constructed to explain the transport and elimination of drug molecules in the posterior segment of the eye (Tojo & Isowaki, 2001). The drug concentration in the vitreous body is given by

$$\frac{\partial c}{\partial t} = \eta\left(\frac{1}{r}\frac{\partial}{\partial r}\left(r\frac{\partial c}{\partial r}\right) + \frac{\partial^2 c}{\partial z^2}\right) - kc + q \tag{13.5}$$

where η is the diffusion coefficient, k is a degradation and metabolism rate constant and q is the drug release rate from the intravitreal device (i.e., source term) defined as

$$q = \begin{cases} S_0 & 0 \le r \le b_0 \\ 0 & b_0 < r \le R \end{cases} \tag{13.6}$$

In this equation, R and b_0 are the radii of the vitreous body and lens, respectively (Fig. 13.1). The initial concentration is

$$c(r, z, 0) = \begin{cases} c_0 & 0 \le r \le b_0 \\ 0 & b_0 < r \le R \end{cases}. \tag{13.7}$$

A symmetry condition is imposed at $r = 0$:

$$\left.\frac{\partial c(r, z, t)}{\partial r}\right|_{r=0} = 0 \tag{13.8}$$

The two following equations describe elimination of the drug through the retina/choroids/sclera (RCS) membrane:

$$\eta\left.\frac{\partial c(r, z, t)}{\partial r}\right|_{r=R} = -\frac{\eta_\rho K_\rho}{\delta_\rho}c(R, z, t) \tag{13.9}$$

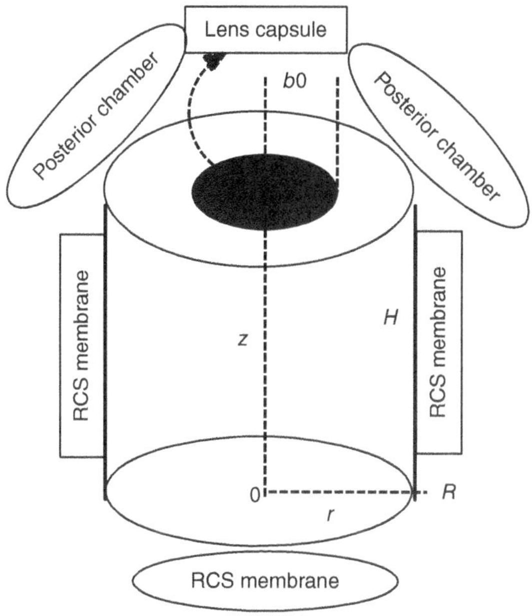

Figure 13.1 Vitreous body model. The three elimination pathways are the retina/choroids/sclera (RCS) membrane, the posterior chamber and the posterior lens capsule.

$$\eta \left. \frac{\partial c(r, z, t)}{\partial z} \right|_{z=0} = \frac{\eta_\rho K_\rho}{\delta_\rho} c(r, 0, t) \tag{13.10}$$

Release from the vitreous body into the lens is represented by

$$\eta \left. \frac{\partial c(r, z, t)}{\partial z} \right|_{z=H} = -\frac{\eta_1 K_1}{\delta_1} c(r, H, t); \qquad 0 \le r \le b_0 \tag{13.11}$$

while the equation

$$\eta \left. \frac{\partial c(r, z, t)}{\partial z} \right|_{z=H} = -\frac{\eta_p K_p}{\delta_p} c(r, H, t); \qquad b_0 < r \le R \tag{13.12}$$

specifies elimination of the drug through the posterior aqueous humor. η_i and δ_i are the diffusion coefficient through the medium i and the thickness of the layer i, respectively. The subscript i represents the RCS membrane (ρ), the lens tissue (l) or the diffusion boundary layer in the posterior aqueous chamber (p) (Tojo & Isowaki, 2001). The mean concentration of the drug in the vitreous body is

$$c_{ave} = \frac{\int_0^H \int_0^R c(r, z, t)\, r dr dz}{\int_0^H \int_0^R r dr dz} \tag{13.13}$$

13.2.2 Mathematical Solution

It is helpful to use the time constant to estimate the time it takes to reach c_{ave}. A solution strategy is first proposed for $k = 0$ to describe the concentration profile in the vitreous body. Equation (13.6) is written as

$$q = S_0 (Heaviside(r) - Heaviside(r - b_0)) \tag{13.14}$$

Replacing q in Eq. (13.5) and setting k equal to zero, we obtain

$$\frac{\partial}{\partial t}c(r, z, t) = \frac{\eta\left(\frac{\partial}{\partial r}c(r, z, t) + r\frac{\partial^2}{\partial r^2}c(r, z, t) + r\frac{\partial^2}{\partial z^2}c(r, z, t)\right)}{r} + S_0(Heaviside(r) - Heaviside(r - b_0)) \tag{13.15}$$

The initial condition is

$$c(r, z, 0) = c_0(Heaviside(r) - Heaviside(r - b_0)) \tag{13.16}$$

The Laplace transform of Eq. (13.15) gives

$$sC(r, z) - c_0(Heaviside(r) - Heaviside(r - b_0)) = \frac{S_0(Heaviside(r) - Heaviside(r - b_0))}{s} + \frac{\eta\frac{\partial}{\partial r}C(r, z)}{r} + \eta\frac{\partial^2}{\partial r^2}C(r, z) + \eta\frac{\partial^2}{\partial z^2}C(r, z) \tag{13.17}$$

where

$$C(r, z) = \int_0^\infty c(r, z, t)e^{-st}\,dt \tag{13.18}$$

We look for a solution to Eq. (13.17) of the form $C(r, z) = C(r) + f(r)g(z)$ where

$$sC(r) - c_0(Heaviside(r) - Heaviside(r - b_0)) = \frac{S_0(Heaviside(r) - Heaviside(r - b_0))}{s} + \frac{\eta\frac{d}{dr}C(r)}{r} + \eta\frac{d^2}{dr^2}C(r) \tag{13.19}$$

$$\frac{d^2}{dr^2}f(r) = c_1 f(r) - \frac{\frac{d}{dr}f(r)}{r} \tag{13.20}$$

and

$$\frac{d^2}{dz^2}g(z) = -c_1 g(z) + \frac{sg(z)}{\eta} \tag{13.21}$$

The solution of Eq. (13.19) is

$$C(r) = C_2 J_0\left(r\sqrt{-\frac{s}{\eta}}\right) + C_1 Y_0\left(r\sqrt{-\frac{s}{\eta}}\right) + \frac{1}{2}\frac{c_0 s + S_0}{s^2}\left(\begin{array}{l} -\pi b_0\sqrt{-\frac{s}{\eta}}Y_0\left(r\sqrt{-\frac{s}{\eta}}\right)(-1 + Heaviside(b_0 - r))J_1\left(b_0\sqrt{-\frac{s}{\eta}}\right) + \\ \pi b_0\sqrt{-\frac{s}{\eta}}J_0\left(r\sqrt{-\frac{s}{\eta}}\right)(-1 + Heaviside(b_0 - r))Y_1\left(b_0\sqrt{-\frac{s}{\eta}}\right) + \\ 2Heaviside(r) - 2 + 2Heaviside(b_0 - r) \end{array}\right) \tag{13.22}$$

In order to satisfy the boundary condition (13.8), let us consider Eq. (13.22) in the region $0 < r < b_0$:

$$C(r) = C_2 J_0\left(r\sqrt{-\frac{s}{\eta}}\right) + C_1 Y_0\left(r\sqrt{-\frac{s}{\eta}}\right) + \frac{c_0 s + S_0}{s^2} \tag{13.23}$$

We need to set $C_1 = 0$ to avoid the singularity around $r = 0$. Therefore, Eq. (13.22) becomes

$$C(r) = C_2 J_0\left(r\sqrt{-\frac{s}{\eta}}\right) + \frac{1}{2}\frac{c_0 s + S_0}{s^2}\left(\begin{array}{l} -\pi b_0\sqrt{-\frac{s}{\eta}}Y_0\left(r\sqrt{-\frac{s}{\eta}}\right)(-1 + Heaviside(b_0 - r))J_1\left(b_0\sqrt{-\frac{s}{\eta}}\right) + \\ \pi b_0\sqrt{-\frac{s}{\eta}}J_0\left(r\sqrt{-\frac{s}{\eta}}\right)(-1 + Heaviside(b_0 - r))Y_1\left(b_0\sqrt{-\frac{s}{\eta}}\right) + \\ 2Heaviside(r) - 2 + 2Heaviside(b_0 - r) \end{array}\right) \tag{13.24}$$

To address the boundary condition (13.9), consider Eq. (13.24) in the region $b_0 < r < R$, which leads to

$$C(r) = C_2 J_0\left(\sqrt{-\frac{s}{\eta}}r\right) + \frac{1}{2}\frac{c_0 s + S_0}{s^2}\left(\begin{array}{l} \pi b_0\sqrt{-\frac{s}{\eta}}Y_0\left(r\sqrt{-\frac{s}{\eta}}\right)J_1\left(b_0\sqrt{-\frac{s}{\eta}}\right) \\ -\pi b_0\sqrt{-\frac{s}{\eta}}J_0\left(r\sqrt{-\frac{s}{\eta}}\right)Y_1\left(b_0\sqrt{-\frac{s}{\eta}}\right) \end{array}\right) \tag{13.25}$$

Now, the boundary condition (13.9) gives

$$\eta\left(-C_2 J_1\left(R\sqrt{-\frac{s}{\eta}}\right)\sqrt{-\frac{s}{\eta}} + \frac{1}{2}\frac{c_0 s + S_0}{s^2}\left(\begin{array}{l} \frac{\pi b_0 s}{\eta}Y_1\left(R\sqrt{-\frac{s}{\eta}}\right)J_1\left(b_0\sqrt{-\frac{s}{\eta}}\right) - \\ \frac{\pi b_0 s}{\eta}J_1\left(R\sqrt{-\frac{s}{\eta}}\right)Y_1\left(b_0\sqrt{-\frac{s}{\eta}}\right) \end{array}\right)\right) = \\ -\frac{\eta_p K_p}{\delta_p}\left(C_2 J_0\left(R\sqrt{-\frac{s}{\eta}}\right) + \frac{1}{2}\frac{c_0 s + S_0}{s^2}\left(\begin{array}{l} \pi b_0\sqrt{-\frac{s}{\eta}}Y_0\left(R\sqrt{-\frac{s}{\eta}}\right)J_1\left(b_0\sqrt{-\frac{s}{\eta}}\right) - \\ \pi b_0\sqrt{-\frac{s}{\eta}}J_0\left(R\sqrt{-\frac{s}{\eta}}\right)Y_1\left(b_0\sqrt{-\frac{s}{\eta}}\right) \end{array}\right)\right) \tag{13.26}$$

The parameter C_2 is obtained from Eq. (13.26):

$$C_2 = \frac{\frac{1}{2}\frac{c_0 s + S_0}{s^2}\left[\begin{array}{l} -\frac{\eta_p K_p}{\delta_p}\left(\left(\begin{array}{l} \pi b_0\sqrt{-\frac{s}{\eta}}Y_0\left(R\sqrt{-\frac{s}{\eta}}\right)J_1\left(b_0\sqrt{-\frac{s}{\eta}}\right) - \\ \pi b_0\sqrt{-\frac{s}{\eta}}J_0\left(R\sqrt{-\frac{s}{\eta}}\right)Y_1\left(b_0\sqrt{-\frac{s}{\eta}}\right) \end{array}\right)\right) \\ -\eta\left(\left(\begin{array}{l} \frac{\pi b_0 s}{\eta}Y_1\left(R\sqrt{-\frac{s}{\eta}}\right)J_1\left(b_0\sqrt{-\frac{s}{\eta}}\right) - \\ \frac{\pi b_0 s}{\eta}J_1\left(R\sqrt{-\frac{s}{\eta}}\right)Y_1\left(b_0\sqrt{-\frac{s}{\eta}}\right) \end{array}\right)\right) \end{array}\right]}{-\eta J_1\left(R\sqrt{-\frac{s}{\eta}}\right)\sqrt{-\frac{s}{\eta}} + \frac{\eta_p K_p}{\delta_p}J_0\left(R\sqrt{-\frac{s}{\eta}}\right)} \tag{13.27}$$

Equation (13.24) becomes

$$C(r) = \frac{\frac{1}{2}\frac{c_0 s+S_0}{s^2}\left[\begin{array}{l} -\frac{\eta_\rho K_\rho}{\delta_\rho}\left(\left(\begin{array}{l}\pi b_0\sqrt{-\frac{s}{\eta}}Y_0\left(R\sqrt{-\frac{s}{\eta}}\right)J_1\left(b_0\sqrt{-\frac{s}{\eta}}\right)- \\ \pi b_0\sqrt{-\frac{s}{\eta}}J_0\left(R\sqrt{-\frac{s}{\eta}}\right)Y_1\left(b_0\sqrt{-\frac{s}{\eta}}\right)\end{array}\right)\right) \\ -\eta\left(\left(\begin{array}{l}\frac{\pi b_0 s}{\eta}Y_1\left(R\sqrt{-\frac{s}{\eta}}\right)J_1\left(b_0\sqrt{-\frac{s}{\eta}}\right)- \\ \frac{\pi b_0 s}{\eta}J_1\left(R\sqrt{-\frac{s}{\eta}}\right)Y_1\left(b_0\sqrt{-\frac{s}{\eta}}\right)\end{array}\right)\right)\end{array}\right]}{-\eta J_1\left(R\sqrt{-\frac{s}{\eta}}\right)\sqrt{-\frac{s}{\eta}}+\frac{\eta_\rho K_\rho}{\delta_\rho}J_0\left(R\sqrt{-\frac{s}{\eta}}\right)} J_0\left(r\sqrt{-\frac{s}{\eta}}\right) +$$

$$\frac{1}{2}\frac{c_0 s+S_0}{s^2}\left[\begin{array}{l} -\pi b_0\sqrt{-\frac{s}{\eta}}Y_0\left(r\sqrt{-\frac{s}{\eta}}\right)(-1+Heaviside(b_0-r))J_1\left(b_0\sqrt{-\frac{s}{\eta}}\right)+ \\ \pi b_0\sqrt{-\frac{s}{\eta}}J_0\left(r\sqrt{-\frac{s}{\eta}}\right)(-1+Heaviside(b_0-r))Y_1\left(b_0\sqrt{-\frac{s}{\eta}}\right)+ \\ 2Heaviside(r)-2+2Heaviside(b_0-r)\end{array}\right] \tag{13.28}$$

Now, after solving Eqs. (13.20) and (13.21), we have

$$f(r)g(z) =$$

$$(C_1 J_0(r\sqrt{-c_1})+C_2 Y_0(r\sqrt{-c_1}))\left(C_3\sin\left(z\sqrt{\frac{c_1\eta-s}{\eta}}\right)+C_4\cos\left(z\sqrt{\frac{c_1\eta-s}{\eta}}\right)\right) \tag{13.29}$$

To satisfy the boundary condition (13.8), set $C_2 = 0$ and, without loss of generality, $C_1 = 1$. Then, Eq. (13.29) reduces to

$$f(r)g(z) = J_0(r\sqrt{-c_1})\left(C_3\sin\left(z\sqrt{\frac{c_1\eta-s}{\eta}}\right)+C_4\cos\left(z\sqrt{\frac{c_1\eta-s}{\eta}}\right)\right) \tag{13.30}$$

Applying the boundary condition (13.9) to Eq. (13.30) results in

$$\frac{(-\eta J_1(R\sqrt{-c_1})\delta_\rho\sqrt{-c_1}+\eta_\rho K_\rho R J_0(\sqrt{-c_1}))}{\delta_\rho}\times$$
$$\left(C_3\sin\left(\frac{\sqrt{-c_1\eta-s}z}{\sqrt{\eta}}\right)+C_4\cos\left(\frac{\sqrt{-c_1\eta-s}z}{\sqrt{\eta}}\right)\right)=0 \tag{13.31}$$

From Eq. (13.31), we write

$$-\eta J_1(R\sqrt{-c_1})\delta_\rho\sqrt{-c_1}+\eta_\rho K_\rho R J_0(\sqrt{-c_1})=0 \tag{13.32}$$

Equation (13.32) implies that

$$c_1 = -\frac{{\alpha_n}^2}{R^2} \tag{13.33}$$

where α_n are the roots of

$$J_1(\alpha_n) = \frac{\eta_\rho K_\rho J_0(\alpha_n)R}{\eta\alpha_n\delta_\rho} \tag{13.34}$$

This leads to

$$f(r)g(z) = J_0\left(\frac{\alpha_n r}{R}\right)\left(iC_3 \sin\text{h}\left(\frac{z}{R}\sqrt{\frac{sR^2+\alpha_n{}^2\eta}{\eta}}\right) + C_4 \cosh\left(\frac{z}{R}\sqrt{\frac{sR^2+\alpha_n{}^2\eta}{\eta}}\right)\right) \tag{13.35}$$

Application of the **principle of superposition** leads to the following expression for $C(r, z) = C(r) + f(r)g(z)$:

$$C(r,z) =$$

$$\frac{\frac{1}{2}\frac{c_0 s+S_0}{s^2}\left[\begin{array}{l} -\frac{\eta_p K_p}{\delta_p}\left(\left(\begin{array}{l}\pi b_0\sqrt{-\frac{s}{\eta}}Y_0\left(R\sqrt{-\frac{s}{\eta}}\right)J_1\left(b_0\sqrt{-\frac{s}{\eta}}\right)-\\ \pi b_0\sqrt{-\frac{s}{\eta}}J_0\left(R\sqrt{-\frac{s}{\eta}}\right)Y_1\left(b_0\sqrt{-\frac{s}{\eta}}\right)\end{array}\right)\right)\\ -\eta\left(\left(\begin{array}{l}\frac{\pi b_0 s}{\eta}Y_1\left(R\sqrt{-\frac{s}{\eta}}\right)J_1\left(b_0\sqrt{-\frac{s}{\eta}}\right)-\\ \frac{\pi b_0 s}{\eta}J_1\left(R\sqrt{-\frac{s}{\eta}}\right)Y_1\left(b_0\sqrt{-\frac{s}{\eta}}\right)\end{array}\right)\right)\end{array}\right]}{-\eta J_1\left(R\sqrt{-\frac{s}{\eta}}\right)\sqrt{-\frac{s}{\eta}}+\frac{\eta_p K_p}{\delta_p}J_0\left(R\sqrt{-\frac{s}{\eta}}\right)} J_0\left(r\sqrt{-\frac{s}{\eta}}\right)+$$

$$\frac{1}{2}\frac{c_0 s+S_0}{s^2}\left[\begin{array}{l}-\pi b_0\sqrt{-\frac{s}{\eta}}Y_0\left(r\sqrt{-\frac{s}{\eta}}\right)(-1+Heaviside(b_0-r))J_1\left(b_0\sqrt{-\frac{s}{\eta}}\right)+\\ \pi b_0\sqrt{-\frac{s}{\eta}}J_0\left(r\sqrt{-\frac{s}{\eta}}\right)(-1+Heaviside(b_0-r))Y_1\left(b_0\sqrt{-\frac{s}{\eta}}\right)+\\ 2Heaviside(r)-2+2Heaviside(b_0-r)\end{array}\right]$$

$$+\sum_{n=1}^{\infty}J_0\left(\frac{\alpha_n r}{R}\right)\left(iA\sinh\left(\frac{z}{R}\sqrt{\frac{sR^2+\alpha_n{}^2\eta}{\eta}}\right)+B_n\cosh\left(\frac{z}{R}\sqrt{\frac{sR^2+\alpha_n{}^2\eta}{\eta}}\right)\right) \tag{13.36}$$

The remaining boundary conditions (13.10), (13.11) and (13.12) must be satisfied using the solution (13.36) to determine the integration constants A_n and B_n. In its actual form, Eq. (13.36) is too complex and approximations are necessary. A first-order approximation is implemented keeping only the first term in the summation of Eq. (13.36):

$$C(r,z) =$$

$$\frac{\frac{1}{2}\frac{c_0 s+S_0}{s^2}\left[\begin{array}{l} -\frac{\eta_p K_p}{\delta_p}\left(\left(\begin{array}{l}\pi b_0\sqrt{-\frac{s}{\eta}}Y_0\left(R\sqrt{-\frac{s}{\eta}}\right)J_1\left(b_0\sqrt{-\frac{s}{\eta}}\right)-\\ \pi b_0\sqrt{-\frac{s}{\eta}}J_0\left(R\sqrt{-\frac{s}{\eta}}\right)Y_1\left(b_0\sqrt{-\frac{s}{\eta}}\right)\end{array}\right)\right)\\ -\eta\left(\left(\begin{array}{l}\frac{\pi b_0 s}{\eta}Y_1\left(R\sqrt{-\frac{s}{\eta}}\right)J_1\left(b_0\sqrt{-\frac{s}{\eta}}\right)-\\ \frac{\pi b_0 s}{\eta}J_1\left(R\sqrt{-\frac{s}{\eta}}\right)Y_1\left(b_0\sqrt{-\frac{s}{\eta}}\right)\end{array}\right)\right)\end{array}\right]}{-\eta J_1\left(R\sqrt{-\frac{s}{\eta}}\right)\sqrt{-\frac{s}{\eta}}+\frac{\eta_p K_p}{\delta_p}J_0\left(R\sqrt{-\frac{s}{\eta}}\right)} J_0\left(r\sqrt{-\frac{s}{\eta}}\right)+$$

$$\frac{1}{2}\frac{c_0 s+S_0}{s^2}\left[\begin{array}{l}-\pi b_0\sqrt{-\frac{s}{\eta}}Y_0\left(r\sqrt{-\frac{s}{\eta}}\right)(-1+Heaviside(b_0-r))J_1\left(b_0\sqrt{-\frac{s}{\eta}}\right)+\\ \pi b_0\sqrt{-\frac{s}{\eta}}J_0\left(r\sqrt{-\frac{s}{\eta}}\right)(-1+Heaviside(b_0-r))Y_1\left(b_0\sqrt{-\frac{s}{\eta}}\right)+\\ 2Heaviside(r)-2+2Heaviside(b_0-r)\end{array}\right]$$

$$+J_0\left(\frac{\alpha_1 r}{R}\right)\left(iA_1\sinh\left(\frac{z}{R}\sqrt{\frac{sR^2+\alpha_1{}^2\eta}{\eta}}\right)+B_1\cosh\left(\frac{z}{R}\sqrt{\frac{sR^2+\alpha_1{}^2\eta}{\eta}}\right)\right) \tag{13.37}$$

Based on the boundary condition (13.10), the following equation holds:

$$\eta J_0\left(\frac{\alpha_1 r}{R}\right)\left(iA_1\frac{1}{R}\sqrt{\frac{sR^2+\alpha_1{}^2\eta}{\eta}}\right)=$$

$$\frac{\frac{\eta_\rho K_\rho}{2\delta_\rho}\frac{c_0 s+S_0}{s^2}\left[\begin{array}{l}-\frac{\eta_\rho K_\rho}{\delta_\rho}\left(\left(\begin{array}{l}\pi b_0\sqrt{-\frac{s}{\eta}}Y_0\left(R\sqrt{-\frac{s}{\eta}}\right)J_1\left(b_0\sqrt{-\frac{s}{\eta}}\right)-\\ \pi b_0\sqrt{-\frac{s}{\eta}}J_0\left(R\sqrt{-\frac{s}{\eta}}\right)Y_1\left(b_0\sqrt{-\frac{s}{\eta}}\right)\end{array}\right)\right)\\ -\eta\left(\left(\begin{array}{l}\frac{\pi b_0 s}{\eta}Y_1\left(R\sqrt{-\frac{s}{\eta}}\right)J_1\left(b_0\sqrt{-\frac{s}{\eta}}\right)-\\ \frac{\pi b_0 s}{\eta}J_1\left(R\sqrt{-\frac{s}{\eta}}\right)Y_1\left(b_0\sqrt{-\frac{s}{\eta}}\right)\end{array}\right)\right)\end{array}\right]}{-\eta J_1\left(R\sqrt{-\frac{s}{\eta}}\right)\sqrt{-\frac{s}{\eta}}+\frac{\eta_\rho K_\rho}{\delta_\rho}J_0\left(R\sqrt{-\frac{s}{\eta}}\right)}J_0\left(r\sqrt{-\frac{s}{\eta}}\right)+$$

$$\frac{\eta_\rho K_\rho}{2\delta_\rho}\frac{c_0 s+S_0}{s^2}\left[\begin{array}{l}-\pi b_0\sqrt{-\frac{s}{\eta}}Y_0\left(r\sqrt{-\frac{s}{\eta}}\right)(-1+Heaviside(b_0-r))J_1\left(b_0\sqrt{-\frac{s}{\eta}}\right)+\\ \pi b_0\sqrt{-\frac{s}{\eta}}J_0\left(r\sqrt{-\frac{s}{\eta}}\right)(-1+Heaviside(b_0-r))Y_1\left(b_0\sqrt{-\frac{s}{\eta}}\right)+\\ 2Heaviside(r)-2+2Heaviside(b_0-r)\end{array}\right]$$

$$+\frac{\eta_\rho K_\rho}{\delta_\rho}J_0\left(\frac{\alpha_1 r}{R}\right)iB_1 \tag{13.38}$$

The boundary conditions (13.11) and (13.12) are combined into a single equation and implemented using the Laplace transformed variable:

$$\eta\left.\frac{\partial C(r,z)}{\partial z}\right|_{z=H}+\frac{\eta_1 K_1}{\delta_1}C(r,H)(Heaviside(r)-Heaviside(r-b_0))+$$

$$\frac{\eta_p K_p}{\delta_p}C(r,H)(Heaviside(r-b_0)-Heaviside(r-R))=0 \tag{13.39}$$

After replacing $C(r, z)$ in Eq. (13.39), we multiply both sides of Eqs. (13.38) and (13.39) by $rJ_0\left(\frac{\alpha_1 r}{R}\right)$ and integrate with respect to r from 0 to R. Two very large expressions, which will not be presented here, are obtained. These equations are solved for A_1 and B_1, which are later used in Eq. (13.37) to yield the concentration in the Laplace domain. Numerical inversion techniques can be implemented to derive the expression for $c(r, z, t)$.

13.2.3 Derivation of the Time Constant t_{eff}

The first-order approximation of $C(r, z)$ can be used in the calculation of the time constant. However, the expression could not be reported here because of page limitation. A zero-order approximation of $C(r, z)$ is considered, instead, by setting $A_1 = B_1 = 0$. Under these conditions, the concentration is

$$C(r, z) =$$

$$\frac{\frac{1}{2}\frac{c_0 s+S_0}{s^2}\left[\begin{array}{l} -\frac{\eta_\rho K_\rho}{\delta_\rho}\left(\left(\begin{array}{l}\pi b_0\sqrt{-\frac{s}{\eta}}Y_0\left(R\sqrt{-\frac{s}{\eta}}\right)J_1\left(b_0\sqrt{-\frac{s}{\eta}}\right)- \\ \pi b_0\sqrt{-\frac{s}{\eta}}J_0\left(R\sqrt{-\frac{s}{\eta}}\right)Y_1\left(b_0\sqrt{-\frac{s}{\eta}}\right)\end{array}\right)\right) \\ -\eta\left(\left(\begin{array}{l}\frac{\pi b_0 s}{\eta}Y_1\left(R\sqrt{-\frac{s}{\eta}}\right)J_1\left(b_0\sqrt{-\frac{s}{\eta}}\right)- \\ \frac{\pi b_0 s}{\eta}J_1\left(R\sqrt{-\frac{s}{\eta}}\right)Y_1\left(b_0\sqrt{-\frac{s}{\eta}}\right)\end{array}\right)\right)\end{array}\right]}{-\eta J_1\left(R\sqrt{-\frac{s}{\eta}}\right)\sqrt{-\frac{s}{\eta}}+\frac{\eta_\rho K_\rho}{\delta_\rho}J_0\left(R\sqrt{-\frac{s}{\eta}}\right)}J_0\left(r\sqrt{-\frac{s}{\eta}}\right)+$$

$$\frac{1}{2}\frac{c_0 s+S_0}{s^2}\left[\begin{array}{l} -\pi b_0\sqrt{-\frac{s}{\eta}}Y_0\left(r\sqrt{-\frac{s}{\eta}}\right)(-1+Heaviside(b_0-r))J_1\left(b_0\sqrt{-\frac{s}{\eta}}\right)+ \\ \pi b_0\sqrt{-\frac{s}{\eta}}J_0\left(r\sqrt{-\frac{s}{\eta}}\right)(-1+Heaviside(b_0-r))Y_1\left(b_0\sqrt{-\frac{s}{\eta}}\right)+ \\ 2Heaviside(r)-2+2Heaviside(b_0-r)\end{array}\right] \quad (13.40)$$

As a result, the average concentration, defined in Eq. (13.13), takes the form

$$C_{av}(s) = 2\left[\begin{array}{l} c_0 s^2 b_0{}^2\delta_\rho J_1\left(R\sqrt{-\frac{s}{\eta}}\right)+c_0 s\sqrt{-\frac{s}{\eta}}b_0{}^2\eta_\rho K_\rho J_0\left(R\sqrt{-\frac{s}{\eta}}\right)+ \\ b_0\pi\sqrt{-\frac{s}{\eta}}J_1\left(b_0\sqrt{-\frac{s}{\eta}}\right)c_0 sY_1\left(R\sqrt{-\frac{s}{\eta}}\right)R\eta_\rho K_\rho J_0\left(R\sqrt{-\frac{s}{\eta}}\right)- \\ \eta_\rho K_\rho\pi b_0 Y_0\left(R\sqrt{-\frac{s}{\eta}}\right)J_1\left(b_0\sqrt{-\frac{s}{\eta}}\right)S_0 RJ_1\left(R\sqrt{-\frac{s}{\eta}}\right)\sqrt{-\frac{s}{\eta}}+ \\ S_0\sqrt{-\frac{s}{\eta}}b_0{}^2\eta_\rho K_\rho J_0\left(R\sqrt{-\frac{s}{\eta}}\right)- \\ \eta_\rho K_\rho\pi b_0 Y_0\left(R\sqrt{-\frac{s}{\eta}}\right)J_1\left(b_0\sqrt{-\frac{s}{\eta}}\right)c_0 sRJ_1\left(R\sqrt{-\frac{s}{\eta}}\right)\sqrt{-\frac{s}{\eta}}+ \\ b_0\pi\sqrt{-\frac{s}{\eta}}J_1\left(b_0\sqrt{-\frac{s}{\eta}}\right)S_0 Y_1\left(R\sqrt{-\frac{s}{\eta}}\right)R\eta_\rho K_\rho J_0\left(R\sqrt{-\frac{s}{\eta}}\right)+ \\ S_0 s b_0{}^2\delta_\rho J_1\left(R\sqrt{-\frac{s}{\eta}}\right)\end{array}\right]\div$$

$$\left(\left(2s^3\delta_\rho J_1\left(R\sqrt{-\frac{s}{\eta}}\right)+2s^2\sqrt{-\frac{s}{\eta}}\eta_\rho K_\rho J_0\left(R\sqrt{-\frac{s}{\eta}}\right)\right)R^2\right) \quad (13.41)$$

in the Laplace domain. From Eq. (13.41), we deduce that the steady-state concentration is

$$\lim_{s\to 0} sC_{av}(s) = c_{ave,ss} = \frac{1}{8}\frac{2S_0 b_0{}^2R^2\eta_\rho K_\rho - b_0{}^4 S_0\eta_\rho K_\rho + 4S_0 b_0{}^2\delta_\rho R\eta}{R^2\eta_\rho K_\rho\eta} \quad (13.42)$$

which results in

$$\frac{C_{av}(s)}{c_{ave,\,ss}} = 16R^2\eta_\rho K_\rho\eta\left[\begin{array}{l} c_0 s^2 b_0{}^2\delta_\rho J_1\left(R\sqrt{-\frac{s}{\eta}}\right)+c_0 s\sqrt{-\frac{s}{\eta}}b_0{}^2\eta_\rho K_\rho J_0\left(R\sqrt{-\frac{s}{\eta}}\right)+ \\ b_0\pi\sqrt{-\frac{s}{\eta}}J_1\left(b_0\sqrt{-\frac{s}{\eta}}\right)c_0 sY_1\left(R\sqrt{-\frac{s}{\eta}}\right)R\eta_\rho K_\rho J_0\left(R\sqrt{-\frac{s}{\eta}}\right)- \\ \eta_\rho K_\rho\pi b_0 Y_0\left(R\sqrt{-\frac{s}{\eta}}\right)J_1\left(b_0\sqrt{-\frac{s}{\eta}}\right)S_0 RJ_1\left(R\sqrt{-\frac{s}{\eta}}\right)\sqrt{-\frac{s}{\eta}}+ \\ S_0\sqrt{-\frac{s}{\eta}}b_0{}^2\eta_\rho K_\rho J_0\left(R\sqrt{-\frac{s}{\eta}}\right)- \\ \eta_\rho K_\rho\pi b_0 Y_0\left(R\sqrt{-\frac{s}{\eta}}\right)J_1\left(b_0\sqrt{-\frac{s}{\eta}}\right)c_0 sRJ_1\left(R\sqrt{-\frac{s}{\eta}}\right)\sqrt{-\frac{s}{\eta}}+ \\ b_0\pi\sqrt{-\frac{s}{\eta}}J_1\left(b_0\sqrt{-\frac{s}{\eta}}\right)S_0 Y_1\left(R\sqrt{-\frac{s}{\eta}}\right)R\eta_\rho K_\rho J_0\left(R\sqrt{-\frac{s}{\eta}}\right)+ \\ S_0 s b_0{}^2\delta_\rho J_1\left(R\sqrt{-\frac{s}{\eta}}\right)\end{array}\right]\div$$

$$\left(\begin{array}{l}\left(2s^3\delta_\rho J_1\left(R\sqrt{-\frac{s}{\eta}}\right)+2s^2\sqrt{-\frac{s}{\eta}}\eta_\rho K_\rho J_0\left(R\sqrt{-\frac{s}{\eta}}\right)\right)R^2\times\\(2S_0b_0{}^2R^2\eta_\rho K_\rho-b_0{}^4S_0\eta_\rho K_\rho+4S_0b_0{}^2\delta_\rho R\eta)\end{array}\right) \tag{13.43}$$

Finally, the effective time constant for $c_{ave}(t)$ to reach a steady-state value $c_{ave,ss}$ is

$$\tau_{eff}=$$

$$\frac{1}{48}\frac{\left(\begin{array}{l}76S_0\eta_\rho{}^3K_\rho{}^3R^6-54S_0\eta_\rho{}^3K_\rho{}^3R^4b_0{}^2+12S_0\eta_\rho{}^3K_\rho{}^3R^2b_0{}^4-S_0\eta_\rho{}^3K_\rho{}^3b_0{}^6+\\456S_0R^5K_\rho{}^2\eta_\rho{}^2\delta_\rho\eta-432c_0\eta_\rho{}^3K_\rho{}^3R^4\eta-216S_0R^3K_\rho{}^2\eta_\rho{}^2b_0{}^2\delta_\rho\eta+\\288c_0\eta_\rho{}^3K_\rho{}^3R^2\eta b_0{}^2+24S_0R\eta_\rho{}^2K_\rho{}^2b_0{}^4\delta_\rho\eta-48c_0\eta_\rho{}^3K_\rho{}^3\eta b_0{}^4+\\1152S_0R^4K_\rho\eta_\rho\delta_\rho{}^2\eta^2-1728c_0\eta_\rho{}^2K_\rho{}^2R^3\eta^2\delta_\rho-288S_0R^2b_0{}^2\eta_\rho K_\rho\delta_\rho{}^2\eta^2+\\576Rc_0\eta_\rho{}^2K_\rho{}^2\eta^2b_0{}^2\delta_\rho+1152S_0R^3\delta_\rho{}^3\eta^3-2304c_0\delta_\rho{}^2\eta^3R^2K_\rho\eta_\rho\end{array}\right)}{\eta K_\rho\eta_\rho\left(\begin{array}{l}9S_0\eta_\rho{}^2K_\rho{}^2R^4-6S_0\eta_\rho{}^2K_\rho{}^2R^2b_0{}^2+S_0\eta_\rho{}^2K_\rho{}^2b_0{}^4+\\36S_0R^3\delta_\rho\eta\eta_\rho K_\rho-48c_0\eta_\rho{}^2K_\rho{}^2R^2\eta-12S_0Rb_0{}^2\eta_\rho K_\rho\delta_\rho\eta+\\24c_0\eta_\rho{}^2K_\rho{}^2\eta b_0{}^2+48S_0R^2\delta_\rho{}^2\eta^2-96c_0\delta_\rho\eta^2RK_\rho\eta_\rho\end{array}\right)} \tag{13.44}$$

Two limiting cases are derived from Eq. (13.44):

$$\lim_{b_0\to R}\tau_{eff}=\frac{R\left(\begin{array}{l}11S_0\eta_\rho{}^3K_\rho{}^3R^4+88S_0R^3K_\rho{}^2\eta_\rho{}^2\delta_\rho\eta-64c_0\eta_\rho{}^3K_\rho{}^3R^2\eta+\\288S_0R^2K_\rho\eta_\rho\delta_\rho{}^2\eta^2-384Rc_0\eta_\rho{}^2K_\rho{}^2\eta^2\delta_\rho+384S_0R\delta_\rho{}^3\eta^3\\-768c_0\delta_\rho{}^2\eta^3K_\rho\eta_\rho\end{array}\right)}{64\eta K_\rho\eta_\rho\left(\begin{array}{l}S_0R^3K_\rho{}^2\eta_\rho{}^2+6S_0R^2\delta_\rho\eta\eta_\rho K_\rho-\\6Rc_0\eta_\rho{}^2K_\rho{}^2\eta+12S_0R\delta_\rho{}^2\eta^2-24c_0\delta_\rho\eta^2K_\rho\eta_\rho\end{array}\right)} \tag{13.45}$$

and

$$\lim_{b_0\to 0}\tau_{eff}=\frac{1}{36}\frac{R\left(\begin{array}{l}19S_0\eta_\rho{}^3K_\rho{}^3R^4+114S_0R^3K_\rho{}^2\eta_\rho{}^2\delta_\rho\eta-\\108c_0\eta_\rho{}^3K_\rho{}^3R^2\eta+288S_0R^2K_\rho\eta_\rho\delta_\rho{}^2\eta^2-\\432Rc_0\eta_\rho{}^2K_\rho{}^2\eta^2\delta_\rho+288S_0R\delta_\rho{}^3\eta^3-576c_0\delta_\rho{}^2\eta^3K_\rho\eta_\rho\end{array}\right)}{\eta K_\rho\eta_\rho\left(\begin{array}{l}3S_0R^3K_\rho{}^2\eta_\rho{}^2+12S_0R^2\delta_\rho\eta\eta_\rho K_\rho-\\16Rc_0\eta_\rho{}^2K_\rho{}^2\eta+16S_0R\delta_\rho{}^2\eta^2-32c_0\delta_\rho\eta^2K_\rho\eta_\rho\end{array}\right)} \tag{13.46}$$

As an example, consider the model parameters for ganciclovir (Tojo & Isowaki, 2001): $S_0 = 163.74\,\mu g/d/cm^3$, $H = 0.72\,cm$, $R = 0.72\,cm$, $b_0 = 0.36\,cm$, $K_1 = 0.324$, $K_p = 1.11$, $K_\rho = 0.74$, $c_0 = 25\,\mu g/cm^3$, $\eta = 0.854\,cm^2/d$, $\eta_1 = 0.158\,cm^2/d$, $\eta_p = 0.864\,cm^2/d$, $\eta_\rho = 0.0476\,cm^2/d$, $\delta_1 = 0.38\,cm$, $\delta_p = 0.0100\,cm$ and $\delta_\rho = 0.0040\,cm$. With these data, the steady-state average concentration, based on Eq. (13.42), is $c_{ave,ss} = 7.10\,\mu g/cm^3$ and the effective time constant is $\tau_{eff} = 0.208\,d$ (Eq. (13.44)).

13.3 ANALYSIS OF A RECTANGULAR PARALLELEPIPED-SHAPED MATRIX WITH A RELEASE AREA

The release kinetics of a polymeric device is described by (Fig. 13.2)

$$\frac{\partial c}{\partial t}=\eta\left(\frac{\partial^2 c}{\partial^2 x^2}+\frac{\partial^2 c}{\partial^2 y^2}+\frac{\partial^2 c}{\partial^2 y^2}\right);\quad -\frac{a}{2}\le x\le\frac{a}{2},\quad -\frac{b}{2}\le y\le\frac{b}{2},\quad -\frac{c}{2}\le z\le\frac{c}{2} \tag{13.47}$$

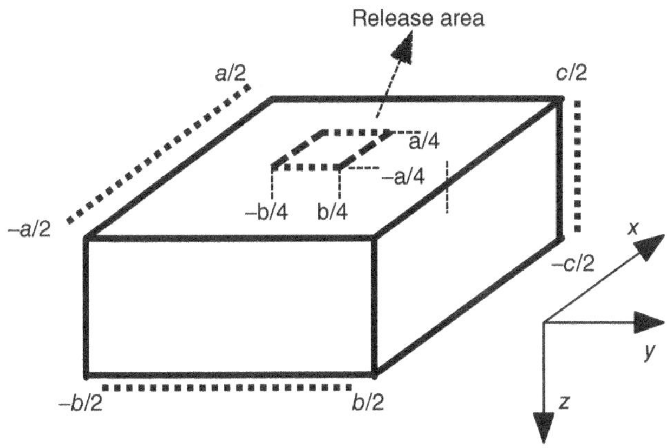

Figure 13.2 Rectangular parallelepiped-shaped device with a release area.

where η is the diffusion coefficient of the API in the polymer. The initial drug concentration is

$$c(x, y, z, 0) = c_0 \tag{13.48}$$

There is no mass transfer at the faces $x = -\frac{a}{2}$, $x = \frac{a}{2}$, $y = -\frac{b}{2}$, $y = \frac{b}{2}$ and $z = -\frac{c}{2}$:

$$\left.\frac{\partial c(x, y, z, t)}{\partial x}\right|_{x=-\frac{a}{2}} = \left.\frac{\partial c(x, y, z, t)}{\partial x}\right|_{x=\frac{a}{2}} = 0 \tag{13.49}$$

$$\left.\frac{\partial c(x, y, z, t)}{\partial y}\right|_{y=-\frac{b}{2}} = \left.\frac{\partial c(x, y, z, t)}{\partial y}\right|_{y=\frac{b}{2}} = 0 \tag{13.50}$$

and

$$\left.\frac{\partial c(x, y, z, t)}{\partial z}\right|_{z=-\frac{c}{2}} = 0 \tag{13.51}$$

At $z = \frac{c}{2}$ we have

$$c\left(x, y, \frac{c}{2}, t\right) = c_b \text{ with } \begin{cases} x \in \left[-\frac{a}{4}, \frac{a}{4}\right] \\ y \in \left[-\frac{b}{4}, \frac{b}{4}\right] \end{cases}$$

$$\left.\frac{\partial c(x, y, z)}{\partial z}\right|_{z=\frac{c}{2}} = 0 \text{ with } \begin{cases} x \in \left[-\frac{a}{2}, -\frac{a}{4}\right) \cup \left(\frac{a}{4}, \frac{a}{2}\right] \\ y \in \left[-\frac{b}{2}, -\frac{b}{4}\right) \cup \left(\frac{b}{4}, \frac{b}{2}\right] \end{cases} \tag{13.52}$$

The amount of drug that leaves the device and the percentage of drug released are

$$M(t) = c_0 abc - \int_{-\frac{a}{2}}^{\frac{a}{2}} \int_{-\frac{b}{2}}^{\frac{b}{2}} \int_{-\frac{c}{2}}^{\frac{c}{2}} c(x, y, z, t)\, dxdydz \tag{13.53}$$

and

$$\frac{M(t)}{M(\infty)} = 1 - \frac{\int_{-\frac{a}{2}}^{\frac{a}{2}}\int_{-\frac{b}{2}}^{\frac{b}{2}}\int_{-\frac{c}{2}}^{\frac{c}{2}} c(x, y, z, t)\,dxdydz}{c_0 abc} \tag{13.54}$$

respectively. The effective time constant for delivering the drug from the matrix will be computed.

13.3.1 Derivation of the Cumulative Amount of Drug Released

The Laplace transform of Eq. (13.47) with condition (13.48) is

$$sC(x, y, z) - c_0 = \eta\frac{\partial^2}{\partial x^2}C(x, y, z) + \eta\frac{\partial^2}{\partial y^2}C(x, y, z) + \eta\frac{\partial^2}{\partial z^2}C(x, y, z) \tag{13.55}$$

where

$$C(x, y, z) = \int_0^\infty c(x, y, z, t)e^{-st}\,dt \tag{13.56}$$

We assume a solution to Eq. (13.55) of the form $C(x, y, z) = c_0/s + f(x)g(y)h(z)$ where

$$\frac{d^2}{dx^2}f(x) = c_1 f(x) \tag{13.57}$$

$$\frac{d^2}{dy^2}g(y) = c_2 g(y) \tag{13.58}$$

and

$$\frac{d^2}{dz^2}h(z) = -c_1 h(z) - c_2 h(z) + \frac{sh(z)}{\eta} \tag{13.59}$$

After solving Eqs. (13.57), (13.58) and (13.59), we obtain

$$f(x)g(y)h(z) = (C_3\,e^{y\sqrt{c_2}} + C_4\,e^{-y\sqrt{c_2}})(C_1\,e^{x\sqrt{c_1}} + C_2\,e^{-x\sqrt{c_1}})\times \left(C_5 \sin\left(z\sqrt{\frac{c_1\eta + c_2\eta - s}{\eta}}\right) + C_6 \cos\left(z\sqrt{\frac{c_1\eta + c_2\eta - s}{\eta}}\right)\right) \tag{13.60}$$

Applying the first boundary condition of Eq. (13.49) to Eq. (13.60) leads to

$$(C_3\,e^{y\sqrt{c_2}} + C_4\,e^{-y\sqrt{c_2}})\left(C_1\sqrt{c_1}\,e^{-\frac{a}{2}\sqrt{c_1}} - C_2\sqrt{c_1}\,e^{\frac{a}{2}\sqrt{c_1}}\right)\times \left(C_5 \sin\left(z\sqrt{\frac{c_1\eta + c_2\eta - s}{\eta}}\right) + C_6 \cos\left(z\sqrt{\frac{c_1\eta + c_2\eta - s}{\eta}}\right)\right) = 0 \tag{13.61}$$

From Eq. (13.61), we derive that

$$C_2 = \frac{C_1\,e^{-\frac{a}{2}\sqrt{c_1}}}{e^{\frac{a}{2}\sqrt{c_1}}} \tag{13.62}$$

As a result, Eq. (13.60) becomes

$$f(x)g(y)h(z) = (C_3 e^{y\sqrt{c_2}} + C_4 e^{-y\sqrt{c_2}}) \left(e^{x\sqrt{c_1}} + \frac{e^{-\frac{a}{2}\sqrt{c_1}} e^{-x\sqrt{c_1}}}{e^{\frac{a}{2}\sqrt{c_1}}} \right) \times \left(C_5 \sin\left(z\sqrt{\frac{c_1\eta + c_2\eta - s}{\eta}} \right) + C_6 \cos\left(z\sqrt{\frac{c_1\eta + c_2\eta - s}{\eta}} \right) \right) \tag{13.63}$$

after replacing C_1 by 1. Application of the second boundary condition of Eq. (13.49) to Eq. (13.63) gives

$$(C_3 e^{y\sqrt{c_2}} + C_4 e^{-y\sqrt{c_2}}) \left(\sqrt{c_1} e^{\frac{a}{2}\sqrt{c_1}} - \frac{\sqrt{c_1} e^{-a\sqrt{c_1}}}{e^{\frac{a}{2}\sqrt{c_1}}} \right) \times \left(C_5 \sin\left(z\sqrt{\frac{c_1\eta + c_2\eta - s}{\eta}} \right) + C_6 \cos\left(z\sqrt{\frac{c_1\eta + c_2\eta - s}{\eta}} \right) \right) = 0 \tag{13.64}$$

Therefore,

$$-2i \sin(ai\sqrt{c_1}) = 0 \tag{13.65}$$

or

$$c_1 = -\frac{n^2\pi^2}{a^2} \tag{13.66}$$

Similarly, after using the two boundary conditions of Eq. (13.50) in Eq. (13.63), we have

$$C_4 = \frac{C_3 e^{-\frac{b}{2}\sqrt{c_2}}}{e^{\frac{b}{2}\sqrt{c_2}}} \tag{13.67}$$

and

$$c_2 = -\frac{m^2\pi^2}{b^2} \tag{13.68}$$

Replacing c_1, c_2 and C_4 in Eq. (13.63) and setting $C_3 = 1$, we get

$$f(x)\,g(y)\,h(z) = \frac{4}{e^{\frac{\pi mi}{2}} e^{\frac{\pi ni}{2}}} \cos\left(\frac{\pi m\,(b+2y)}{2b} \right) \cos\left(\frac{\pi n\,(a+2x)}{2a} \right) \times \begin{pmatrix} iA_{n,m} \sinh\left(\frac{z}{ab}\sqrt{\frac{m^2\pi^2\eta a^2 + n^2\pi^2\eta b^2 + sa^2b^2}{\eta}} \right) + \\ B_{n,m} \cosh\left(\frac{z}{ab}\sqrt{\frac{m^2\pi^2\eta a^2 + n^2\pi^2\eta b^2 + sa^2b^2}{\eta}} \right) \end{pmatrix} \tag{13.69}$$

Application of the boundary condition (13.51) to Eq. (13.69) yields

$$B_{n,m} = \frac{iA_{n,m} \cosh\left(\frac{c}{2ab}\sqrt{\frac{m^2\pi^2\eta a^2 + n^2\pi^2\eta b^2 + sa^2b^2}{\eta}} \right)}{\sinh\left(\frac{c}{2ab}\sqrt{\frac{m^2\pi^2\eta a^2 + n^2\pi^2\eta b^2 + sa^2b^2}{\eta}} \right)} \tag{13.70}$$

Consequently, Eq. (13.69) is

$$f(x)g(y)h(z) = \frac{4iA_{n,m}}{e^{\frac{\pi mi}{2}} e^{\frac{\pi ni}{2}}} \cos\left(\frac{\pi m (b+2y)}{2b}\right) \cos\left(\frac{\pi n (a+2x)}{2a}\right) \times$$
$$\left(\begin{array}{l} \sinh\left(\frac{z}{ab}\sqrt{\frac{m^2\pi^2\eta a^2+n^2\pi^2\eta b^2+sa^2b^2}{\eta}}\right) \times \\ \sinh\left(\frac{c}{2ab}\sqrt{\frac{m^2\pi^2\eta a^2+n^2\pi^2\eta b^2+sa^2b^2}{\eta}}\right) + \\ \cosh\left(\frac{c}{2ab}\sqrt{\frac{m^2\pi^2\eta a^2+n^2\pi^2\eta b^2+sa^2b^2}{\eta}}\right) \times \\ \cosh\left(\frac{z}{ab}\sqrt{\frac{m^2\pi^2\eta a^2+n^2\pi^2\eta b^2+sa^2b^2}{\eta}}\right) \end{array}\right) \div$$
$$\left.\left.\left.\sinh\left(\frac{c}{2ab}\sqrt{\frac{m^2\pi^2\eta a^2+n^2\pi^2\eta b^2+sa^2b^2}{\eta}}\right)\right)\right)\right) \tag{13.71}$$

The principle of superposition gives the following solution

$$C(x,y,z) = \frac{c_0}{s} + \sum_{m=0}^{\infty}\left(\sum_{n=0}^{\infty}\left(\frac{4iA_{n,m}}{e^{\frac{\pi mi}{2}} e^{\frac{\pi ni}{2}}} \cos\left(\frac{\pi m (b+2y)}{2b}\right) \cos\left(\frac{\pi n (a+2x)}{2a}\right) \times\right.\right.$$
$$\left(\begin{array}{l} \sinh\left(\frac{z}{ab}\sqrt{\frac{m^2\pi^2\eta a^2+n^2\pi^2\eta b^2+sa^2b^2}{\eta}}\right) \times \\ \sinh\left(\frac{c}{2ab}\sqrt{\frac{m^2\pi^2\eta a^2+n^2\pi^2\eta b^2+sa^2b^2}{\eta}}\right) + \\ \cosh\left(\frac{c}{2ab}\sqrt{\frac{m^2\pi^2\eta a^2+n^2\pi^2\eta b^2+sa^2b^2}{\eta}}\right) \times \\ \cosh\left(\frac{z}{ab}\sqrt{\frac{m^2\pi^2\eta a^2+n^2\pi^2\eta b^2+sa^2b^2}{\eta}}\right) \end{array}\right) \div$$
$$\left.\left.\left.\sinh\left(\frac{c}{2ab}\sqrt{\frac{m^2\pi^2\eta a^2+n^2\pi^2\eta b^2+sa^2b^2}{\eta}}\right)\right)\right)\right) \tag{13.72}$$

The two conditions in Eq. (13.52) are combined into a single equation:

$$C\left(x,y,\frac{c}{2}\right)\alpha(x,y) + (1-\alpha(x,y))\frac{\partial}{\partial z}C(x,y,z)\bigg|_{z=\frac{c}{2}} = c_b\alpha(x,y) \tag{13.73}$$

where

$$\alpha(x,y) = \left(Heaviside\left(x+\frac{a}{4}\right) - Heaviside\left(x-\frac{a}{4}\right)\right) \times$$
$$\left(Heaviside\left(y+\frac{b}{4}\right) - Heaviside\left(y-\frac{b}{4}\right)\right) \tag{13.74}$$

Before using the boundary condition (13.73), it is convenient to limit the number of terms in Eq. (13.72). A first-order approximation is implemented:

$$
\begin{aligned}
C(x,y,z) &= \frac{c_0}{s} + \frac{4iA_{0,0}\left(\sinh\left(z\sqrt{\frac{s}{\eta}}\right)\sinh\left(\frac{c}{2}\sqrt{\frac{s}{\eta}}\right)+\cosh\left(\frac{c}{2}\sqrt{\frac{s}{\eta}}\right)\cosh\left(z\sqrt{\frac{s}{\eta}}\right)\right)}{\sinh\left(\frac{c}{2}\sqrt{\frac{s}{\eta}}\right)} + \\
&4A_{1,0}\cos\left(\frac{\pi(a+2x)}{2a}\right)\frac{\begin{pmatrix}\sinh\left(\frac{z}{a}\sqrt{\frac{\pi^2\eta+sa^2}{\eta}}\right)\sinh\left(\frac{c}{2a}\sqrt{\frac{\pi^2\eta+sa^2}{\eta}}\right)+\\ \cosh\left(\frac{c}{2a}\sqrt{\frac{\pi^2\eta+sa^2}{\eta}}\right)\cosh\left(\frac{z}{a}\sqrt{\frac{\pi^2\eta+sa^2}{\eta}}\right)\end{pmatrix}}{\sinh\left(\frac{c}{2a}\sqrt{\frac{\pi^2\eta+sa^2}{\eta}}\right)} + \\
&4A_{0,1}\cos\left(\frac{\pi(b+2y)}{2b}\right)\frac{\begin{pmatrix}\sinh\left(\frac{z}{b}\sqrt{\frac{\pi^2\eta+sb^2}{\eta}}\right)\sinh\left(\frac{c}{2b}\sqrt{\frac{\pi^2\eta+sb^2}{\eta}}\right)+\\ \cosh\left(\frac{c}{2b}\sqrt{\frac{\pi^2\eta+sb^2}{\eta}}\right)\cosh\left(\frac{z}{b}\sqrt{\frac{\pi^2\eta+sb^2}{\eta}}\right)\end{pmatrix}}{\sinh\left(\frac{c}{2b}\sqrt{\frac{\pi^2\eta+sb^2}{\eta}}\right)} \\
&- 4iA_{1,1}\cos\left(\frac{\pi(b+2y)}{2b}\right)\cos\left(\frac{\pi(a+2x)}{2a}\right)\times \\
&\frac{\begin{pmatrix}\sinh\left(\frac{z}{ab}\sqrt{\frac{\pi^2\eta\, a^2+\pi^2\eta\, b^2+sa^2b^2}{\eta}}\right)\sinh\left(\frac{c}{2ab}\sqrt{\frac{\pi^2\eta\, a^2+\pi^2\eta\, b^2+sa^2b^2}{\eta}}\right)+\\ \cosh\left(\frac{c}{2ab}\sqrt{\frac{\pi^2\eta\, a^2+\pi^2\eta\, b^2+sa^2b^2}{\eta}}\right)\cosh\left(\frac{z}{ab}\sqrt{\frac{\pi^2\eta\, a^2+\pi^2\eta\, b^2+sa^2b^2}{\eta}}\right)\end{pmatrix}}{\sinh\left(\frac{c}{2ab}\sqrt{\frac{\pi^2\eta\, a^2+\pi^2\eta\, b^2+sa^2b^2}{\eta}}\right)}
\end{aligned}
\tag{13.75}
$$

The boundary condition (13.73) leads to

$$
\begin{aligned}
&C\left(x,y,\frac{c}{2}\right)\left(Heaviside\left(x+\frac{a}{4}\right)-Heaviside\left(x-\frac{a}{4}\right)\right)\times \\
&\left(Heaviside\left(y+\frac{b}{4}\right)-Heaviside\left(y-\frac{b}{4}\right)\right)+ \\
&\begin{pmatrix}1-\left(Heaviside\left(x+\frac{a}{4}\right)-Heaviside\left(x-\frac{a}{4}\right)\right)\times\\ \left(Heaviside\left(y+\frac{b}{4}\right)-Heaviside\left(y-\frac{b}{4}\right)\right)\end{pmatrix}\left.\frac{\partial C(x,y,z)}{\partial z}\right|_{z=\frac{c}{2}} = \\
&c_b\left(Heaviside\left(x+\frac{a}{4}\right)-Heaviside\left(x-\frac{a}{4}\right)\right)\times \\
&\left(Heaviside\left(y+\frac{b}{4}\right)-Heaviside\left(y-\frac{b}{4}\right)\right)
\end{aligned}
\tag{13.76}
$$

with $C(x, y, z)$ defined by Eq. (13.75). We integrate both sides of Eq. (13.76) with respect to x from $-a/2$ to $a/2$ and with respect to y from $-b/2$ to $b/2$:

$$\frac{1}{4}\frac{\begin{pmatrix} 8iabA_{0,0}\left(\cosh\left(\frac{c}{2}\sqrt{\frac{s}{\eta}}\right)\right)^2 s\sqrt{\eta} + abc_0\sinh\left(\frac{c}{2}\sqrt{\frac{s}{\eta}}\right)\sqrt{\eta} - 4iabA_{0,0}s\sqrt{\eta}+ \\ 8iabs^{3/2}A_{0,0}\cosh\left(\frac{c}{2}\sqrt{\frac{s}{\eta}}\right)\sinh\left(\frac{c}{2}\sqrt{\frac{s}{\eta}}\right) \end{pmatrix}}{s\sqrt{\eta}\sinh\left(\frac{c}{2}\sqrt{\frac{s}{\eta}}\right)}+$$

$$4iA_{0,0}ba\sqrt{\frac{s}{\eta}}\cosh\left(\frac{c}{2}\sqrt{\frac{s}{\eta}}\right) = \frac{1}{4}c_b ab \tag{13.77}$$

The constant $A_{0,0}$ is obtained from Eq. (13.77):

$$A_{0,0} = \frac{-\frac{1}{4}i\sqrt{\eta}(c_b s - c_0)\sinh\left(\frac{c}{2}\sqrt{\frac{s}{\eta}}\right)}{2\left(\cosh\left(\frac{c}{2}\sqrt{\frac{s}{\eta}}\right)\right)^2 s\sqrt{\eta} - s\sqrt{\eta} + 6s^{3/2}\cosh\left(\frac{c}{2}\sqrt{\frac{s}{\eta}}\right)\sinh\left(\frac{c}{2}\sqrt{\frac{s}{\eta}}\right)} \tag{13.78}$$

Using the Laplace transform of Eq. (13.53), in addition to Eqs. (13.75) and (13.78), we have

$$M(s) = \frac{abc_0c}{s} - ab\frac{\begin{pmatrix} -c_b s\eta\cosh\left(\frac{c}{2}\sqrt{\frac{s}{\eta}}\right)e^{-\frac{c}{2}\sqrt{\frac{s}{\eta}}} - c_0\eta\cosh\left(\frac{c}{2}\sqrt{\frac{s}{\eta}}\right)e^{\frac{c}{2}\sqrt{\frac{s}{\eta}}}+ \\ c_b s\eta\cosh\left(\frac{c}{2}\sqrt{\frac{s}{\eta}}\right)e^{\frac{c}{2}\sqrt{\frac{s}{\eta}}} + 2c_0\left(\cosh\left(\frac{c}{2}\sqrt{\frac{s}{\eta}}\right)\right)^2 c\sqrt{\eta s}+ \\ 6sc_0\cosh\left(\frac{c}{2}\sqrt{\frac{s}{\eta}}\right)\sinh\left(\frac{c}{2}\sqrt{\frac{s}{\eta}}\right)c - c_0c\sqrt{\eta s} + c_0\eta\cosh\left(\frac{c}{2}\sqrt{\frac{s}{\eta}}\right)e^{-\frac{c}{2}\sqrt{\frac{s}{\eta}}} \end{pmatrix}}{2s^{3/2}\left(\cosh\left(\frac{c}{2}\sqrt{\frac{s}{\eta}}\right)\right)^2\sqrt{\eta} - s^{3/2}\sqrt{\eta} + 6s^2\cosh\left(\frac{c}{2}\sqrt{\frac{s}{\eta}}\right)\sinh\left(\frac{c}{2}\sqrt{\frac{s}{\eta}}\right)} \tag{13.79}$$

13.3.2 Derivation of the Percentage Amount of Drug Released and t_{eff}

The percentage of drug release is derived from Eq. (13.54) as

$$\frac{M(s)}{M(\infty)} = \frac{\eta\cosh\left(\frac{c}{2}\sqrt{\frac{s}{\eta}}\right)\left(c_b s e^{-\frac{c}{2}\sqrt{\frac{s}{\eta}}} + c_0 e^{\frac{c}{2}\sqrt{\frac{s}{\eta}}} - c_b s e^{\frac{c}{2}\sqrt{\frac{s}{\eta}}} - c_0 e^{-\frac{c}{2}\sqrt{\frac{s}{\eta}}}\right)}{c_0 cs\left(2\sqrt{s}\left(\cosh\left(\frac{c}{2}\sqrt{\frac{s}{\eta}}\right)\right)^2\sqrt{\eta} - \sqrt{s\eta} + 6s\cosh\left(\frac{c}{2}\sqrt{\frac{s}{\eta}}\right)\sinh\left(\frac{c}{2}\sqrt{\frac{s}{\eta}}\right)\right)} \tag{13.80}$$

and the effective time constant is

$$t_{eff} = \frac{1}{5}\frac{c(2c_0c^3 + 30c_0c^2 + 5c_b\eta c + 135c_0c + 45c_b\eta)}{\eta(c_0c^2 + 9c_0c + 3c_b\eta)} \tag{13.81}$$

Finally, the inverse Laplace transform of Eq. (13.80) is

$$\frac{M(t)}{M(\infty)} = 1 - \sum_{n=1}^{\infty} \frac{24(\cos(\beta_n))^3 \sin(\beta_n)(4c_b\beta_n{}^2\eta + c_0c^2)e^{-\frac{4\beta_n{}^2\eta t}{c^2}}}{cc_0\beta_n \begin{pmatrix} 24\left(\cos\left(\beta_n\right)\right)^4 c - 12(\cos(\beta_n))^2 c + \\ 4c^2(\cos(\beta_n))^4 - c^2 + 144\beta_n{}^2(\cos(\beta_n))^4 \end{pmatrix}} \tag{13.82}$$

where

$$\sin(\beta_n) = -\frac{1}{12}\frac{c - 2(\cos(\beta_n))^2 c}{\beta_n \cos(\beta_n)} \tag{13.83}$$

13.4 SUMMARY

A method for calculating the effective time constant (t_{eff}) for controlled-release devices was outlined. The technique could be implemented in the time or the Laplace domain. This parameter was calculated to estimate the time it took to reach an average drug concentration in a cylindrical vitreous body model following application of a ganciclovir- release system. Calculations were performed for limiting cases, as well. The second example dealt with the release of a medication through a small area of a parallelepiped. We obtained closed-form expressions for t_{eff} and the fraction of drug released when a first-order approximation was considered.

13.5 APPENDIX: MAPLE AND MATHEMATICA CODE LISTINGS

13.5.1 Maple Code: 2Dvitreous.mws (Section 13.2.2)

```
> restart:with(VectorCalculus):with(inttrans):with(PDETools):
> eq:=diff(c(r,z,t),t)=eta*Laplacian(c(r,z,t),cylindrical[r,theta,z])-
0*c(r,z,t)+S[0]*(Heaviside(r)-Heaviside(r-b[0]));
```

$$eq := \frac{\partial}{\partial t}c(r,z,t) = \frac{\eta\left(\frac{\partial}{\partial r}c(r,z,t) + r\left(\frac{\partial^2}{\partial r^2}c(r,z,t)\right) + r\left(\frac{\partial^2}{\partial z^2}c(r,z,t)\right)\right)}{r} + S_0(\text{Heaviside}(r) - \text{Heaviside}(r - b_0))$$

```
> q=S[0]*(Heaviside(r)-Heaviside(r-b[0])):
> eq1:=c(r,z,0)=c[0]*(Heaviside(r)-Heaviside(r-b[0])):
>
>
> eq2:=laplace(eq,t,s):
> eq3:=subs(eq1,eq2):
> eq4:=subs(laplace(c(r,z,t),t,s)=C(r,z),eq3):
> eq5:=eval(subs(C(r,z)=C(r),eq4)):
> C(r,z)=C(r)+f(r)*g(z):
> C(r,z)=Int(c(r,z,t)*exp(-s*t),t=0..infinity):
> eq6:=eval(subs(Int=int,dsolve(eq5))):
>
> eq9:=eq6 assuming r>0 and r<b[0]:
> eq9a:=series(rhs(eq9),r=0,1):
```

```
> eq9A:=subs(_C1=0,eq6):
> eq10:=eq9A assuming r>b[0] and r<0:
>
> eq11:=eta*subs(r=R,diff(rhs(eq10),r))=-(eta[rho]*K[rho]/delta[rho])
*subs(r=R,rhs(eq10)):
> eq12:=isolate(eq11,_C2):
> eq13:=subs(eq12,eq9A):
>
> eq14:=subs(c[0]=0,S[0]=0,eq4):
> eq15:=pdsolve(eq14,HINT=f(r)*g(z)):
>
> eq16:=factor(build(eq15)):
> eq17:=subs(_C2=0,eq16):
> C(r,z) = BesselJ(0,(-_c[1])^(1/2)*r)*(_C3*sin(1/eta^(1/2)*(_c[1]*
eta-s)^(1/2)*z)+_C4*cos(1/eta^(1/2)*(_c[1]*eta-s)^(1/2)*z)):
> eq18:=factor(eta*subs(r=R,diff(rhs(eq17),r))+(eta[rho]*K[rho]/delta
[rho])*subs(r=R,rhs(eq17)))=0:
> -(_C3*sin(1/eta^(1/2)*(_c[1]*eta-s)^(1/2)*z)+_C4*cos(1/eta^(1/2)*
(_c[1]*eta-s)^(1/2)*z))*(eta*BesselJ(1,(-_c[1])^(1/2)*R)*(-_c[1])^
(1/2)*delta[rho]-eta[rho]*K[rho]*BesselJ(0,(-_c[1])^(1/2)*R))/
delta[rho] =0:
> eq19:=-eta*BesselJ(1,(-_c[1])^(1/2)*R)*(-_c[1])^(1/2)*delta[rho]+
eta[rho]*K[rho]*BesselJ(0,(-_c[1])^(1/2)*R)=0:
> eq20:=(-_c[1])^(1/2)*R=alpha[n]:
> eq21:=isolate(eq20,_c[1]):
> eq22:=simplify(subs(eq21,eq19),power,symbolic):
> eq22A:=isolate(eq22,BesselJ(1,alpha[n])):
> eq23:=simplify(subs(eq21,eq17),power,symbolic):
> eq24:=subs(_C1=1,eq23):
>
> eq25:=subs(C=C[n],_C3=A[n],_C4=B[n],eq24):
> eq26:=C(r,z)=C(r)+Sum(C[n](r,z),n=1..infinity):
> eq27:=subs(eq13,eq25,eq26):
```

First-order approximation

```
> eq27A:=eval(subs(Sum=sum,subs(infinity=1,eq27))):
> eq28:=eval(eta*subs(z=0,diff(rhs(eq27A),z))-(eta[rho]*K[rho]/delta
[rho])*subs(z=0,rhs(eq27A)))=0:
> eq29:=eta*subs(z=H,diff(rhs(eq27A),z))+(eta[1]*K[1]/delta[1])*subs
(z=H,rhs(eq27A))*(Heaviside(r)-Heaviside(r-b[0]))+(eta[p]*K[p]/delta
[p])*subs(z=H,rhs(eq27A))*(Heaviside(r-b[0])-Heaviside(r-R))=0:
> r*BesselJ(0,alpha[1]/R*r):
> eq30:=expand(int(expand(lhs(eq28)*r*BesselJ(0,alpha[1]/R*r)),r=0..R))
=0 assuming R>0 and b[0]>0 and b[0]<R:
> eq31:=expand(int(expand(lhs(eq29)*r*BesselJ(0,alpha[1]/R*r)),r=0..R))
=0 assuming R>0 and b[0]>0 and b[0]<R:
> eq32 := solve({eq30, eq31}, {A[1], B[1]}):
> eq33:=subs(eq32,eq27A):
>
>
```

```
> aux1:=c[av](steady)=s*series(int(int(rhs(eq13)*r,r=0..R),z=
0..H)/int(int(1*r,r=0..R),z=0..H),s=0,3) assuming R>0 and s>0:
> aux3:=convert(simplify(((aux1)),power,symbolic),polynom) assum-
ing R>b[0] and b[0]>0:
>
>
> aux4A:=c[av](s)/c[av](steady)=int(int(rhs(eq13)*r,r=0..R),
z=0..H)/int(int(1*r,r=0..R),z=0..H)/rhs(aux3) assuming R>0
and b[0]>0 and R>b[0]:
> aux4B:=simplify(convert(series(rhs(aux4A),s=0,6),polynom),power,
symbolic) assuming R>b[0] and b[0]>0 and s>0:
> aux4C:=(diff(aux4B,s)):
> aux5:=tau[eff]=factor(simplify(limit(simplify(1/s^2+aux4C),s=0),
power,symbolic)/simplify(limit(simplify(1/s-aux4B),s=0),power,
symbolic));
```

$$aux5 := \tau_{eff} = \frac{1}{48}(76R^6K_\rho^3S_0\eta_\rho^3 - 54R^4K_\rho^3S_0b_0^2\eta_\rho^3 + 12R^2K_\rho^3S_0b_0^4\eta_\rho^3 - K_\rho^3S_0b_0^6\eta_\rho^3 + 456R^5\eta K_\rho^2S_0\delta_\rho\eta_\rho^2 - 432R^4\eta K_\rho^3c_0\eta_\rho^3 - 216R^3\eta K_\rho^2S_0b_0^2\delta_\rho\eta_\rho^2 + 288R^2\eta K_\rho^3b_0^2c_0\eta_\rho^3 + 24R\eta K_\rho^2S_0b_0^4\delta_\rho\eta_\rho^2 - 48\eta K_\rho^3b_0^4c_0\eta_\rho^3 + 1152R^4\eta^2K_\rho S_0\delta_\rho^2\eta_\rho - 1728R^3\eta^2K_\rho^2c_0\delta_\rho\eta_\rho^2 - 288R^2\eta^2K_\rho S_0b_0^2\delta_\rho^2\eta_\rho + 576R\eta^2K_\rho^2b_0^2c_0\delta_\rho\eta_\rho^2 + 1152R^3\eta^3S_0\delta_\rho^3 - 2304R^2\eta^3K_\rho c_0\delta_\rho^2\eta_\rho)/(\eta K_\rho\eta_\rho(9R^4K_\rho^2S_0\eta_\rho^2 - 6R^2K_\rho^2S_0b_0^2\eta_\rho^2 + K_\rho^2S_0b_0^4\eta_\rho^2 + 36R^3\eta K_\rho S_0\delta_\rho\eta_\rho - 48R^2\eta K_\rho^2c_0\eta_\rho^2 - 12R\eta K_\rho S_0b_0^2\delta_\rho\eta_\rho + 24\eta K_\rho^2b_0^2c_0\eta_\rho^2 + 48R^2\eta^2S_0\delta_\rho^2 - 96R\eta^2K_\rho c_0\delta_\rho\eta_\rho))$$

```
> Limit(lhs(aux5),b[0]=R)=factor(subs(b[0]=R,rhs(aux5)));
```

$$\lim_{b_0\to R}\tau_{eff} = \frac{1}{64}(R(11R^4K_\rho^3S_0\eta_\rho^3 + 88R^3\eta K_\rho^2S_0\delta_\rho\eta_\rho^2 - 64R^2\eta K_\rho^3c_0\eta_\rho^3 + 288R^2\eta^2K_\rho S_0\delta_\rho^2\eta_\rho - 384R\eta^2K_\rho^2c_0\delta_\rho\eta_\rho^2 + 384R\eta^3S_0\delta_\rho^3 - 768\eta^3K_\rho c_0\delta_\rho^2\eta_\rho)) /(\eta K_\rho\eta_\rho(R^3K_\rho^2S_0\eta_\rho^2 + 6R^2\eta K_\rho S_0\delta_\rho\eta_\rho - 6R\eta K_\rho^2c_0\eta_\rho^2 + 12R\eta^2S_0\delta_\rho^2 - 24\eta^2K_\rho c_0\delta_\rho\eta_\rho))$$

```
> Limit(lhs(aux5),b[0]=0)=factor(subs(b[0]=0,rhs(aux5)));
```

$$\lim_{b_0\to 0}\tau_{eff} = \frac{1}{36}(R(19R^4K_\rho^3S_0\eta_\rho^3 + 114R^3\eta K_\rho^2S_0\delta_\rho\eta_\rho^2 - 108R^2\eta K_\rho^3c_0\eta_\rho^3 + 288R^2\eta^2K_\rho S_0\delta_\rho^2\eta_\rho - 432R\eta^2K_\rho^2c_0\delta_\rho\eta_\rho^2 + 288R\eta^3S_0\delta_\rho^3 - 576\eta^3K_\rho c_0\delta_\rho^2\eta_\rho)) /(\eta K_\rho\eta_\rho(3R^3K_\rho^2S_0\eta_\rho^2 + 12R^2\eta K_\rho S_0\delta_\rho\eta_\rho - 16R\eta K_\rho^2c_0\eta_\rho^2 + 16R\eta^2S_0\delta_\rho^2 - 32\eta^2K_\rho c_0\delta_\rho\eta_\rho))$$

```
>
> eq13:
> aux7:=c[av](s)=int(int(rhs(eq13)*r,r=0..R),z=0..H)/int(int(1*r,r=
0..R),z=0..H) assuming R>0 and b[0]>0 and R>b[0]:
> aux9:=Limit(s*lhs(aux7),s=0)=limit(s*rhs(aux7),s=0);
```

$$aux9 := \lim_{s\to 0} sc_{av}(s) = \frac{1}{8}\,\frac{2R^2K_\rho S_0b_0^2\eta_\rho - K_\rho S_0b_0^4\eta_\rho + 4R\eta S_0b_0^2\delta_\rho}{R^2\eta K_\rho\eta_\rho}$$

```
>
```

13.5.2 Maple Code: Parallelepiped.mws (Section 13.3)

```
> restart:with(VectorCalculus):with(inttrans):with(PDETools):
> eq:=diff(C(x,y,z,t),t)=eta*Laplacian(C(x,y,z,t),cartesian[x,y,z]);
```

$$eq := \frac{\partial}{\partial t}C(x,y,z,t) = \eta\left(\frac{\partial^2}{\partial x^2}C(x,y,z,t) + \frac{\partial^2}{\partial y^2}C(x,y,z,t) + \frac{\partial^2}{\partial z^2}C(x,y,z,t)\right)$$

```
> eq1:=C(x,y,z,0)=c[0]:
> eq2:=laplace(eq,t,s):
```

```
> eq3:=subs(eq1,eq2):
>
> eq4:=subs(laplace(C(x,y,z,t),t,s)=C(x,y,z),eq3):
>
> eq5:=subs(c[0]=0,eq4):
> eq6:=pdsolve(eq5,HINT=f(x)*g(y)*h(z)):
> eq7:=factor(build(eq6)):
> eq8:=_C1*exp(_c[1]^(1/2)*x)^2+_C2=exp(_c[1]^
(1/2)*x)*(_C1*exp(_c[1]^(1/2)*x)+_C2*exp(-_c[1]^(1/2)*x)):
> eq9:=exp(_c[2]^(1/2)*y)^2*_C3+_C4=exp(_c[2]^
(1/2)*y)*(exp(_c[2]^(1/2)*y)*_C3+_C4*exp(-_c[2]^(1/2)*y)):
> eq10:=subs(eq8,eq9,eq7):
> eq11:=subs(x=-a/2,diff(rhs(eq10),x))=0:
> eq12:=isolate(eq11,_C2):
> eq13:=subs(eq12,eq10):
> eq13A:=subs(_C1=1,eq13):
> eq13B:=(subs(x=a/2,diff(rhs(eq13A),x)))=0:
> eq13C:=convert(simplify(exp(1/2*_c[1]^(1/2)*a)*(exp(1/2*_c[1]^(1/2)
*a)-exp(-1/2*_c[1]^(1/2)*a)^2/exp(1/2*_c[1]^(1/2)*a))),sin)=0:
> eq13D:=_c[1]^(1/2)*a*I=n*Pi:
>
> eq13E:=isolate(eq13D,_c[1]):
> eq13Ea:=subs(y=-b/2,diff(rhs(eq13A),y))=0:
> eq13Eb:=isolate(eq13Ea,_C4):
> eq13Ec:=subs(_C3=1,subs(eq13Eb,eq13A)):
> eq13F:=_c[2]=-m^2*Pi^2/b^2:
> eq14:=simplify(subs(eq13E,eq13F,eq13Ec),power,symbolic):
>
> eq15:=subs(C=C[n,m],_C5=A[n,m],_C6=B[n,m],eq14):
> eq15A:=exp(Pi*n/a*x*I)+exp(-1/2*I*Pi*n)^2*exp(-I*Pi*n/a*x)=simplify
(exp(I*n*Pi/2)*(exp(Pi*n/a*x*I)+exp(-1/2*I*Pi*n)^2*exp(-I*Pi*n/a*x)))/
(exp(I*n*Pi/2)):
> eq15B:=exp(Pi*m/b*y*I)+exp(-1/2*I*Pi*m)^2*exp(-I*Pi*m/b*y)=simplify
(exp(I*m*Pi/2)*(exp(Pi*m/b*y*I)+exp(-1/2*I*Pi*m)^2*exp(-I*Pi*m/b*y)))/
(exp(I*m*Pi/2)):
> eq15C:=subs(eq15A,eq15B,eq15):
> eq15D:=subs(z=-c/2,diff(rhs(eq15C),z))=0:
> eq15E:=isolate(eq15D,B[n,m]):
> eq15F:=factor(subs(eq15E,eq15C)):
> eq16:=C(x,y,z)=(c[0])/s+Sum(Sum(rhs(eq15F),n=0..
infinity),m=0..infinity):
> eq16A:=simplify(eval(subs(Sum=sum,subs(infinity=1,eq16))),power,
symbolic):
>
> eq17A:=C(x,y,c/2)*alpha(x,y)+(1-alpha(x,y))*Eval(diff(C(x,y,z),z),
z=c/2)=c[b]*alpha(x,y):
> eq17B:=alpha(x,y)=(Heaviside(x+a/4)-Heaviside(x-a/4))*(Heaviside
(y+b/4)-Heaviside(y-b/4)):
> eq17C:=subs(eq17B,eq17A):
```

```
>
> eq17E:=subs(z=c/2,eq16A):
> eq17F:=eval(subs(Eval=eval,eval(subs(eq17E,eq16A,eq17C)))):
>
> eq19:=int(int(lhs(eq17F),x=-a/2..a/2),y=-b/2..b/2)=int(int(rhs
(eq17F),x=-a/2..a/2),y=-b/2..b/2) assuming a>0 and b>0:
> eq19A:=factor(isolate(eq19,A[0,0])):
> eq20:=subs(eq19A,eq16A):
> eq21:=Q(s)=factor(int(int(int(rhs(eq20),y=-b/2..b/2),x=-a/2..a/2),
z=-c/2..c/2)):
> eq22:=M(s)=c[0]*a*b*c/s-Q(s):
>
> eq23:=subs(eq21,eq22):
>
> eq24:=M(s)/M(infinity)=simplify(rhs(eq23)/(c[0]*a*b*c));
```

$$eq24 := \frac{M(s)}{M(\infty)} = \left(\cosh\left(\frac{1}{2}\frac{\sqrt{s}\,c}{\sqrt{\eta}}\right)\eta\left(e^{-\frac{1}{2}\frac{\sqrt{s}\,c}{\sqrt{\eta}}}s\,c_b - e^{\frac{1}{2}\frac{\sqrt{s}\,c}{\sqrt{\eta}}}s\,c_b - e^{-\frac{1}{2}\frac{\sqrt{s}\,c}{\sqrt{\eta}}}c_0 + e^{\frac{1}{2}\frac{\sqrt{s}\,c}{\sqrt{\eta}}}c_0\right)\right) \Big/ \left(c_0\,c\,s\left(2\cosh\left(\frac{1}{2}\frac{\sqrt{s}\,c}{\sqrt{\eta}}\right)^2\sqrt{\eta}\sqrt{s} + 6s\sinh\left(\frac{1}{2}\frac{\sqrt{s}\,c}{\sqrt{\eta}}\right)\cosh\left(\frac{1}{2}\frac{\sqrt{s}\,c}{\sqrt{\eta}}\right) - \sqrt{\eta}\sqrt{s}\right)\right)$$

```
> eq25:=t[eff]=factor(simplify(limit(simplify(1/s^2+diff(rhs(eq24),
s)),s=0),power,symbolic)/simplify(limit(simplify(1/s-rhs(eq24)),s=0),
power,symbolic));
```

$$eq25 := t_{eff} = \frac{1}{5}\,\frac{c(2c^3c_0+30c^2c_0+5c\,\eta\,c_b+135\,c\,c_0+45\,\eta\,c_b)}{\eta(c^2c_0+9\,c\,c_0+3\,\eta\,c_b)}$$

```
> eq26:=residue(rhs(eq24),s=0):
> eq27:=denom(rhs(eq24))/s=0:
> eq28:=(1/2/eta^(1/2)*s^(1/2)*c)=I*beta[n]:
> eq29:=isolate(eq28,s):
> eq30:=factor(simplify(subs(eq29,eq27),power,symbolic)):
> eq31:=isolate(eq30,sin(beta[n])):
> eq32:=numer(rhs(eq24))*exp(s*t)/diff(denom(rhs(eq24)),s):
> eq33:=(factor(evalc(subs(eq31,simplify(subs(eq29,eq32),power,
symbolic))))):
> eq34:=M(t)/M(infinity)=eq26-Sum(-eq33,n=1..infinity);
```

$$eq34 := \frac{M(t)}{M(\infty)} = 1 - \left(\sum_{n=1}^{\infty}\frac{24\cos(\beta_n)^3\sin(\beta_n)(4\,\eta\,\beta_n^2c_b+c^2c_0)\,e^{-\frac{4\beta_n^2\eta t}{c^2}}}{c\,c_0\,\beta_n(4\cos(\beta_n)^4c^2+144\cos(\beta_n)^4\beta_n^2+24\cos(\beta_n)^4c-12\cos(\beta_4)^2c-c^2)}\right)$$

```
>
```

13.5.3 Maple Code: 2Dvitreousmat.nb (Section 13.2)

$\partial_t c[r,z,t] = \eta$Laplacian$[c[r,z,t],\{r,\theta,z\},$ "Cylindrical"$]+$
$S[0](1-$UnitStep$[r-b[0]])$//Expand

$$c^{(0,0,1)}[r,z,t] = S[0] - S[0]\text{UnitStep}[r-b[0]] + \eta c^{(0,2,0)}[r,z,t] + \frac{\eta c^{(1,0,0)}[r,z,t]}{r} + \eta c^{(2,0,0)}[r,z,t]$$

eq0A $= c[r,z,0] = c[0](1-$UnitStep$[r-b[0]])$;
eq $= sc[r,z]-c[r,z,0] =$
$S[0](1-$UnitStep$[r-b[0]])/s+$
$\frac{1}{r}(\eta(rc^{(0,2)}[r,z]+c^{(1,0)}[r,z]+rc^{(2,0)}[r,z]))$

$$sc[r,z]-c[r,z,0] = \frac{S[0](1-\text{UnitStep}[r-b[0]])}{s} + \frac{\eta(rc^{(0,2)}[r,z]+c^{(1,0)}[r,z]+rc^{(2,0)}[r,z])}{r}$$

eq1 $=$ eq/.eq0A[[1]]−> eq0A[[2]];
eq1A $= sC[r]-c[0](1-$UnitStep$[r-b[0]]) =$
$\frac{S[0](1-\text{UnitStep}[r-b[0]])}{s} + \frac{\eta(C^{(1)}[r]+rC^{(2)}[r])}{r}$;
eq1B $=$ Factor[DSolve[eq1A, $C[r]$, r]];
eq1C $= (C[r]$/.eq1B)[[1]];
eq1D $=$ Simplify[eq1C, $r>0\&\&r<b[0]$]

$$\frac{1}{4s^{3/2}\eta}\left(\text{BesselJ}\left[0,\frac{ir\sqrt{s}}{\sqrt{\eta}}\right]\left(4s^{3/2}\eta C[1]+2i\pi r\sqrt{\eta}\text{BesselY}\left[1,-\frac{ir\sqrt{s}}{\sqrt{\eta}}\right](sc[0]+S[0])\right)+\sqrt{s}\text{BesselY}\left[0,-\frac{ir\sqrt{s}}{\sqrt{\eta}}\right]\left(4s\eta C[2]-\pi r^2\text{Hypergeometric0F1Regularized}\left[2,\frac{r^2 s}{4\eta}\right](sc[0]+S[0])\right)\right)$$

eq1E $=$ Series[eq1D, $\{r,0,1\}$];
eq1F $=$ eq1C/.$C[2]\to 0$;
eq1G $=$ Simplify[eq1F, $r>0\&\&r>b[0]$]

$$\frac{1}{4s^{3/2}\eta}\left(-\pi\sqrt{s}b[0]^2\text{BesselY}\left[0,-\frac{ir\sqrt{s}}{\sqrt{\eta}}\right]\text{Hypergeometric0F1Regularized}\left[2,\frac{sb[0]^2}{4\eta}\right](sc[0]+S[0]+\text{BesselJ}\left[0,\frac{ir\sqrt{s}}{\sqrt{\eta}}\right]\left(4s^{3/2}\eta C[1]+2i\pi\sqrt{\eta}b[0]\text{BesselY}\left[1,-\frac{i\sqrt{s}b[0]}{\sqrt{\eta}}\right](sc[0]+S[0])\right)\right)$$

eq1H $= \eta((\partial_r\text{eq1G})/.r\to R) = -(\eta[\text{rho}]K[\text{rho}]/\delta[\text{rho}])((\text{eq1G})/.r\to R)$;
eq1I $=$ Solve[eq1H, $C[1]$];
eq1J $=$ eq1F/.eq1I;
eq2A $=$ (eq1/.$c[0]\to 0$)/.$S[0]\to 0$

$$sc[r,z] = \frac{\eta(rc^{(0,2)}[r,z]+c^{(1,0)}[r,z]+rc^{(2,0)}[r,z])}{r}$$

eq2 $=$ eq2A/.$c\to$ Function[$\{r,z\}, f[r]*g[z]$];
eq3 $=$ Expand[eq2[[1]]/$(f[r]*g[z])$] $=$ Expand[eq2[[2]]/$(f[r]*g[z])$]

$$s = \frac{\eta f'[r]}{rf[r]} + \frac{\eta f''[r]}{f[r]} + \frac{\eta g''[z]}{g[z]}$$

eq3A $= \frac{f'[r]}{rf[r]} + \frac{f''[r]}{f[r]} = c[1]$

$$\frac{f'[r]}{rf[r]} + \frac{f''[r]}{f[r]} = c[1]$$

eq4 $= s = c[1] + \frac{\eta g''[z]}{g[z]}$

$$s = c[1] + \frac{\eta g''[z]}{g[z]}$$

eq5 $=$ DSolve[eq3A, $f[r]$, r];

eq5A $=$ (DSolve[eq4, $g[z]$, z]/.$C[1]$−> $A[1]$)/.$C[2]\to B[1]$;
eq6 $= (f[r]$/.eq5)$(g[z]$/.eq5A)[[1]];

```
eq6A = ((eq6/.C[2] → 0)/.C[1] → 1)[[1]];
eq6C =
Factor[η * ((∂_r eq6A)/.r → R)+
(η[rho] * K[rho]/δ[rho]) * ((eq6A)/.r → R)] = 0
(1/δ[rho]) e^(-(z Sqrt[s-c[1]])/Sqrt[η]) (e^((2z Sqrt[s-c[1]])/Sqrt[η]) A[1] + B[1])
(-iηBesselJ[1, iR Sqrt[c[1]]] Sqrt[c[1]] δ[rho] + BesselJ[0, iR Sqrt[c[1]]] K[rho] η[rho]) = 0
eq6D = iηBesselJ[1, -iR Sqrt[c[1]]] Sqrt[c[1]] δ[rho]+
BesselJ[0, -iR Sqrt[c[1]]] K[rho] η[rho] = 0;
eq6E = -iR Sqrt[c[1]] = α[n]
-iR Sqrt[c[1]] = α[n]
eq6F = Solve[eq6E, c[1]]
{{c[1] → -α[n]^2/R^2}}
eq6G = Simplify[eq6D/.eq6F, α[n] > 0&&R > 0];
eq6H = Solve[eq6G, BesselJ[1, α[n]]]
{{BesselJ[1, α[n]] → (R BesselJ[0, α[n]] K[rho] η[rho])/(η α[n] δ[rho])}}
eq6J = Simplify[eq6A/.eq6F, α[n] > 0&&R > 0];
eq6J1 = (eq6J/.A[1] → A[n])/.B[1]-> B[n];
eq6K =
C[r, z] =
Sum[BesselJ[0, r α[n]/R]
(A[n] Cos[(z Sqrt[-s α[n]^2/R^2])/Sqrt[η]] + B[n] Sin[(z Sqrt[-s - α[n]^2/R^2])/Sqrt[η]]), {n, 1, ∞}];
(* Computing the effective time using an approximation of
zerothorder :*)
eq7 = Normal[(Simplify[Series[r * eq1J, {s, 0, 3}], r > 0&&eta > 0])];
eq7A = Simplify[Integrate[Integrate[eq7, {r, 0, R}], {z, 0, H}],
r > 0&&b[0] > 0&&R > 0&&b[0] < R];
eq7B = Simplify[Limit[s * eq7A, s → 0], η > 0]
{(H b[0]^2 S[0] (4RηΔ[rho] + (2R^2 - b[0]^2) K[rho] η[rho]))/(16η K[rho] η[rho])}
eq7C = Simplify[eq7A/eq7B];
eq7D = Simplify[Limit[s * eq7C, s → 0], η > 0]
{1}
eq7E = Simplify[Normal[Series[eq7C, {s, 0, 2}]]][[1]];
eq7F = Limit[FullSimplify[(∂_s eq7E + 1/s^2)], s → 0];
eq7G = Limit[FullSimplify[(-eq7E + 1/s)], s → 0]
(48R^2 η^2 S[0] δ[rho]^2 + 12RηK[rho](-8ηc[0] + (3R^2 - b[0]^2)S[0])δ[rho]η[rho] +
K[rho]^2 (24ηb[0]^2 c[0] + 9R^4 S[0] + b[0]^4 S[0] - 6R^2 (8ηc[0] + b[0]^2 S[0]))η[rho]^2) /
(24ηK[rho]S[0]η[rho](4RηΔ[rho] + (2R^2 - b[0]^2)K[rho]η[rho]))
eq8 = tau[eff] = Factor[eq7F/eq7G];
eq8A = Factor[(eq8/.b[0] → R)];
eq8B = Factor[(eq8/.b[0] → 0)];
```

PROBLEMS

13.1. Considering the model parameters for ganciclovir (Tojo & Isowaki, 2001): $S_0 = 163.74\,\mu g/d/cm^3$, $H = 0.72\,cm$, $R = 0.72\,cm$, $b_0 = 0.36\,cm$, $K_1 = 0.324$, $K_p = 1.11$, $K_\rho = 0.74$, $c_0 = 25\,\mu g/cm^3$, $\eta = 0.854\,cm^2/d$, $\eta_1 = 0.158\,cm^2/d$, $\eta_p = 0.864\,cm^2/d$, $\eta_\rho = 0.0476\,cm^2/d$, $\delta_1 = 0.38\,cm$, $\delta_p = 0.0100\,cm$ and $\delta_\rho = 0.0040\,cm$, plot $\frac{c_{av}(t)}{c_{ave,ss}}$ from Eq. (13.43). Show the effective time constant.

13.2. Given the following parameters: $S_0 = 163.74\,\mu g/d/cm^3$, $H = 0.72\,cm$, $R = 0.72$ cm, $K_1 = 0.324$, $K_p = 1.11$, $K_\rho = 0.74$, $c_0 = 25\,\mu g/cm^3$, $\eta = 0.854\,cm^2/d$, $\eta_1 = 0.158\,cm^2/d$, $\eta_p = 0.864\,cm^2/d$, $\eta_\rho = 0.0476\,cm^2/d$, $\delta_1 = 0.38\,cm$, $\delta_p = 0.0100\,cm$ and $\delta_\rho = 0.0040\,cm$, calculate $\lim_{b_0 \to R} \tau_{eff}$ using Eq. (13.45).

13.3. Considering the following constants: $S_0 = 163.74\,\mu g/d/cm^3$, $H = 0.72\,cm$, $R = 0.72\,cm$, $K_1 = 0.324$, $K_p = 1.11$, $K_\rho = 0.74$, $c_0 = 25\,\mu g/cm^3$, $\eta = 0.854\,cm^2/d$, $\eta_1 = 0.158\,cm^2/d$, $\eta_p = 0.864\,cm^2/d$, $\eta_\rho = 0.0476\,cm^2/d$, $\delta_1 = 0.38\,cm$, $\delta_p = 0.0100\,cm$ and $\delta_\rho = 0.0040\,cm$, calculate $\lim_{b_0 \to 0} \tau_{eff}$ using Eq. (13.46).

13.4. Calculate t_{eff} using $a = 85\,\mu m$, $b = 170\,\mu m$, $c = 340\,\mu m$, $\eta = 8.2 \times 10^{-10}\,cm^2/s$, $c_b = 0$ and Eq.(13.81).

13.5. Plot $M(t)/M(\infty)$ using $a = 85\,\mu m$, $b = 170\,\mu m$, $c = 340\,\mu m$, $\eta = 8.2 \times 10^{-10}\,cm^2/s$, $c_b = 0$ and Eq. (13.82).

REFERENCES

Collins R. The choice of an effective time constant for diffusive processes in finite systems (Thermal conduction and sputtering examples). *J Phys D: Appl Phys* 1980;13(11):1935.

Fernandes M, Simon L, Loney NW. Mathematical modeling of transdermal drug-delivery systems: analysis and applications. *J Membr Sci* 2005;256(1–2):184–192.

Ferreira JA, de Oliveira P, da Silva P, Simon L. Flux tracking in drug delivery. *Appl Math Model* 2011;35(10):4684–4696.

Simon L. Timely drug delivery from controlled-release devices: dynamic analysis and novel design concepts. *Math Biosci* 2009;217(2):151–158.

Simon L. Graphical process design tools for iontophoretic transdermal drug-delivery devices. *Comput Methods Programs Biomed* 2012;107(3):447–455.

Simon L, Ospina J. Two-dimensional solution and analysis of a cylindrical matrix device with a circular release area. *Chem Eng Commun* 2012;200(1):115–138.

Tojo K, Isowaki A. Pharmacokinetic model for in vivo/in vitro correlation of intravitreal drug delivery. *Adv Drug Deliv Rev* 2001;52(1):17–24.

Wei R, Simon L, Hu L, Michniak-Kohn B. Effects of iontophoresis and chemical enhancers on the transport of lidocaine and nicotine across the oral mucosa. *Pharm Res* 2012;29(4):961–971.

14

DATA FITTING FOR TWO- AND THREE-DIMENSIONAL CONTROLLED-RELEASE DRUG-DELIVERY MODELS

14.1 DATA FITTING IN CONTROLLED-RELEASE DRUG-DELIVERY SYSTEMS

Parameter estimation methods are applied in many fields to help researchers understand and control processes (Wang & Lou, 2010). A study was conducted to assess the effects of physical parameters on the cumulative amount of drug collected (Simon et al., 2006). Based on the results, it was determined that the enhanced permeation rate was due to an increase in the drug concentration at the membrane surface and not because of a rise in the diffusion coefficient. A similar technique was implemented to shed light on the key mechanisms controlling the transport of lidocaine and nicotine across the oral mucosa (Wei et al., 2012). Physical and chemical penetration enhancers were employed in the study. In addition to allowing researchers to simulate the process at different conditions, dosage regimens can be developed to maintain plasma drug concentration within a desired range.

Computation of the parameter values provides a way of determining whether a mathematical model is correct (Schittkowski, 2002). After deriving the governing equations that describe the system, researchers need to compare the experimental data to the model predictions. This step is only possible if parameter values are available. The estimates are often computed by solving a problem of the form

$$\min_{\mathbf{p}_l \le \mathbf{p} \le \mathbf{p}_u} \sum_{i=1}^{N} [X_{i,2} - Q_e(X_{i,1}, \mathbf{p})]^2 \tag{14.1}$$

where $X_{i,1}$ and $X_{i,2}$ represent the time and a process variable, respectively; N is the number of data points and $\mathbf{p}$ is a vector of parameters. For example, $X_{i,2}$ denotes the experimental

Closed-form Solutions for Drug Transport through Controlled-Release Devices in Two and Three Dimensions, First Edition. Laurent Simon and Juan Ospina.

cumulative amount of drug released in Simon et al. (2006) while Q_e is the predicted value. The solution to the minimization problem yields estimates of $\mathbf{p}$ that lie between $\mathbf{p_l}$ and $\mathbf{p_u}$. Other residual forms, such as the sum of absolute errors, can be selected instead of the sum of squared errors. To solve the problem (14.1), a local optimization algorithm, based on Newton's method can be implemented in Mathematica using the command "FindMinimum" (Simon et al., 2006).

Eq. (14.1) assumes that a solution of the process mathematical model exists. It is customary to use the method of lines to discretize the spatial variables, first. The resulting system of differential equations is then integrated in time. In this chapter, we present an analytical expression for the variable of interest (e.g., Q_e) before solving the minimization problem.

14.2 ESTIMATION OF DIFFUSION COEFFICIENT IN A SOLID CYLINDER OF FINITE LENGTH

The model of a cylindrical dosage form is described by (Fig. 14.1)

$$\frac{\partial c}{\partial t} = \eta\left(\frac{1}{r}\frac{\partial}{\partial r}r\frac{\partial c}{\partial r} + \frac{\partial^2 c}{\partial z^2}\right); \quad 0 \le r \le R, \quad -H \le z \le H \tag{14.2}$$

$$c(r, z, 0) = c_o \tag{14.3}$$

$$c(R, z, t) = c_b \tag{14.4}$$

$$\left.\frac{\partial c(r, z, t)}{\partial r}\right|_{r=0} = 0 \tag{14.5}$$

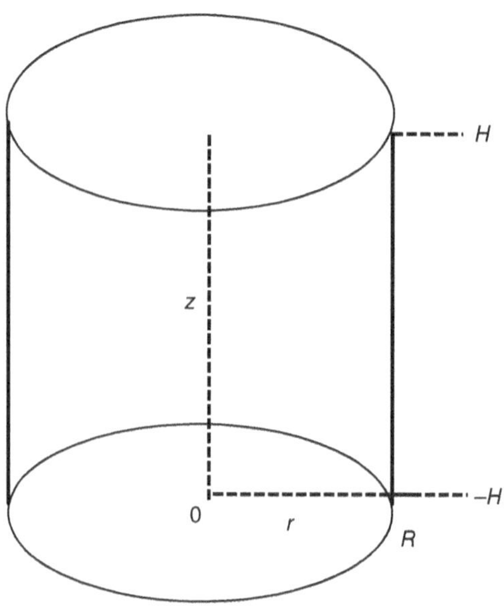

Figure 14.1 Controlled-release cylindrical dosage form.

$$c(r, -H, t) = c_b \tag{14.6}$$

and

$$c(r, H, t) = c_b \tag{14.7}$$

The goal is to derive an expression for the cumulative amount of drug released:

$$M(t) = c_0 \pi R^2 (2H) - 2\pi \int_{-H}^{H} \int_{0}^{R} c(r, z, t)\, r dr dz \tag{14.8}$$

We first take the Laplace transform of Eq. (14.2):

$$sC(r, z) - c_0 = \frac{\eta \frac{\partial}{\partial r} C(r, z)}{r} + \eta \frac{\partial^2}{\partial r^2} C(r, z) + \eta \frac{\partial^2}{\partial z^2} C(r, z) \tag{14.9}$$

where

$$C(r, z) = \int_0^\infty C(r, z, t) e^{-st}\, dt \tag{14.10}$$

A solution to Eq. (14.9) is of the form $C(r, z) = C(r) + f(r)g(z)$. Therefore,

$$sC(r) - c_0 = \frac{\eta \frac{d}{dr} C(r)}{r} + \eta \frac{d^2}{dr^2} C(r) \tag{14.11}$$

$$\frac{d^2}{dr^2} f(r) = c_1\, f(r) - \frac{\frac{d}{dr} f(r)}{r} \tag{14.12}$$

$$\frac{d^2}{dr^2} g(z) = -c_1 g(z) + \frac{sg(z)}{\eta} \tag{14.13}$$

Now, the solution of Eq. (14.11) is

$$C(r) = C_2 J_0 \left(r\sqrt{-\frac{s}{\eta}} \right) + C_1 Y_0 \left(r\sqrt{-\frac{s}{\eta}} \right) + \frac{c_0}{s} \tag{14.14}$$

Applying the boundary condition (14.5) to Eq. (14.14), we obtain $C_1 = 0$ resulting in the following equation:

$$C(r) = C_2 J_0 \left(r\sqrt{-\frac{s}{\eta}} \right) + \frac{c_0}{s} \tag{14.15}$$

Similarly, after using the Laplace transform of Eq. (14.4) in Eq.(14.15), we have

$$C_2 J_0 \left(R\sqrt{-\frac{s}{\eta}} \right) + \frac{c_0}{s} = \frac{c_b}{s} \tag{14.16}$$

which gives

$$C_2 = \frac{\left(\frac{c_b}{s} - \frac{c_0}{s} \right)}{J_0 \left(R\sqrt{-\frac{s}{\eta}} \right)} \tag{14.17}$$

Equation (14.15) becomes

$$C(r) = \frac{\left(\frac{c_b}{s} - \frac{c_0}{s}\right) J_0\left(r\sqrt{-\frac{s}{\eta}}\right)}{J_0\left(R\sqrt{-\frac{s}{\eta}}\right)} + \frac{c_0}{s} \tag{14.18}$$

The solution of Eqs. (14.12) and (14.13) gives

$$f(r)g(z) = \left(C_3 \sin\left(z\sqrt{\frac{c_1\eta - s}{\eta}}\right) + C_4 \cos\left(z\sqrt{\frac{c_1\eta - s}{\eta}}\right)\right) \times (C_1 J_0(r\sqrt{-c_1}) + C_2 Y_0(r\sqrt{-c_1})) \tag{14.19}$$

Based on Eq. (14.5), we have $C_2 = 0$, leading to

$$f(r)g(z) = \left(C_3 \sin\left(z\sqrt{\frac{c_1\eta - s}{\eta}}\right) + C_4 \cos\left(z\sqrt{\frac{c_1\eta - s}{\eta}}\right)\right) C_1 J_0(r\sqrt{-c_1}) \tag{14.20}$$

Without loss of generality, C_1 is set to 1. Therefore,

$$f(r)g(z) = \left(C_3 \sin\left(z\sqrt{\frac{c_1\eta - s}{\eta}}\right) + C_4 \cos\left(z\sqrt{\frac{c_1\eta - s}{\eta}}\right)\right) J_0(r\sqrt{-c_1}) \tag{14.21}$$

The boundary condition $f(R)g(z) = 0$ yields

$$\left(C_3 \sin\left(z\sqrt{\frac{c_1\eta - s}{\eta}}\right) + C_4 \cos\left(z\sqrt{\frac{c_1\eta - s}{\eta}}\right)\right) J_0(R\sqrt{-c_1}) = 0 \tag{14.22}$$

or

$$J_0\,(R\sqrt{-c_1}) = 0 \tag{14.23}$$

which gives

$$c_1 = -\frac{{\alpha_{0,n}}^2}{R^2} \tag{14.24}$$

where $\alpha_{0,n}$ are the zeroes of $J_0(x)$. Then, Eq. (14.21) is

$$f(r)g(z) = \left(iC_3 \sinh\left(\frac{z}{R}\sqrt{\frac{sR^2 + {\alpha_{0,n}}^2\eta}{\eta}}\right) + C_4 \cosh\left(\frac{z}{R}\sqrt{\frac{sR^2 + {\alpha_{0,n}}^2\eta}{\eta}}\right)\right) J_0\left(\frac{\alpha_{0,n} r}{R}\right) \tag{14.25}$$

Finally, the general solution to Eq. (14.9) is obtained by applying Eq. (14.18) and the **principle of superposition:**

$$C(r, z) = \frac{\left(\frac{c_b}{s} - \frac{c_0}{s}\right) J_0\left(r\sqrt{-\frac{s}{\eta}}\right)}{J_0\left(R\sqrt{-\frac{s}{\eta}}\right)} + \frac{c_0}{s} + \sum_{n=1}^{\infty} \left(iA_n \sinh\left(\frac{z}{R}\sqrt{\frac{sR^2 + {\alpha_{0,n}}^2\eta}{\eta}}\right) + B_n \cosh\left(\frac{z}{R}\sqrt{\frac{sR^2 + {\alpha_{0,n}}^2\eta}{\eta}}\right)\right) J_0\left(\frac{\alpha_{0,n} r}{R}\right) \tag{14.26}$$

The boundary conditions (14.6) and (14.7) become

$$\frac{c_b}{s} = \frac{\left(\frac{c_b}{s} - \frac{c_0}{s}\right) J_0\left(r\sqrt{-\frac{s}{\eta}}\right)}{J_0\left(R\sqrt{-\frac{s}{\eta}}\right)} + \frac{c_0}{s}$$
$$+ \sum_{n=1}^{\infty}\left(iA_n \sinh\left(-\frac{H}{R}\sqrt{\frac{sR^2 + \alpha_{0,n}{}^2\eta}{\eta}}\right) + B_n \cosh\left(-\frac{H}{R}\sqrt{\frac{sR^2 + \alpha_{0,n}{}^2\eta}{\eta}}\right)\right) J_0\left(\frac{\alpha_{0,n} r}{R}\right) \tag{14.27}$$

and

$$\frac{c_b}{s} = \frac{\left(\frac{c_b}{s} - \frac{c_0}{s}\right) J_0\left(r\sqrt{-\frac{s}{\eta}}\right)}{J_0\left(R\sqrt{-\frac{s}{\eta}}\right)} + \frac{c_0}{s}$$
$$+ \sum_{n=1}^{\infty}\left(iA_n \sinh\left(\frac{H}{R}\sqrt{\frac{sR^2 + \alpha_{0,n}{}^2\eta}{\eta}}\right) + B_n \cosh\left(\frac{H}{R}\sqrt{\frac{sR^2 + \alpha_{0,n}{}^2\eta}{\eta}}\right)\right) J_0\left(\frac{\alpha_{0,n} r}{R}\right) \tag{14.28}$$

respectively. After multiplying both sides of Eq. (14.27) by $rJ_0\left(\frac{\alpha_{0,n} r}{R}\right)$ and integrating from $r = 0$ to $r = R$, we obtain

$$\frac{1}{2}\frac{R^2}{J_0\left(R\sqrt{-\frac{s}{\eta}}\right) s(sR^2 + \alpha_{0,n}{}^2\eta)\alpha_{0,n}} \times \begin{pmatrix} 2c_b\eta\alpha_{0,n}{}^2 J_0\left(R\sqrt{-\frac{s}{\eta}}\right) J_1(\alpha_{0,n}) - \\ is^2 J_0\left(R\sqrt{-\frac{s}{\eta}}\right) A_n\alpha_{0,n}R^2(J_1(\alpha_{0,n}))^2 \sinh\left(\frac{H}{R}\sqrt{\frac{sR^2+\alpha_{0,n}{}^2\eta}{\eta}}\right) - \\ isJ_0\left(R\sqrt{-\frac{s}{\eta}}\right) A_n\alpha_{0,n}{}^3(J_1(\alpha_{0,n}))^2\eta \sinh\left(\frac{H}{R}\sqrt{\frac{sR^2+\alpha_{0,n}{}^2\eta}{\eta}}\right) + \\ s^2 J_0\left(\sqrt{-\frac{s}{\eta}}R\right) B_n \cosh\left(\frac{H}{R}\sqrt{\frac{sR^2+\alpha_{0,n}{}^2\eta}{\eta}}\right) \alpha_{0,n}R^2(J_1(\alpha_{0,n}))^2 + \\ 2c_0 J_0\left(R\sqrt{-\frac{s}{\eta}}\right) R^2 J_1(\alpha_{0,n})s + \\ sJ_0\left(R\sqrt{-\frac{s}{\eta}}\right) B_n \cosh\left(\frac{H}{R}\sqrt{\frac{sR^2+\alpha_{0,n}{}^2\eta}{\eta}}\right) \alpha_{0,n}{}^3(J_1(\alpha_{0,n}))^2\eta \end{pmatrix} = \frac{R^2 c_b J_1(\alpha_{0,n})}{\alpha_{0,n} s} \tag{14.29}$$

Similarly, we multiply both sides of Eq. (14.28) by $rJ_0\left(\frac{\alpha_{0,n}r}{R}\right)$ and integrate from $r = 0$ to $r = R$ to have

$$\frac{1}{2}\frac{R^2}{J_0\left(\sqrt{-\frac{s}{\eta}}R\right)s(sR^2+\alpha_{0,n}{}^2\eta)\alpha_{0,n}} \times \left(\begin{array}{l} 2c_b\eta\alpha_{0,n}{}^2J_0\left(R\sqrt{-\frac{s}{\eta}}\right)J_1(\alpha_{0,n}) + \\ isJ_0\left(R\sqrt{-\frac{s}{\eta}}\right)A_n\alpha_{0,n}{}^3(J_1(\alpha_{0,n}))^2\eta\sinh\left(\frac{H}{R}\sqrt{\frac{sR^2+\alpha_{0,n}{}^2\eta}{\eta}}\right)+ \\ is^2J_0\left(R\sqrt{-\frac{s}{\eta}}\right)A_n\alpha_{0,n}R^2(J_1(\alpha_{0,n}))^2\sinh\left(\frac{H}{R}\sqrt{\frac{sR^2+\alpha_{0,n}{}^2\eta}{\eta}}\right)+ \\ s^2J_0\left(R\sqrt{-\frac{s}{\eta}}\right)B_n\cosh\left(\frac{H}{R}\sqrt{\frac{sR^2+\alpha_{0,n}{}^2\eta}{\eta}}\right)\alpha_{0,n}R^2(J_1(\alpha_{0,n}))^2+ \\ 2c_0J_0\left(R\sqrt{-\frac{s}{\eta}}\right)R^2J_1(\alpha_{0,n})s+ \\ sJ_0\left(R\sqrt{-\frac{s}{\eta}}\right)B_n\cosh\left(\frac{H}{R}\sqrt{\frac{sR^2+\alpha_{0,n}{}^2\eta}{\eta}}\right)\alpha_{0,n}{}^3(J_1(\alpha_{0,n}))^2\eta \end{array}\right) = \frac{R^2c_bJ_1(\alpha_{0,n})}{\alpha_{0,n}s} \quad (14.30)$$

Equations (14.29) and (14.30) are solved for A_n and B_n:

$$A_n = 0;\ \ B_n = \frac{4R^2(c_b-c_0)e^{\frac{H}{R}\sqrt{\frac{sR^2+\alpha_{0,n}{}^2\eta}{\eta}}}}{\alpha_{0,n}J_1(\alpha_{0,n})(sR^2+\alpha_{0,n}{}^2\eta)\left(\left(e^{\frac{H}{R}\sqrt{\frac{sR^2+\alpha_{0,n}{}^2\eta}{\eta}}}\right)^2+1\right)} \quad (14.31)$$

Therefore, Eq. (14.26) becomes

$$C(r,z) = \frac{\left(\frac{c_b}{s}-\frac{c_0}{s}\right)J_0\left(r\sqrt{-\frac{s}{\eta}}\right)}{J_0\left(R\sqrt{-\frac{s}{\eta}}\right)} + \frac{c_0}{s} + \sum_{n=1}^{\infty}\left(\frac{4R^2\left(c_b-c_0\right)e^{\frac{H}{R}\sqrt{\frac{sR^2+\alpha_{0,n}{}^2\eta}{\eta}}}\cosh\left(\frac{z}{R}\sqrt{\frac{sR^2+\alpha_{0,n}{}^2\eta}{\eta}}\right)J_0\left(\frac{\alpha_{0,n}r}{R}\right)}{\alpha_{0,n}J_1(\alpha_{0,n})(sR^2+\alpha_{0,n}{}^2\eta)\left(\left(e^{\frac{H}{R}\sqrt{\frac{sR^2+\alpha_{0,n}{}^2\eta}{\eta}}}\right)^2+1\right)}\right) \quad (14.32)$$

Equation (14.32) combined with the Laplace transform of Eq. (14.8) give

$$M(s) = \frac{2c_0\pi R^2 H}{s} - \frac{2\pi RH}{sJ_0\left(R\sqrt{-\frac{s}{\eta}}\right)\sqrt{-\frac{s}{\eta}}}\begin{pmatrix} -2J_1\left(R\sqrt{-\frac{s}{\eta}}\right)c_0 + c_0 J_0\left(R\sqrt{-\frac{s}{\eta}}\right)\times \\ R\sqrt{-\frac{s}{\eta}} + 2J_1\left(R\sqrt{-\frac{s}{\eta}}\right)c_b \end{pmatrix}$$

$$+\sum_{n=1}^{\infty}\frac{8\pi R^5\sqrt{\eta}(c_0 - c_b)\left(e^{\frac{2H}{R}\sqrt{\frac{sR^2+\alpha_{0,n}{}^2\eta}{\eta}}} - 1\right)}{\alpha_{0,n}{}^2(sR^2 + \alpha_{0,n}{}^2\eta)^{3/2}\left(e^{\frac{2H}{R}\sqrt{\frac{sR^2+\alpha_{0,n}{}^2\eta}{\eta}}} + 1\right)} \tag{14.33}$$

The inverse Laplace transform of Eq. (14.33) is computed using the **Bromwich integral** and the residue theorem (see Problem 14.1). Finally, the fraction of drug released at time t is

$$\frac{M(t)}{M(\infty)} = 1 - \sum_{n=1}^{\infty}\left(\sum_{m=0}^{\infty}\frac{32e^{-\frac{\eta t\pi^2(2m+1)^2}{4H^2}-\frac{\alpha_{0,n}{}^2\eta t}{R^2}}}{\pi^2\alpha_{0,n}{}^2(2m+1)^2}\right) \tag{14.34}$$

where

$$M(\infty) = 2H\pi R^2(c_0 - c_b) \tag{14.35}$$

Let us test the algorithm described in Section 14.1 on a polymeric-controlled release device of height 180 μm and radius 90 μm. The diffusion coefficient of the active agent in the polymer is 144 μm²/h. Equation (14.34) is used to generate the following data (Fig. 14.2):

$$\begin{pmatrix} 0 & 0.0796902 \\ 1 & 0.336727 \\ 2 & 0.453583 \\ 3 & 0.534762 \\ 4 & 0.59753 \\ 5 & 0.648592 \\ 6 & 0.691369 \\ 7 & 0.727879 \\ 8 & 0.759428 \\ 9 & 0.78692 \\ 10 & 0.811014 \\ 11 & 0.832217 \\ 12 & 0.850928 \\ 13 & 0.867475 \\ 14 & 0.882129 \end{pmatrix}$$

The goal is to estimate the diffusion coefficient η from the above data. To this end, the following lines are written in Mathematica to declare the parameters and create the function $M(t)/M(\infty)$: "infinity=7;=180;*R*=90;etaf=144.;" and "M1Minf[t_, eta_]:=1-(Sum[Sum[32*Exp[−eta*t*Pi^2*(2*m+1)^2/(4*H^2)-alpha[[n]]^2*eta*t/R^2]/(Pi^2*alpha[[n]]^2*(2*m+1)^2), {*m*, 0, *infinity*}], {*n*, 1, *infinity*}])"

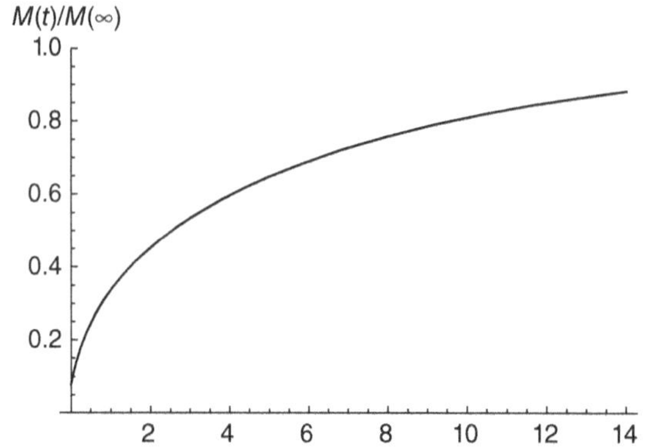

Figure 14.2 Cumulative fraction of drug released from a cylindrical dosage form of height 180 μm and radius 90 μm.

In Mathematica, the size of the data is "n=Length[data];" and the objective function is written as "$sterror[eta\_] := \left(\sum_{i=1}^{n} (data\,[[i,\,2]] - M1Minf[data[[i,\,1]],\,eta])^2\right)$;" The diffusion coefficient is estimated from the data by using the following code: "*FindMinimum*[*sterror*[*eta*], *eta*]", which yields the function evaluated at the optimum η and the parameter η: $\{3.8518 \times 10^{-32}, \{eta \rightarrow 144.\}\}$.

14.3 ESTIMATION OF DIFFUSION COEFFICIENT IN A RECTANGULAR PARALLELEPIPED-SHAPED MATRIX

The release kinetics of a polymeric device is given by (Fig. 14.3)

$$\frac{\partial c}{\partial t} = \eta \left(\frac{\partial^2 c}{\partial^2 x^2} + \frac{\partial^2 c}{\partial^2 y^2} + \frac{\partial^2 c}{\partial^2 y^2} \right); \quad -\frac{a}{2} \le x \le \frac{a}{2}, \quad -\frac{b}{2} \le y \le \frac{b}{2}, \quad -\frac{c}{2} \le z \le \frac{c}{2} \tag{14.36}$$

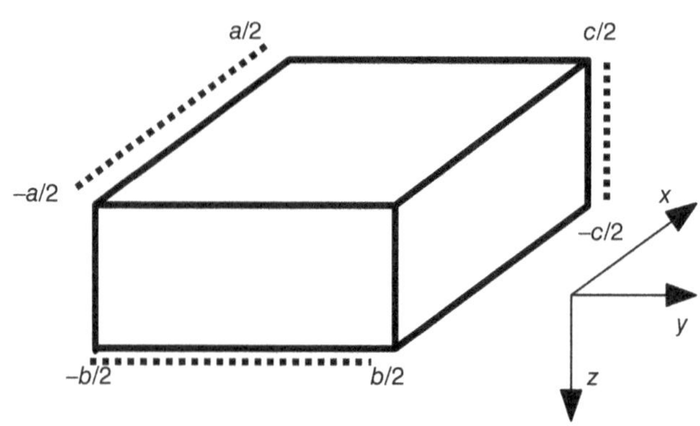

Figure 14.3 Drug released from a rectangular parallelepiped-shaped matrix.

where η is the drug diffusion coefficient in the polymer. The drug concentration at $t = 0$ is

$$c(x,\ y,\ z,\ 0) = c_0 \tag{14.37}$$

The device is in contact with a medium where the drug concentration is c_b:

$$c\left(-\frac{a}{2},\ y,\ z,\ t\right) = c\left(\frac{a}{2},\ y,\ z,\ t\right) = c_b \tag{14.38}$$

$$c\left(x, -\frac{b}{2},\ z,\ t\right) = c\left(x,\ \frac{b}{2},\ z,\ t\right) = c_b \tag{14.39}$$

$$c\left(x,\ y,\ -\frac{c}{2},\ t\right) = c\left(x,\ y,\ \frac{c}{2},\ t\right) = c_b \tag{14.40}$$

The amount of drug released at time t is

$$M(t) = c_0abc - \int_{-\frac{a}{2}}^{\frac{a}{2}} \int_{-\frac{b}{2}}^{\frac{b}{2}} \int_{-\frac{c}{2}}^{\frac{c}{2}} c(x,\ y,\ z,\ t)\ dxdydz \tag{14.41}$$

To solve the above system, the corresponding one-dimensional problem will be solved using the Laplace transform technique. The one-dimensional equation derived from Eq. (14.36) reads

$$\frac{\partial c}{\partial t} = \eta \frac{\partial^2 c}{\partial^2 x^2} \tag{14.42}$$

and the initial condition is

$$c(x,\ 0) = c_0 \tag{14.43}$$

The Laplace transform of Eq. (14.42) is given by

$$sC(x) - c_0 = \eta \frac{d^2}{d^2x^2} C(x) \tag{14.44}$$

where

$$C(x) = \int_0^{\infty} c(x,\ t)e^{-st}\ dt \tag{14.45}$$

The general solution of Eq. (14.44) is

$$C(x) = C_2 e^{x\sqrt{\frac{s}{\eta}}} + C_1 e^{-x\sqrt{\frac{s}{\eta}}} + \frac{c_0}{s} \tag{14.46}$$

Applying the Laplace transform of the boundary condition (14.38) to Eq. (14.46), we obtain

$$C_2 e^{L\sqrt{\frac{s}{\eta}}} + C_1 e^{-L\sqrt{\frac{s}{\eta}}} + \frac{c_0}{s} = \frac{c_b}{s} \tag{14.47}$$

and

$$C_2 e^{-L\sqrt{\frac{s}{\eta}}} + C_1 e^{L\sqrt{\frac{s}{\eta}}} + \frac{c_0}{s} = \frac{c_b}{s} \tag{14.48}$$

which give

$$C_1 = \frac{c_b - c_0}{2\cosh\left(L\sqrt{\frac{s}{\eta}}\right)}, \quad C_2 = \frac{c_b - c_0}{2\cosh\left(L\sqrt{\frac{s}{\eta}}\right)} \tag{14.49}$$

Therefore,

$$C(x) = \frac{c_b - c_0}{2\cosh\left(L\sqrt{\frac{s}{\eta}}\right)}\left(e^{x\sqrt{\frac{s}{\eta}}} + e^{-x\sqrt{\frac{s}{\eta}}}\right) + \frac{c_0}{s} \tag{14.50}$$

The inverse Laplace transform of Eq. (14.50) is computed using the Bromwich integral and the residue theorem (see Problem 14.4). The result is

$$c(x, t) = c_b + \sum_{n=0}^{\infty} \frac{2(-1)^{1+n}}{\pi(2n+1)}\left(-e^{\frac{i(2n+1)}{2L}} c_0 + e^{\frac{i(2n+1)\pi x}{2L}} c_b - e^{\frac{-i(2n+1)\pi x}{2L}} c_0 + e^{\frac{-i(2n+1)\pi x}{2L}} c_b\right) e^{-\frac{(2n+1)^2\pi^2\eta t}{4L^2}} \tag{14.51}$$

Therefore, we can calculate $M(t)$ for the one-dimensional problem as

$$M(t) = 2Lc_0 - 2c_bL - \sum_{n=0}^{\infty} \frac{8(-1)^n L(c_0 - c_b + c_0 e^{2i\pi n} - c_b e^{2i\pi n}) e^{-\frac{\pi(4\pi\eta t n^2 + 4\pi\eta t n + \pi\eta t + 4inL^2)}{4L^2}}}{\pi^2(4n^2 + 4n + 1)} \tag{14.52}$$

The cumulative fraction of drug released is calculated from Eq. (14.52):

$$\frac{M(t)}{M(\infty)} = 1 - \sum_{n=0}^{\infty} \frac{8e^{-\frac{(2n+1)^2\pi^2\eta t}{4L^2}}}{\pi^2(2n^2+1)^2} \tag{14.53}$$

Finally, Eq. (14.53) can be generalized for the original problem

$$\frac{M(t)}{M(\infty)} = 1 - \frac{512}{\pi^6}\left(\sum_{n=0}^{\infty} \frac{e^{-\frac{(2n+1)^2\pi^2\eta t}{a^2}}}{(2n+1)^2}\right)\left(\sum_{m=0}^{\infty} \frac{e^{-\frac{(2m+1)^2\pi^2\eta t}{b^2}}}{(2m+1)^2}\right)\left(\sum_{k=0}^{\infty} \frac{e^{-\frac{(2k+1)^2\pi^2\eta t}{c^2}}}{(2k+1)^2}\right) \tag{14.54}$$

We implemented the algorithm outlined in Section 14.1 on a rectangular parallelepiped -shaped device with the following dimension: $a = 85\,\mu m$, $b = 170\,\mu m$ and $c = 340\,\mu m$. The diffusion coefficient of the drug in the polymer is $144\,\mu m^2/h$. After applying Eq. (14.54), the

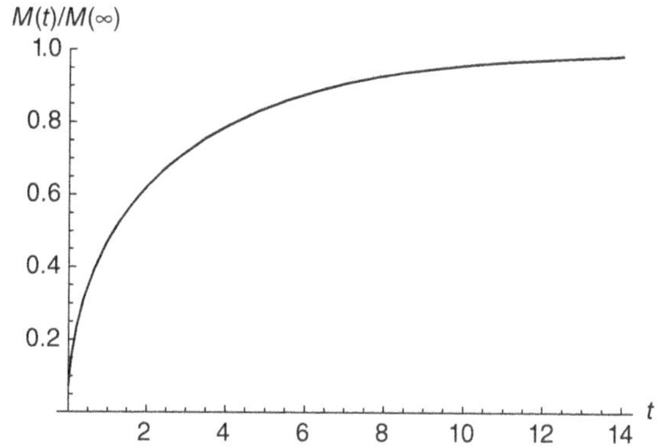

Figure 14.4 Cumulative fraction of drug released from a rectangular parallelepiped-shaped matrix with the following dimension: $a = 85\ \mu$m, $b = 170\ \mu$m and $c = 340\ \mu$m.

following data are generated (Fig. 14.4):

$$\begin{pmatrix} 0 & 0.0739888 \\ 1 & 0.472844 \\ 2 & 0.622219 \\ 3 & 0.719297 \\ 4 & 0.788552 \\ 5 & 0.839592 \\ 6 & 0.877779 \\ 7 & 0.906597 \\ 8 & 0.928469 \\ 9 & 0.945134 \\ 10 & 0.957867 \\ 11 & 0.967616 \\ 12 & 0.975091 \\ 13 & 0.98083 \\ 14 & 0.98524 \end{pmatrix}$$

By using the following command: "Find Minimum [*sterror* [*eta*], *eta*]", the function value at the optimum η and the parameter η are $\{2.31978 \times 10^{-31}\}$, $\{eta \rightarrow 144.\}$.

14.4 SUMMARY

Optimal model parameters were estimated in Mathematica using analytical solutions of two and three-dimensional transport models. The two case studies involve the fractions of drug released from a cylindrical controlled-release dosage form and a rectangular parallelepiped-shaped matrix. For the latter device, it was possible to solve a one-dimensional problem and to extend the results to three dimensions. The governing equations were solved using Laplace transform-based methods. The function "*FindMinimum*" successfully calculated the diffusion coefficient of the drug in the polymer. No constraints were placed on the parameter.

14.5 APPENDIX: MAPLE AND MATHEMATICA CODE LISTINGS

14.5.1 Maple Code: cyl2D.mws (Section 14.2)

```
> restart:with(VectorCalculus):with(inttrans):with(PDETools):
> eq:=diff(C(r,z,t),t)=eta*Laplacian(C(r,z,t),cylindrical[r,theta,z]);
```

$$eq := \frac{\partial}{\partial t}\ C(r, z, t) = \frac{\eta\ \left(\frac{\partial}{\partial r}\ C(r, z, t) + r\left(\frac{\partial^2}{\partial r^2} C(r, z, t)\right) + r\left(\frac{\partial^2}{\partial z^2}\ C(r, z, t)\right)\right)}{r}$$

```
> eq1:=C(r,z,0)=c[0]:
>
> eq2:=laplace(eq,t,s):
> eq3:=subs(eq1,eq2):
> eq4:=subs(laplace(C(r,z,t),t,s)=C(r,z),eq3):
> eq5:=subs(C(r,z)=C(r),eq4):
> eq6:=dsolve(eq5):
> eq7:=subs(_C1=0,eq6):
> eq8:=subs(r=R,rhs(eq7))=c[b]/s:
> eq9:=isolate(eq8,_C2):
>
> eq10:=subs(eq9,eq7):
> eq11:=subs(c[0]=0,eq4):
> eq12:=pdsolve(eq11,HINT=f(r)*g(z)):
> eq13:=factor(build(eq12)):
> eq14:=subs(_C2=0,eq13):
> eq15:=subs(_C1=1,eq14):
> eq16:=subs(r=R,rhs(eq15))=0:
> eq17:=BesselJ(0,(-_c[1])^(1/2)*R)=0:
> eq18:=(-_c[1])^(1/2)*R=alpha[0,n]:
> eq19:=isolate(eq18,_c[1]):
> eq20:=simplify(subs(eq19,eq15),power,symbolic):
> eq21:=subs(C=C[n],_C3=A[n],_C4=B[n],eq20):
> eq22:=C(r,z)=rhs(eq10)+Sum(rhs(eq21),n=1..infinity):
> eq23:=subs(z=-H,rhs(eq22))=c[b]/s:
> eq23A:=subs(z=H,rhs(eq22))=c[b]/s:
> eq24:=subs(BesselJ(0,alpha[0,n])=0,int(subs(z=-H,rhs(eq10)
   +rhs(eq21))*BesselJ(0,alpha[0,n]/R*r)*r,r=0..R))=int(BesselJ
   (0,alpha[0,n]/R*r)*r*c[b]/s,r=0..R):
> eq24A:=subs(BesselJ(0,alpha[0,n])=0,int(subs(z=H,rhs(eq10)
   +rhs(eq21))*BesselJ(0,alpha[0,n]/R*r)*r,r=0..R))=int(BesselJ
   (0,alpha[0,n]/R*r)*r*c[b]/s,r=0..R):
> eq25:=factor(solve(eq24,eq24A, , {A[n], B[n]})):
> eq26:=subs(eq25,eq22):
> eq27:=Q(s)=2*Pi*int(int(rhs(eq10)*r,r=0..R),z=-H..H)+2*Pi*Sum
   (factor(int(int(subs(eq25,rhs(eq21))*r,r=0..R),z=-H..H)),
   n=1..infinity):
> eq28:=M(s)=c[0]*Pi*R^2*(2*H)/s-Q(s):
> eq29:=subs(eq27,eq28):
> aux1:=2*Pi*int(int(rhs(eq10)*r,r=0..R),z=-H..H):
> eq30:=coeff(convert(series(aux1,s=0,2),polynom),s,-1):
> eq31:=BesselJ(0,(-s/eta)^(1/2)*R)=0:
```

```
>
> eq32:=(-s/eta)^(1/2)*R=alpha[0,n]:
> eq33:=isolate(eq32,s):
> eq34:=numer(aux1)*exp(s*t)/diff(denom(aux1),s):
> eq35:=factor(subs(BesselJ(0,alpha[0,n])=0,simplify(subs(eq33,eq34),
   power,symbolic))):
> aux2:=-4*R^5*eta^(1/2)*(c[0]-c[b])*(-1+exp(2/eta^(1/2)*(alpha[0,n]
   ^2*eta+s*R^2)^(1/2)/R*H))/alpha[0,n]^2/(exp(2/eta^(1/2)*(alpha
   [0,n]^2*eta+s*R^2)^(1/2)/R*H)+1)/(alpha[0,n]^2*eta+s*R^2)^(3/2):
> aux3:=exp(2/eta^(1/2)*(alpha[0,n]^2*eta+s*R^2)^(1/2)/R*H)+1=exp
   (1/eta^(1/2)*(alpha[0,n]^2*eta+s*R^2)^(1/2)/R*H)*convert(exp(1/eta
   ^(1/2)*(alpha[0,n]^2*eta+s*R^2)^(1/2)/R*H)+exp(-1/eta^(1/2)
   *(alpha[0,n]^2*eta+s*R^2)^(1/2)/R*H),cosh):
>
> aux4:=-1+exp(2/eta^(1/2)*(alpha[0,n]^2*eta+s*R^2)^(1/2)/R*H)=exp
   (1/eta^(1/2)*(alpha[0,n]^2*eta+s*R^2)^(1/2)/R*H)*convert(-exp
   (-1/eta^(1/2)*(alpha[0,n]^2*eta+s*R^2)^(1/2)/R*H)+exp(1/eta^(1/2)
   *(alpha[0,n]^2*eta+s*R^2)^(1/2)/R*H),sinh):
> aux5:=subs(aux3,aux4,aux2):
> eq36:=coeff(convert(series(aux5*exp(s*t),isolate(alpha[0,n]^2*eta+s*
   R^2=0,s),2),polynom),s+alpha[0,n]^2/R^2*eta,-1):
> eq37:=cosh(1/eta^(1/2)*(alpha[0,n]^2*eta+s*R^2)^(1/2)/R*H)=0:
> eq38:=(1/eta^(1/2)*(alpha[0,n]^2*eta+s*R^2)^(1/2)/R*H)
   =I*(2*m+1)*Pi/2:
> eq39:=isolate(eq38,s):
> eq40:=numer(aux5)*exp(s*t)/diff(denom(aux5),s):
> eq41:=factor(simplify(subs(eq39,eq40),power,symbolic)) assum-
ing m::integer:
> eq42:=M(t)=simplify(2*c[0]*Pi*R^2*H-eq30-Sum(eq35,n=1..infinity)
   -Sum(2*Pi*eq36,n=1..infinity)-Sum(Sum(2*Pi*eq41,m
   =0..infinity),n=1..infinity)):
> eq43:=M(infinity)=factor(2*c[0]*Pi*R^2*H-2*Pi*R^2*c[b]*H):
> eq44:=M(t)/M(infinity)=1-Sum(Sum(factor(2*Pi*eq41/(rhs(eq43))),
   m=0..infinity),n=1..infinity):
> eq45:=-1/4*eta*(4*Pi^2*R^2*m^2+4*Pi^2*R^2*m+Pi^2*R^2+4*alpha[0,n]
   ^2*H^2)*t/H^2/R^2=factor(-eta*t/H^2*Pi^2*m^2-eta*t/
   H^2*Pi^2*m-1/4*eta*t/H^2*Pi^2)-alpha[0,n]^2/R^2*eta*t:
> eq46:=subs(eq45,eq44);
>
```

$$eq46 := \frac{M(t)}{M(\infty)} = 1 - \left(\sum_{n=1}^{\infty}\left(\sum_{m=0}^{\infty} \frac{32\ e^{-\frac{1}{4}\frac{\eta\, t\, \pi^2 (2m+1)^2}{H^2} - \frac{\alpha_{0,n}^2\, \eta\, t}{R^2}}}{\pi^2\, \alpha_{0,n}^2\, (2m+1)^2}\right)\right)$$

```
>
```

14.5.2 Maple Code: rect3D.mws (Section 14.3)

```
> restart:with(inttrans):
> eq:=diff(c(x,t),t)=eta*diff(c(x,t),x,x);
```

$$eq := \frac{\partial}{\partial t}\, c(x,t) = \eta\, \left(\frac{\partial^2}{\partial x^2}\, c(x,t)\right)$$

```
> eq1:=c(x,0)=C[ini]:
>
> eq2:=laplace(eq,t,s):
> eq3:=subs(eq1,eq2):
> eq4:=subs(laplace(c(x,t),t,s)=c(x),eq3):
> eq5:=dsolve(eq4):
> eq6:=subs(x=-L,rhs(eq5))=C[infinity]/s:
> eq7:=subs(x=L,rhs(eq5))=C[infinity]/s:
> eq8:=convert(solve({eq6, eq7}, {_C1, _C2}),cosh):
> eq9:=simplify(subs(eq8,eq5)):
> eq10:=coeff(convert(series(rhs(eq9),s=0,2),polynom),s,-1):
> eq11:=cosh(1/eta^(1/2)*s^(1/2)*L)=0:
> eq12:=1/eta^(1/2)*s^(1/2)*L=(2*n+1)*I*Pi/2:
> eq13:=isolate(eq12,s):
> eq14:=numer(rhs(eq9))*exp(s*t)/diff(denom(rhs(eq9)),s):
> eq15:=simplify(subs(eq13,eq14),power,symbolic) assuming n::integer:
> eq16:=c(x,t)=eq10+Sum(eq15,n=0..infinity):
> eq17:=M(t)=C[ini]*2*L-C[infinity]*2*L-Sum(int(eq15,x=-L..L),
   n=0..infinity):
> eq18:=M(t)/M(infinity)=1-Sum(int(eq15/(C[ini]*2*L-C[infinity]*2*L),
   x=-L..L),n=0..infinity):
> aux:=(1+exp(2*I*Pi*n))*expand(exp(-1/4*Pi*(4*Pi*eta*t*n^2+4*Pi*eta
   *t*n+Pi*eta*t+4*I*n*L^2)/L^2)) assuming n::integer:
> aux1:=factor(simplify(aux, power,symbolic)):
> aux2:=(1+exp(2*I*Pi*n))=aux1/exp(-1/4*Pi*(4*Pi*eta*t*n^2+4*Pi*eta
   *t*n+Pi*eta*t+4*I*n*L^2)/L^2):
> eq19:=simplify(factor(subs(aux2,eq18)),power,symbolic);
```

$$eq19 := \frac{M(t)}{M(\infty)} = 1 - \left(\sum_{n=0}^{\infty} \frac{8 e^{-\frac{1}{4} \frac{(2n+1)^2 \pi^2 \eta t}{L^2}}}{\pi^2 (2n+1)^2} \right)$$

```
> aux3:=Sum(8*exp(-1/4*(2*n+1)^2*Pi^2*eta/L^2*t)/Pi^2/(2*n+1)^2,
   n =0 .. infinity):
> aux4:=expand(subs(L=a/2,aux3)):
> aux5:=expand(subs(L=b/2,n=m,aux3)):
>
> aux6:=expand(subs(L=c/2,n=k,aux3)):
> eq20:=M(t)=1-factor(simplify(aux4*aux5*aux6,power,symbolic));
```

$$eq20 := M(t) = 1 - \frac{512 \left(\sum_{n=0}^{\infty} \frac{e^{-\frac{(2n+1)^2 \pi^2 \eta t}{a^2}}}{(2n+1)^2} \right) \left(\sum_{m=0}^{\infty} \frac{e^{-\frac{(2m+1)^2 \pi^2 \eta t}{b^2}}}{(2m+1)^2} \right) \left(\sum_{k=0}^{\infty} \frac{e^{-\frac{(2k+1)^2 \pi^2 \eta t}{c^2}}}{(2k+1)^2} \right)}{\pi^6}$$

```
>
>
```

14.5.3 Mathematica Code: cyl2Dmat.nb (Section 14.2)

```
eq0 = ∂_t c[r, z, t] = η Laplacian [c[r, z, t], {r, θ, z}, "Cylindrical"]//Expand
```

$$c^{(0,0,1)}[r, z, t] = \eta c^{(0,2,0)}[r, z, t] + \frac{\eta c^{(1,0,0)}[r, z, t]}{r} + \eta c^{(2,0,0)}[r, z, t]$$

eq0A = $c[r, z, 0] = c[0]$

$$c[r, z, 0] = c[0]$$

eq = $s^*c[r.z] - c[r, z, 0] = \frac{1}{r}(\eta(rc^{(0,2)}[r, z] + c^{(1,0)}[r, z] + rc^{(2,0)}[r, z]))$

$$sc[r, z] - c[r, z, 0] = \frac{\eta(rc^{(0,2)}[r, z]+c^{(1,0)}[r, z]+rc^{(2,0)}[r, z])}{r}$$

eq1 = eq/.eq0A[[1]]- > eq0A[[2]]

$$-c[0] + sc[r, z] = \frac{\eta(rc^{(0,2)}[r, z]+c^{(1,0)}[r, z]+rc^{(2,0)}[r, z])}{r}$$

eq1A = $sC[r] - c[0] = \frac{\eta(C^{(1)}[r]+rC^{(2)}[r])}{r}$

$$-c[0] + sC[r] = \frac{\eta(C'[r]+rC''[r])}{r}$$

eq1B = `Factor[DSolve[`eq1A, $C[r]$, r`]]`

$$\left\{\left\{C[r] \to \frac{c[0]+s\text{BesselJ}\left[0, \frac{ir\sqrt{s}}{\sqrt{\eta}}\right]c[1]+s\text{BesselY}\left[0, -\frac{ir\sqrt{s}}{\sqrt{\eta}}\right]C[2]}{s}\right\}\right\}$$

eq1C = $((C[r]/.\text{eq1B}[[1]])/.C[2] \to 0$

$$\frac{c[0]+s\text{BesselJ}\left[0, \frac{ir\sqrt{s}}{\sqrt{\eta}}\right]C[1]}{s}$$

eq1D = (eq1C/.$r \to R$) = $c[b]/s$

$$\frac{c[0]+s\text{BesselJ}\left[0, \frac{iR\sqrt{s}}{\sqrt{\eta}}\right]C[1]}{s} = \frac{c[b]}{s}$$

eq1E = (`Solve[`eq1D, $C[1]$`])[[1]]`

$$\left\{C[1] \to \frac{-c[0]+c[b]}{s\text{BesselJ}\left[0, \frac{iR\sqrt{s}}{\sqrt{\eta}}\right]}\right\}$$

eq1F = eq1C/.eq1E

$$\frac{c[0]+\frac{\text{BesselJ}\left[0, \frac{ir\sqrt{s}}{\sqrt{\eta}}\right](-c[0]+c[b])}{\text{BesselJ}\left[0, \frac{iR\sqrt{s}}{\sqrt{\eta}}\right]}}{s}$$

eq2A = (eq1/.$c[0] \to 0$)

$$sc[r, z] = \frac{\eta(rc^{(0,2)}[r, z]+c^{(1,0)}[r, z]+rc^{(2,0)}[r, z])}{r}$$

eq2 = eq2A/.$c \to$ `Function[`$\{r, z\}$, $f[r]^*g[z]$`]`

$$sf[r]g[z] = \frac{\eta(g[z]f'[r]+rg[z]f''[r]+rf[r]g''[z])}{r}$$

eq3 = `Expand[`eq2[[1]]/($f[r]^*g[z]$)`]` = `Expand[`eq2[[2]]/($f[r]^*g[z]$)`]`

$$s = \frac{\eta f'[r]}{rf[r]} + \frac{\eta f''[r]}{f[r]} + \frac{\eta g''[z]}{g[z]}$$

eq3A = $\frac{f'[r]}{rf[r]} + \frac{f''[r]}{f[r]} = c[1]$

$$\frac{f'[r]}{rf[r]} + \frac{f''[r]}{f[r]} = c[1]$$

eq4 = $s = c[1]^*\eta + \frac{\eta g''[z]}{g[z]}$

$$s = \eta c[1] + \frac{\eta g''[z]}{g[z]}$$

eq5 = `DSolve[`eq3A, $f[r]$, r`]`

$\{\{f[r] \to \texttt{BesselJ}[0,\ ir\sqrt{c[1]}]C[1] + \texttt{BesselY}\ [0,\ -ir\sqrt{c[1]}]C[2]\}\}$

eq5A = (Dsolve[eq4, g[z], z]/.C[1]−> A[1])/.C[2] → B[1]

$$\left\{\left\{g[z] \to e^{\frac{z\sqrt{s-\eta c[1]}}{\sqrt{\eta}}} A[1] + e^{-\frac{z\sqrt{s-\eta c[1]}}{\sqrt{\eta}}} B[1]\right\}\right\}$$

eq6 = (f[r]/.eq5) * (g[z]/.eq5A)[[1]]

$$\left\{\left(e^{\frac{z\sqrt{s-\eta c[1]}}{\sqrt{\eta}}} A[1] + e^{-\frac{z\sqrt{s-\eta c[1]}}{\sqrt{\eta}}} B[1]\ (\texttt{BesselJ}\ [0,\ ir\sqrt{c[1]}]\ C[1]) + \texttt{BesselY}\left[0,\ -ir\sqrt{c[1]}\right] C[2]\right)\right\}$$

eq6A = ((eq6/.C[2] → 0)/.C[1] → 1)[[1]]

$$\left(e^{\frac{z\sqrt{s-\eta c[1]}}{\sqrt{\eta}}} A[1] + e^{-\frac{z\sqrt{s-\eta c[1]}}{\sqrt{\eta}}} B[1]\right)\texttt{BesselJ}[0,\ ir\sqrt{c[1]}]$$

eq6C = (eq6A/.r → R) = 0

$$\left(e^{\frac{z\sqrt{s-\eta c[1]}}{\sqrt{\eta}}} A[1] + e^{-\frac{z\sqrt{s-\eta c[1]}}{\sqrt{\eta}}} B[1]\right)\texttt{BesselJ}[0,\ iR\sqrt{c[1]}] = 0$$

eq6D = BesselJ[0, $iR\sqrt{c[1]}$] = 0

BesselJ[0, $iR\sqrt{c[1]}$] = 0

eq6E = $iR\sqrt{c[1]} = \alpha[0,\ n]$

$iR\sqrt{c[1]} = \alpha[0,\ n]$

eq6F = Solve[eq6E, c[1]]

$$\left\{\left\{c[1] \to -\frac{\alpha[0,n]^2}{R^2}\right\}\right\}$$

eq6J = Simplify[eq6A/.eq6F, α[0, n] > 0 & & R > 0]

$$\left\{e^{-\frac{\sqrt{s+\frac{\eta\alpha[0,n]^2}{R^2}}}{\sqrt{\eta}}}\left(e^{\frac{2z\sqrt{s+\frac{\eta\alpha[0,n]^2}{R^2}}}{\sqrt{\eta}}} A[1] + B[1]\right)\texttt{BesselJ}\left[0,\ \frac{r\alpha[0,n]}{R}\right]\right\}$$

eq6J1 = (eq6J/.A[1] → A[n])/.B[1]− > B[n]

$$\left\{e^{-\frac{z\sqrt{s+\frac{\eta\alpha[0,n]^2}{R^2}}}{\sqrt{\eta}}}\left(e^{\frac{2z\sqrt{s+\frac{\eta\alpha[0,n]^2}{R^2}}}{\sqrt{\eta}}} A[n] + B[n]\right)\texttt{BesselJ}\left[0,\ \frac{r\alpha[0,n]}{R}\right]\right\}$$

eq6K =
C[r, z] =
eq1F+

$$\sum_{n=1}^{\infty}\texttt{BesselJ}\left[0,\ \frac{r\alpha[0,n]}{R}\right]\left(A[n]\,\mathrm{Cos}\left[\frac{z\sqrt{-s-\frac{\eta\alpha[0,n]^2}{R^2}}}{\sqrt{\eta}}\right] + B[n]\,Sin\left[\frac{z\sqrt{-s-\frac{\eta\alpha[0,n]^2}{R^2}}}{\sqrt{\eta}}\right]\right)$$

$$C[r, z] = \frac{c[0] + \frac{\text{BesselJ}\left[0, \frac{ir\sqrt{s}}{\sqrt{\eta}}\right](-c[0]+c[b])}{\text{BesselJ}\left[0, \frac{iR\sqrt{s}}{\sqrt{\eta}}\right]}}{s} + \sum_{n=1}^{\infty} \text{BesselJ}\left[0, \frac{r\alpha[0,n]}{R}\right] \times$$

$$\left(A[n]\text{Cos}\left[\frac{z\sqrt{-s-\frac{\eta\alpha[0,n]^2}{R^2}}}{\sqrt{n}}\right] + B[n]Sin\left[\frac{z\sqrt{-s-\frac{\eta\alpha[0,n]^2}{R^2}}}{\sqrt{n}}\right]\right)$$

eq7 = ((eq6K[[2]])/.$z \to -H$) = $c[b]/s$

$$\frac{c[0] + \frac{\text{BesselJ}\left[0, \frac{ir\sqrt{s}}{\sqrt{\eta}}\right](-c[0]+c[b])}{\text{BesselJ}\left[0, \frac{iR\sqrt{s}}{\sqrt{\eta}}\right]}}{s} + \sum_{n=1}^{\infty} \text{BesselJ}\left[0, \frac{r\alpha[0,n]}{R}\right] \times$$

$$\left(A[n]\text{Cos}\left[\frac{H\sqrt{-s-\frac{\eta\alpha[0,n]^2}{R^2}}}{\sqrt{\eta}}\right] - B[n]Sin\left[\frac{H\sqrt{-s-\frac{\eta\alpha[0,n]^2}{R^2}}}{\sqrt{\eta}}\right]\right) = \frac{c[b]}{s}$$

eq7A = ((eq6K[[2]])/.$z \to H$) = $c[b]/s$

$$\frac{c[0] + \frac{\text{BesselJ}\left[0, \frac{ir\sqrt{s}}{\sqrt{\eta}}\right](-c[0]+c[b])}{\text{BesselJ}\left[0, \frac{iR\sqrt{s}}{\sqrt{\eta}}\right]}}{s} +$$

$$\sum_{n=1}^{\infty} \text{BesselJ}\left[0, \frac{r\alpha[0,n]}{R}\right]\left(A[n]\text{Cos}\left[\frac{H\sqrt{-s-\frac{\eta\alpha[0,n]^2}{R^2}}}{\sqrt{\eta}}\right] + B[n]Sin\left[\frac{H\sqrt{-s-\frac{\eta\alpha[0,n]^2}{R^2}}}{\sqrt{\eta}}\right]\right) = \frac{c[b]}{s}$$

eq8 =

$$\left(\left(\text{Integrate}\left[((\text{eq1F} + \text{eq6J1}[[1]])/.z \to -H)*r*\text{BesselJ}\left[0, \frac{r\alpha[0,n]}{R}\right], \{r, 0, R\}\right]\right)/.$$

$$\text{BesselJ}[0, \alpha[0, n]] \to 0\Big) =$$

$$\text{Integrate}\left[r*\text{BesselJ}\left[0, \frac{r\alpha[0,n]}{R}\right] * \frac{c[b]}{s}, \{r, 0, R\}\right]$$

$$\frac{1}{2}R^2\left(e^{-\frac{H\sqrt{s+\frac{\eta\alpha[0,n]^2}{R^2}}}{\sqrt{\eta}}} A[n]\text{BesselJ}[1, \alpha[0,n]]^2 + e^{\frac{H\sqrt{s+\frac{\eta\alpha[0,n]^2}{R^2}}}{\sqrt{\eta}}} B[n]\text{BesselJ}[1, \alpha[0,n]]^2 + \left(2\left(R^2 s\text{BesselJ}[1,\alpha[0,n]]c[0]\text{Hypergeometric0F1Regularized}\left[1, \frac{R^2 s}{4\eta}\right] + \eta\ \text{BesselJ}[1,\alpha[0,n]]c[b]\text{Hypergeometric0F1Regularized}\left[1, \frac{R^2 s}{4\eta}\right]\alpha[0,n]^2\right)\right)\Big/ s\text{Hypergeometric0F1Regularized}\left[1, \frac{R^2 s}{4\eta}\right]\alpha[0,n](R^2 s + \eta\alpha[0,n]^2))\right) =$$

$$\frac{R^2\text{BesselJ}[1,\alpha[0,n]]c[b]}{s\alpha[0,n]}$$

eq9 =

$$((\text{Integrate}[((\text{eq1F} + \text{eq6J1}[[1]]])/.z \to H)*r*\text{BesselJ}\left[0, \frac{r\alpha[0,n]}{R}\right],$$

$$\{r, 0, R\}])/.\text{BesselJ}[0, \alpha[0,n]] \to 0) =$$

$$\text{Integrate}\left[r*\text{BesselJ}\left[0, \frac{r\alpha[0,n]}{R}\right] * \frac{c[b]}{s}, \{r, 0, R\}\right]$$

$$\frac{1}{2}R^2\left(e^{\frac{H\sqrt{s+\eta\alpha[0,n]^2}}{\frac{R^2}{\sqrt{\eta}}}}A[n]\texttt{BesselJ}[1,\alpha[0,n]]^2+e^{-\frac{H\sqrt{s+\eta\alpha[0,n]^2}}{\frac{R^2}{\sqrt{\eta}}}}B[n]\texttt{BesselJ}[1,\alpha[0,n]]^2+\left(2\left(R^2 s\texttt{BesselJ}[1,\alpha[0,n]]c[0]\texttt{Hypergeometric0F1Regularized}\left[1,\frac{R^2 s}{4\eta}\right]+\eta\texttt{BesselJ}[1,\alpha[0,n]]c[b]\texttt{Hypergeometric0F1Regularized}\left[1,\frac{R^2 s}{4\eta}\right]\alpha[0,n]^2\right)\right)\Big/\left(s\texttt{Hypergeometric0F1Regularized}\left[1,\frac{R^2 s}{4\eta}\right]\alpha[0,n](R^2 s+\eta\alpha[0,n]^2)\right)\right)=$$

$$\frac{R^2\texttt{BesselJ}[1,\alpha[0,n]]c[b]}{s\alpha[0,n]}$$

eq10 = `Solve[{eq8, eq9}, {A[n], B[n]}]`

$$\left\{\left\{A[n]\to-\frac{2e^{\frac{H\sqrt{s+\frac{\eta\alpha[0,n]^2}{R^2}}}{\sqrt{\eta}}}R^2(c[0]-c[b])}{\left(1+e^{\frac{2H\sqrt{s+\frac{\eta\alpha[0,n]^2}{R^2}}}{\sqrt{\eta}}}\right)\texttt{BesselJ}[1,\alpha[0,n]]\alpha[0,n](R^2 s+\eta\alpha[0,n]^2)},\right.\right.$$
$$\left.\left.B[n]\to-\frac{2e^{\frac{H\sqrt{s+\frac{\eta\alpha[0,n]^2}{R^2}}}{\sqrt{\eta}}}R^2(c[0]-c[b])}{\left(1+e^{\frac{2H\sqrt{s+\frac{\eta\alpha[0,n]^2}{R^2}}}{\sqrt{\eta}}}\right)\texttt{BesselJ}[1,\alpha[0,n]]\alpha[0,n](R^2 s+\eta\alpha[0,n]^2)}\right\}\right\}$$

eq11 = (eq6K/.eq10)[[1]]

$$C[r,z]=\frac{c[0]+\frac{\texttt{BesselJ}\left[0,\frac{ir\sqrt{s}}{\sqrt{\eta}}\right](-c[0]+c[b])}{\texttt{BesselJ}\left[0,\frac{iR\sqrt{s}}{\sqrt{\eta}}\right]}}{s}+$$
$$\sum_{n=1}^{\infty}\texttt{BesselJ}\left[0,\frac{r\alpha[0,n]}{R}\right]\left(-\frac{2e^{\frac{H\sqrt{s+\frac{\eta\alpha[0,n]^2}{R^2}}}{\sqrt{\eta}}}R^2(c[0]-c[b])\text{Cos}\left[\frac{z\sqrt{-s-\frac{\eta\alpha[0,n]^2}{R^2}}}{\sqrt{\eta}}\right]}{\left(1+e^{\frac{2H\sqrt{\frac{s+\eta\alpha[0,n]^2}{R^2}}}{\sqrt{\eta}}}\right)\texttt{BesselJ}[1,\alpha[0,n]]\alpha[0,n](R^2 s+\eta\alpha[0,n]^2)}-\frac{2e^{\frac{H\sqrt{s+\frac{\eta\alpha[0,n]^2}{R^2}}}{\sqrt{\eta}}}R^2(c[0]-c[b])Sin\left[\frac{z\sqrt{-s-\frac{\eta\alpha[0,n]^2}{R^2}}}{\sqrt{\eta}}\right]}{\left(1+e^{\frac{2H\sqrt{s+\frac{\eta\alpha[0,n]^2}{R^2}}}{\sqrt{\eta}}}\right)\texttt{BesselJ}[1,\alpha[0,n]]\alpha[0,n](R^2 s+\eta\alpha[0,n]^2)}\right)$$

eq12 =
Simplify[
Expand[2 * Pi * Integrate[Integrate[$r *$ eq1F, $\{r,\ 0,\ R\}$], $\{z,\ -H,\ H\}$]]] +
(2 * Pi) *

$$\sum_{n=1}^{\infty}\left(2R^2\texttt{Integrate}\left[r*\texttt{BesselJ}\left[0,\ \frac{r\alpha[0,\ n]}{R}\right],\ \{r,\ 0,\ R\}\right](-c[0]+c[b])\right.$$

$$\texttt{Integrate}\left[\text{Cos}\left[\frac{z\sqrt{-s-\frac{\eta*\alpha[0,\ n]^2}{R^2}}}{\sqrt{\eta}}\right],\ \{z,\ -H,\ H\}\right]$$

$$\left.\text{Sec}\left[\frac{H\sqrt{-s-\frac{\eta*\alpha[0,n]^2}{R^2}}}{\sqrt{\eta}}\right]\right)$$

$$(\texttt{BesselJ}[1,\ \alpha[0,\ n]]\alpha[0,\ n](R^2s+\ \eta\alpha[0,\ n]^2))$$

$$\left(2H\pi R^2\left(c[0]\left(\texttt{Hypergeometric0F1Regularized}\left[1,\ \frac{R^2s}{4\eta}\right]-\texttt{Hypergeometric0F1Regularized}\left[2,\ \frac{R^2s}{4\eta}\right]\right)+c[b]\texttt{Hypergeometric0F1Regularized}\left[2,\ \frac{R^2s}{4\eta}\right]\right)\right)/$$

$$\left(s\texttt{Hypergeometric0F1Regularized}\left[1,\ \frac{R^2s}{4\eta}\right]\right)$$

$$+\ 2\pi\sum_{n=1}^{\infty}\frac{4R^4\ \sqrt{n}(-c[0]+c[b])\text{Tan}\left[\frac{H\sqrt{-s-\frac{\eta\alpha[0,\ n]^2}{R^2}}}{\sqrt{\eta}}\right]}{\alpha[0,\ n]^2(R^2s+\eta\alpha[0,\ n]^2)\sqrt{-s-\frac{\eta\alpha[0,n]^2}{R^2}}}$$

eq13 = $M[s] = 2 *$ Pi $* R^2 * c[0] * H/s -$ eq12

$$M[s]=\frac{2H\ \pi R^2\,c[0]}{s}-\left(2H\pi R^2\left(c[0]\left(\texttt{Hypergeometric0F1Regularized}\left[1,\ \frac{R^2s}{4\eta}\right]-\texttt{Hypergeometric0F1Regularized}\left[2,\ \frac{R^2s}{4\eta}\right]\right)+c[b]\texttt{Hypergeometric0F1Regularized}\left[2,\ \frac{R^2s}{4\eta}\right]\right)\right)/$$

$$\left(s\texttt{Hypergeometric0F1Regularized}\left[1,\ \frac{R^2s}{4\eta}\right]\right)$$

$$-2\pi\sum_{n=1}^{\infty}\frac{4R^4\ \sqrt{n}(-c[0]+c[b])\text{Tan}\left[\frac{H\sqrt{-s-\frac{\eta\alpha[0,\ n]^2}{R^2}}}{\sqrt{\eta}}\right]}{\alpha[0,\ n]^2(R^2s+\ \eta\alpha[0,\ n]^2)\sqrt{-s-\frac{\eta\alpha[0,\ n]^2}{R^2}}}$$

eq14 =

$$(2H\pi R^2$$

$$\left(c[0]\left(\texttt{Hypergeometric0F1Regularized}\left[1,\ \frac{R^2s}{4\eta}\right]-\right.\right.$$

$$\left.\texttt{Hypergeometric0F1Regularized}\left[2,\ \frac{R^2s}{4\eta}\right]\right)+$$

$$\left.c[b]\texttt{Hypergeometric0F1Regularized}\left[2,\ \frac{R^2s}{4\eta}\right]\right))$$

$$\left(s\texttt{Hypergeometric0F1Regularized}\left[1,\ \frac{R^2s}{4\eta}\right]\right)$$

$$\left(2H\pi R^2\left(c[0]\left(\texttt{Hypergeometric0F1Regularized}\left[1,\ \frac{R^2s}{4\eta}\right]-\texttt{Hypergeometric0F1Regularized}\left[2,\ \frac{R^2s}{4\eta}\right]\right)+c[b]\texttt{Hypergeometric0F1Regularized}\left[2,\ \frac{R^2s}{4\eta}\right]\right)\right)/$$

$$\left(s\texttt{Hypergeometric0F1Regularized}\left[1,\ \frac{R^2s}{4\eta}\right]\right)$$

eq15 = Residue[eq14, {s, 0}]

$$2H\pi R^2 c[b]$$

$$\text{eq16} = \left(4R^4\sqrt{\eta}(-c[0]+c[b])\,\text{Sin}\left[\frac{H\sqrt{-s-\frac{\eta*\alpha[0,n]^2}{R^2}}}{\sqrt{\eta}}\right]\right)$$

$$\left(\alpha[0,n]^2\sqrt{-s-\frac{\eta*\alpha[0,n]^2}{R^2}}\,(R^2 s+\eta\alpha[0,n]^2)^*\text{Cos}\left[\frac{H\sqrt{-s-\frac{\eta*\alpha[0,n]^2}{R^2}}}{\sqrt{\eta}}\right]\right)$$

$$\frac{4R^4\sqrt{\eta}(-c[0]+c[b])\text{Tan}\left[\frac{H\sqrt{-s-\frac{\eta\alpha[0,n]^2}{R^2}}}{\sqrt{\eta}}\right]}{\alpha[0,n]^2(R^2 s+\eta\alpha[0,n]^2)\sqrt{-s-\frac{\eta\alpha[0,n]^2}{R^2}}}$$

$$\text{Cos}\left[\frac{H\sqrt{-s-\frac{\eta*\alpha[0,n]^2}{R^2}}}{\sqrt{\eta}}\right]=0$$

$$\text{Cos}\left[\frac{H\sqrt{-s-\frac{\eta\alpha[0,n]^2}{R^2}}}{\sqrt{\eta}}\right]=0$$

$$\text{eq17} = \frac{H\sqrt{-s-\frac{\eta*\alpha[0,n]^2}{R^2}}}{\sqrt{\eta}} == (2*m+1)*\text{Pi}/2$$

$$\frac{H\sqrt{-s-\frac{\eta\alpha[0,n]^2}{R^2}}}{\sqrt{\eta}} == \frac{1}{2}(1+2m)\pi$$

eq18 = Solve[eq17, s]

$$\left\{\left\{s\to\frac{-\pi^2R^2\eta-4m\pi^2R^2\eta-4m^2\pi^2R^2\eta-4H^2\eta\alpha[0,n]^2}{4H^2R^2}\right\}\right\}$$

$$\text{eq19} = \left(4R^4\sqrt{\eta}(-c[0]+c[b])\,\text{Sin}\left[\frac{H\sqrt{-s-\frac{\eta*\alpha[0,n]^2}{R^2}}}{\sqrt{\eta}}\right]\right)*$$

$$\text{Exp}[s*t]/$$

$$\left(\partial_s\left(\alpha[0,n]^2\sqrt{-s-\frac{\eta*\alpha[0,n]^2}{R^2}}(R^2 s+\eta\alpha[0,n]^2)^*\text{Cos}\left[\frac{H\sqrt{-s-\frac{\eta*\alpha[0,n]^2}{R^2}}}{\sqrt{\eta}}\right]\right)\right)$$

$$\left(4e^{st}R^4\sqrt{\eta}(-c[0]+c[b])\,\text{Sin}\left[\frac{H\sqrt{-s-\frac{\eta\alpha[0,n]^2}{R^2}}}{\sqrt{\eta}}\right]\right)\Bigg/$$

$$\left(\frac{H\text{Sin}\left[\frac{H\sqrt{-s-\frac{\eta\alpha[0,n]^2}{R^2}}}{\sqrt{\eta}}\right]\alpha[0,n]^2(R^2 s+\eta\alpha[0,n]^2)}{2\sqrt{\eta}}-\frac{\text{Cos}\left[\frac{H\sqrt{-s-\frac{\eta\alpha[0,n]^2}{R^2}}}{\sqrt{\eta}}\right]\alpha[0,n]^2(R^2 s+\eta\alpha[0,n]^2)}{2\sqrt{-s-\frac{\eta\alpha[0,n]^2}{R^2}}}+\right.$$

$$\left.R^2\text{Cos}\left[\frac{H\sqrt{-s-\frac{\eta\alpha[0,n]^2}{R^2}}}{\sqrt{\eta}}\right]\alpha[0,n]^2\sqrt{-s-\frac{\eta\alpha[0,n]^2}{R^2}}\right)$$

eq20 = Simplify[eq19/.eq18, $m>0$ & & $H>0$ & & $\eta>0$]

$$\left\{\frac{32e^{-\frac{t\eta\left((1+2m)^2\pi^2R^2+4H^2\alpha[0,n]^2\right)}{4H^2R^2}}HR^2(c[0]-c[b])\mathrm{Cos}[m\pi]}{(1+2m)\pi((1+2m)\pi\mathrm{Cos}[m\pi]+6\mathrm{Sin}[m\pi])\alpha[0,n]^2}\right\}$$

eq21 = Simplify[(eq20/.Sin[$m\pi$] → 0)/.Cos[$m\pi$] → $(-1)^m$]

$$\left\{\frac{32e^{-\frac{t\eta\left((1+2m)^2\pi^2R^2+4H^2\alpha[0,n]^2\right)}{4H^2R^2}}HR^2(c[0]-c[b])}{(1+2m)^2\pi^2\alpha[0,n]^2}\right\}$$

eq22 =

$$M[t]=2H\pi R^2c[0]-\mathrm{eq15}-\sum_{n=1}^{\infty}\sum_{m=0}^{\infty}\frac{2^*\mathrm{Pi}^*32e^{-\frac{t\eta((1+2m)^2\pi^2R^2+4H^2\alpha[0,n]^2)}{4H^2R^2}}HR^2(c[0]-c[b])}{(1+2m)^2\pi^2\alpha[0,n]^2}$$

$$M[t]=2H\pi R^2c[0]-2H\pi R^2c[b]-\sum_{n=1}^{\infty}\sum_{m=0}^{\infty}\frac{64e^{-\frac{t\eta((1+2m)^2\pi^2R^2+4H^2\alpha[0,n]^2)}{4H^2R^2}}HR^2(c[0]-c[b])}{(1+2m)^2\pi\alpha[0,n]^2}$$

$$\mathrm{eq23}=\mathrm{Factor}\left[\left(\frac{64e^{-\frac{t\eta\left((1+2m)^2\pi^2R^2+4H^2\alpha[0,n]^2\right)}{4H^2R^2}}HR^2(c[0]-c[b])}{(1+2m)^2\pi\alpha[0,n]^2}\right)\right.$$

$$\left.(2H\pi R^2c[0]-2H\pi R^2c[b])\right]$$

$$\frac{32e^{-\frac{t\eta((1+2m)^2\pi^2R^2+4H^2\alpha[0,n]^2)}{4H^2R^2}}}{(1+2m)^2\pi^2\alpha[0,n]^2}$$

$$\mathrm{eq24}=M[t]/M[\infty]=1-\sum_{n=1}^{\infty}\sum_{m=0}^{\infty}\frac{32e^{-\frac{t\eta((1+2m)^2\pi^2R^2+4H^2\alpha[0,n]^2)}{4H^2R^2}}}{(1+2m)^2\pi^2\alpha[0,n]^2}$$

$$\frac{M[t]}{M[\infty]}=1-\sum_{n=1}^{\infty}\sum_{m=0}^{\infty}\frac{32e^{-\frac{t\eta((1+2m)^2\pi^2R^2+4H^2\alpha[0,n]^2)}{4H^2R^2}}}{(1+2m)^2\pi^2\alpha[0,n]^2}$$

14.5.4 Mathematica Code: rec3Dmat.nb (Section 14.3)

eq0 = $\partial_t c[x,\ t] = \eta * \partial_{x,x}c[x,\ t]$

$c^{(0,1)}[x,\ t] = \eta c^{(2,0)}[x,\ t]$

eq0A = $c[x,\ 0] = C[0]$

$c[x,\ 0] = C[0]$

eq = $s * c[x] - c[x,\ 0] = \eta * \partial_{x,x}c[x]$

$sc[x] - c[x,\ 0] = \eta c''[x]$

eq1 = eq/.eq0A[[1]]− > eq0A[[2]]

$sc[x] - C[0] = \eta c''[x]$

eq1B = (`Factor[DSolve[`eq1, $c[x], x$]])[[1]]

$$\left\{c[x] \to \frac{e^{-\frac{\sqrt{sx}}{\sqrt{\eta}}}\left(e^{\frac{\sqrt{sx}}{\sqrt{\eta}}}C[0]+e^{\frac{2\sqrt{sx}}{\sqrt{\eta}}}sC[1]+sC[2]\right)}{s}\right\}$$

$$\text{eq1C} = \frac{C[0]}{s} + e^{\frac{\sqrt{sx}}{\sqrt{\eta}}}C[1] + e^{-\frac{\sqrt{sx}}{\sqrt{\eta}}}C[2]$$

$$\frac{C[0]}{s} + e^{\frac{\sqrt{sx}}{\sqrt{\eta}}}C[1] + e^{-\frac{\sqrt{sx}}{\sqrt{\eta}}}C[2]$$

eq1D = (eq1C/.$x \to -L$) = $c[b]/s$

$$\frac{C[0]}{s} + e^{-\frac{L\sqrt{s}}{\sqrt{\eta}}}C[1] + e^{\frac{L\sqrt{s}}{\sqrt{\eta}}}C[2] = \frac{c[b]}{s}$$

eq1DA = (eq1C/.$x \to L$) = $c[b]/s$

$$\frac{C[0]}{s} + e^{\frac{L\sqrt{s}}{\sqrt{\eta}}}C[1] + e^{-\frac{L\sqrt{s}}{\sqrt{\eta}}}C[2] = \frac{c[b]}{s}$$

eq1E = `Solve[{`eq1D, eq1DA`}, {`$C[1]$, $C[2]$`}]`

$$\left\{\left\{C[1] \to \frac{e^{\frac{L\sqrt{s}}{\sqrt{\eta}}}(c[b]-C[0])}{\left(1+e^{\frac{2L\sqrt{s}}{\sqrt{\eta}}}\right)s},\ C[2] \to \frac{e^{\frac{L\sqrt{s}}{\sqrt{\eta}}}(c[b]-C[0])}{\left(1+e^{\frac{2L\sqrt{s}}{\sqrt{\eta}}}\right)s}\right\}\right\}$$

eq1F = (`Factor[`eq1C/.eq1E])[[1]]

$$\frac{1}{\left(1+e^{\frac{2L\sqrt{s}}{\sqrt{\eta}}}\right)s}e^{-\frac{\sqrt{sx}}{\sqrt{\eta}}}\left(\begin{array}{l} e^{\frac{L\sqrt{s}}{\sqrt{\eta}}+\frac{2\sqrt{sx}}{\sqrt{\eta}}}c[b] + e^{\frac{L\sqrt{s}}{\sqrt{\eta}}}c[b] + e^{\frac{2L\sqrt{s}}{\sqrt{\eta}}+\frac{\sqrt{sx}}{\sqrt{\eta}}}C[0]- \\ e^{\frac{L\sqrt{s}}{\sqrt{\eta}}+\frac{2\sqrt{sx}}{\sqrt{\eta}}}C[0] - e^{\frac{L\sqrt{s}}{\sqrt{\eta}}}C[0] + e^{\frac{\sqrt{sx}}{\sqrt{\eta}}}C[0]\end{array}\right)$$

eq2A = `Normal[Series[`eq1F, `{`s, 0, −1`}]]`*s

$c[b]$

$$\text{eq2} = \left(1 + e^{\frac{2L\sqrt{s}}{\sqrt{\eta}}}\right) -> 2^{*}e^{\frac{L\sqrt{s}}{\sqrt{\eta}}}{}^{*}\text{Cosh}\left[\frac{L\sqrt{s}}{\sqrt{\eta}}\right]$$

$$1 + e^{\frac{2L\sqrt{s}}{\sqrt{\eta}}} \to 2e^{\frac{L\sqrt{s}}{\sqrt{\eta}}}\text{Cosh}\left(\frac{L\sqrt{s}}{\sqrt{\eta}}\right)$$

eq3 = eq1F/.eq2

$$\frac{1}{2s}e^{-\frac{L\sqrt{s}}{\sqrt{\eta}}-\frac{\sqrt{sx}}{\sqrt{\eta}}}\left(\begin{array}{l} e^{\frac{L\sqrt{s}}{\sqrt{\eta}}+\frac{2\sqrt{sx}}{\sqrt{\eta}}}c[b] + e^{\frac{L\sqrt{s}}{\sqrt{\eta}}}c[b] + e^{\frac{2L\sqrt{s}}{\sqrt{\eta}}+\frac{\sqrt{sx}}{\sqrt{\eta}}}C[0]- \\ e^{\frac{L\sqrt{s}}{\sqrt{\eta}}+\frac{2\sqrt{sx}}{\sqrt{\eta}}}C[0] - e^{\frac{L\sqrt{s}}{\sqrt{\eta}}}C[0] + e^{\frac{\sqrt{sx}}{\sqrt{\eta}}}C[0]\end{array}\right)\text{Sech}\left[\frac{L\sqrt{s}}{\sqrt{\eta}}\right]$$

$$\text{eq3A} = \text{Cosh}\left[\frac{L\sqrt{s}}{\sqrt{\eta}}\right] = 0$$

$$\text{Cosh}\left[\frac{L\sqrt{s}}{\sqrt{\eta}}\right] = 0$$

eq4 = $\frac{L\sqrt{s}}{\sqrt{\eta}}$ = `Sqrt[-1]` ∗ (2 ∗ n + 1) ∗ `Pi/2`

$$\frac{L\sqrt{s}}{\sqrt{\eta}} = \frac{1}{2}i(1 + 2n)\pi$$

eq5 = `Solve[`eq4, s]

$$\left\{\left\{s \to \frac{-\pi^2\eta-4n\pi^2\eta-4n^2\pi^2\eta}{4L^2}\right\}\right\}$$

eq5A =

$$\left(e^{-\frac{L\sqrt{s}}{\sqrt{\eta}}-\frac{\sqrt{sx}}{\sqrt{\eta}}}\left(e^{\frac{L\sqrt{s}}{\sqrt{\eta}}+\frac{2\sqrt{sx}}{\sqrt{\eta}}}c[b]+e^{\frac{L\sqrt{s}}{\sqrt{\eta}}}c[b]+e^{\frac{2L\sqrt{s}}{\sqrt{\eta}}+\frac{\sqrt{sx}}{\sqrt{\eta}}}C[0]-e^{\frac{L\sqrt{s}}{\sqrt{\eta}}+\frac{2\sqrt{sx}}{\sqrt{\eta}}}C[0]-e^{\frac{L\sqrt{s}}{\sqrt{\eta}}}C[0]+e^{\frac{\sqrt{sx}}{\sqrt{\eta}}}C[0]\right)\right)*$$

$$\texttt{Exp}[s*t]\left(\partial_s\left(2*s*\texttt{Cosh}\left[\frac{L\sqrt{s}}{\sqrt{\eta}}\right]\right)\right)$$

$$\left(e^{st-\frac{L\sqrt{s}}{\sqrt{\eta}}-\frac{\sqrt{sx}}{\sqrt{\eta}}}\left(e^{\frac{L\sqrt{s}}{\sqrt{\eta}}+\frac{2\sqrt{sx}}{\sqrt{\eta}}}c[b]+e^{\frac{L\sqrt{s}}{\sqrt{\eta}}}c[b]+e^{\frac{2L\sqrt{s}}{\sqrt{\eta}}+\frac{\sqrt{sx}}{\sqrt{\eta}}}C[0]-e^{\frac{L\sqrt{s}}{\sqrt{\eta}}+\frac{2\sqrt{sx}}{\sqrt{\eta}}}C[0]-e^{\frac{L\sqrt{s}}{\sqrt{\eta}}}C[0]+e^{\frac{\sqrt{sx}}{\sqrt{\eta}}}C[0]\right)\right)\Big/$$

$$\left(2\mathrm{Cosh}\left[\frac{L\sqrt{s}}{\sqrt{\eta}}\right]+\frac{L\sqrt{s}\mathrm{Sinh}\left[\frac{L\sqrt{s}}{\sqrt{\eta}}\right]}{\sqrt{\eta}}\right)$$

eq6 = Simplify[eq5A / .eq5, $n > 0$ & & $L > 0$ & & $\eta > 0$]

$$\left\{\left(e^{-\frac{(1+2n)\pi(2iLx+(1+2n)\pi t\eta)}{4L^2}}\left(-4e^{in\pi}\left(1+e^{\frac{i(1+2n)\pi x}{L}}\right)c[b]+4\left(e^{in\pi}+ie^{\frac{i(1+2n)\pi x}{2L}}+ie^{\frac{i(1+2n)\pi(2L+x)}{2L}}-ie^{\frac{i(1+2n)\pi(L+2x)}{2L}}\right)C[0]\right)\right)\Big/(4i+\pi+2n\pi+e^{2in\pi}(-4i+\pi+2n\pi))\right\}$$

eq7 = (eq6 / .Sin[$n\pi$] → 0) / .Sec[$n\pi$] → (−1) ^ n

$$\left\{\left(e^{-\frac{(1+2n)\pi(2iLx+(1+2n)\pi t\eta)}{4L^2}}\left(-4e^{in\pi}\left(1+e^{\frac{i(1+2n)\pi x}{L}}\right)c[b]+4\left(e^{in\pi}+ie^{\frac{i(1+2n)\pi x}{2L}}+ie^{\frac{i(1+2n)\pi(2L+x)}{2L}}-ie^{\frac{i(1+2n)\pi(L+2x)}{2L}}\right)C[0]\right)\right)\Big/(4i+\pi+2n\pi+e^{2in\pi}(-4i+\pi+2n\pi))\right\}$$

eq8 =

$C[x,t]$ =

$$eq2A+\sum_{n=0}^{\infty}\frac{1}{(1+2n)\pi}\left((-1)^n e^{-\frac{(1+2n)\pi(2iL^2+2iLx+(1+2n)\pi t\eta)}{4L^2}}\left(-2ie^{in\pi}\left(1+e^{\frac{i(1+2n)\pi x}{L}}\right)c[b]+2\left(ie^{in\pi}-e^{\frac{i(1+2n)\pi x}{2L}}-e^{\frac{i(\pi+2n\pi)(2L+x)}{2L}}+e^{\frac{i(\pi+2n\pi)(L+2x)}{2L}}\right)C[0]\right)\right);$$

eq9 =

$$M[t]=2^*L^*c[0]-2^*L^*c[b]-\sum_{n=0}^{\infty}\frac{1}{(1+2n)\pi}$$

$$\texttt{Integrate}\left[\left((-1)^n e^{-\frac{(1+2n)\pi(2iL^2+2iLx+(1+2n)\pi t\eta)}{4L^2}}\left(-2ie^{in\pi}\left(1+e^{\frac{i(1+2n)\pi x}{L}}\right)c[b]+2\left(ie^{in\pi}-e^{\frac{i(1+2n)\pi x}{2L}}-e^{\frac{i(\pi+2n\pi)(2L+x)}{2L}}+e^{\frac{i(\pi+2n\pi)(L+2x)}{2L}}\right)C[0]\right)\right),\{x,\ -L,\ L\}\right]$$

$$M[t]=2Lc[0]-2Lc[b]-\sum_{n=0}^{\infty}-\left(\left(8ie^{\frac{1}{4}\pi\left(2i(-3+2n)-\frac{(1+2n)^2\pi t\eta}{L^2}\right)}L(2(-c[b]+C[0])\mathrm{Cos}[n\pi]+(1+2n)\pi C[0]\mathrm{Sin}[n\pi]\right)\Big/((1+2n)\pi(\pi+2n\pi))\right)$$

eq10 = Factor[(eq9 / .Sin[$n\pi$] → 0) / .Cos[$n\pi$] → (−1) ^ n]

$$M[t] = 2Lc[0] - 2Lc[b] - \sum_{n=0}^{\infty} \frac{-16i(-1)^n e^{\frac{1}{4}\pi\left(2i(-3+2n)-\frac{(1+2n)^2\pi t\eta}{L^2}\right)} L(-c[b]+C[0])}{(1+2n)\pi(\pi+2n\pi)}$$

$$\text{aux0} = \texttt{Factor}\left[\frac{16(-1)^n e^{in\pi-\frac{(\pi+2n\pi)^2 t\eta}{4L^2}} L((-c[b]+C[0])/(2LC[0]-2Lc[b]))}{(1+2n)\pi(\pi+2n\pi)}\right]$$

$$\frac{8(-1)^n e^{in\pi-\frac{(\pi+2n\pi)^2 t\eta}{4L^2}}}{(1+2n)^2\pi^2}$$

$$\text{aux} = e^{in\pi - \frac{(\pi+2n\pi)^2 t\eta}{4L^2}} \to ((-1)\,\hat{}\,n)^* e^{-\frac{\texttt{Factor}[(\pi+2n\pi)^2 t\eta]}{4L^2}}$$

$$e^{in\pi - \frac{(\pi+2n\pi)^2 t\eta}{4L^2}} \to (-1)^n e^{-\frac{(1+2n)^2\pi^2 t\eta}{4L^2}}$$

$$\text{aux1} = \texttt{Factor}[(\text{aux0}/.\text{aux})/.(-1)^{2n} \to 1]$$

aux0

$$\text{eq11} = M[t]/M[\infty] = 1 - \sum_{n=0}^{\infty} \frac{8\ e^{-\frac{(1+2n)^2\pi^2 t\eta}{4L^2}}}{(1+2n)^2\pi^2}$$

$$\frac{M[t]}{M[\infty]} = 1 - \sum_{n=0}^{\infty} \frac{8\ e^{-\frac{(1+2n)^2\pi^2 t\eta}{4L^2}}}{(1+2n)^2\pi^2}$$

$$\text{eq12} = \text{eq11}/.L \to a/2$$

$$\frac{M[t]}{M[\infty]} = 1 - \sum_{n=0}^{\infty} \frac{8\ e^{-\frac{(1+2n)^2\pi^2 t\eta}{a^2}}}{(1+2n)^2\pi^2}$$

eq13 =

$$\frac{M[t]}{M[\infty]} = 1 - \left(\sum_{n=0}^{\infty} \frac{8\ e^{-\frac{(1+2n)^2\pi^2 t\eta}{a^2}}}{(1+2n)^2\pi^2}\right) * \left(\sum_{m=0}^{\infty} \frac{8e^{-\frac{(1+2m)^2\pi^2 t\eta}{b^2}}}{(1+2m)^2\pi^2}\right) * \left(\sum_{k=0}^{\infty} \frac{8e^{-\frac{(1+2k)^2\pi^2 t\eta}{c^2}}}{(1+2k)^2\pi^2}\right)$$

$$\frac{M[t]}{M[\infty]} = 1 - \left(\sum_{k=0}^{\infty} \frac{8e^{-\frac{(1+2k)^2\pi^2 t\eta}{c^2}}}{(1+2k)^2\pi^2}\right)\left(\sum_{m=0}^{\infty} \frac{8e^{-\frac{(1+2m)^2\pi^2 t\eta}{b^2}}}{(1+2m)^2\pi^2}\right)\sum_{n=0}^{\infty} \frac{8\ e^{-\frac{(1+2n)^2\pi^2 t\eta}{a^2}}}{(1+2n)^2\pi^2}$$

PROBLEMS

14.1. Show that the ratio of the inverse Laplace transform of the following equation

$$M(s) = \frac{2c_0\pi R^2 H}{s} - \frac{2\pi RH}{sJ_0\left(R\sqrt{-\frac{s}{\eta}}\right)\sqrt{-\frac{s}{\eta}}}\left(\begin{array}{l}-2J_1\left(R\sqrt{-\frac{s}{\eta}}\right)c_0 + c_0 J_0\left(R\sqrt{-\frac{s}{\eta}}\right)\ \times \\ R\sqrt{-\frac{s}{\eta}} + 2J_1\left(R\sqrt{-\frac{s}{\eta}}\right)c_b\end{array}\right)$$
$$+\sum_{n=1}^{\infty} \frac{8\pi R^5\sqrt{\eta}(c_0-c_b)\left(\mathrm{e}^{\frac{2H}{R}\sqrt{\frac{sR^2+\alpha_{0,n}{}^2\eta}{\eta}}}-1\right)}{\alpha_{0,n}{}^2(sR^2+\alpha_{0,n}{}^2\eta)^{3/2}\left(\mathrm{e}^{\frac{2H}{R}\sqrt{\frac{sR^2+\alpha_{0,n}{}^2\eta}{\eta}}}+1\right)} \tag{1}$$

and $M(\infty)$ is

$$\frac{M(t)}{M(\infty)} = 1 - \sum_{n=1}^{\infty}\left(\sum_{m=0}^{\infty} \frac{32\mathrm{e}^{-\frac{\eta t\pi^2(2m+1)^2}{4H^2}-\frac{\alpha_{0,n}{}^2\eta t}{R^2}}}{\pi^2\alpha_{0,n}{}^2(2m+1)^2}\right) \tag{2}$$

14.2. The $(t[h], M(t))$-data from a cylindrical matrix device (140 μm in height and 85 μm in radius) are given below. Estimate the diffusion coefficient of the drug in the polymer.

$$\begin{pmatrix} 0 & 0.0796902 \\ 1 & 0.297441 \\ 2 & 0.402865 \\ 3 & 0.477135 \\ 4 & 0.535324 \\ 5 & 0.583296 \\ 6 & 0.624054 \\ 7 & 0.659374 \\ 8 & 0.690411 \\ 9 & 0.71796 \\ 10 & 0.742599 \\ 11 & 0.764762 \\ 12 & 0.784786 \\ 13 & 0.802937 \\ 14 & 0.819435 \\ 15 & 0.834461 \\ 16 & 0.848168 \\ 17 & 0.860688 \\ 18 & 0.872136 \\ 19 & 0.882612 \\ 20 & 0.892205 \end{pmatrix}$$

14.3. The $(t[h], M(t))$-data from a cylindrical matrix device (125 μm in height and 70 μm in radius) are given below. Estimate the diffusion coefficient of the drug in the polymer.

$$\begin{pmatrix} 0 & 0.0796902 \\ 2 & 0.488205 \\ 4 & 0.63746 \\ 6 & 0.732194 \\ 8 & 0.798968 \\ 10 & 0.847987 \\ 12 & 0.884617 \\ 14 & 0.912224 \\ 16 & 0.933127 \\ 18 & 0.949 \\ 20 & 0.961075 \end{pmatrix}$$

14.4. Show that the inverse Laplace transform of

$$C(x) = \frac{c_b - c_0}{2\cosh\left(L\sqrt{\frac{s}{\eta}}\right)}\left(e^{x\sqrt{\frac{s}{\eta}}} + e^{-x\sqrt{\frac{s}{\eta}}}\right) + \frac{c_0}{s} \tag{1}$$

is

$$c(x,t) = c_b + \sum_{n=0}^{\infty} \frac{2(-1)^{1+n}}{\pi(2n+1)} \left(-e^{\frac{i(2n+1)\pi x}{2L}} c_0 + e^{\frac{i(2n+1)\pi x}{2L}} c_b - e^{\frac{-i(2n+1)\pi x}{2L}} c_0 + e^{\frac{-i(2n+1)\pi x}{2L}} c_b\right) e^{\frac{-(2n+1)^2\pi^2\eta t}{4L^2}} \quad (2)$$

14.5. The (*t[h], M(t)*)-data from a rectangular parallelepiped-shaped device (Dimension: $a = 75\,\mu\text{m}$, $b = 180\,\mu\text{m}$ and $c = 380\,\mu\text{m}$) are given below. What is the diffusion coefficient of the drug in the polymer?

$$\begin{pmatrix} 0 & 0.0739888 \\ 1 & 0.451386 \\ 2 & 0.599285 \\ 3 & 0.69774 \\ 4 & 0.769248 \\ 5 & 0.822732 \\ 6 & 0.863273 \\ 7 & 0.894243 \\ 8 & 0.918025 \\ 9 & 0.936356 \\ 10 & 0.950525 \\ 11 & 0.961501 \\ 12 & 0.970018 \\ 13 & 0.976635 \\ 14 & 0.981782 \end{pmatrix}$$

REFERENCES

Schittkowski, K. (2002). *Numerical data fitting in dynamical systems: a practical introduction with applications and software*. Dordrecht, The Netherlands: Kluwer Academic Publishers.

Simon, L., Weltner, A. N., Wang, Y., & Michniak, B. (2006). A parametric study of iontophoretic transdermal drug-delivery systems. *J Membr Sci*, 278(1–2), 124–132.

Wang, S, & Lou, X. (2010). Numerical methods for the estimation of effective diffusion coefficients of 2D controlled drug delivery systems. *Optim Eng*, 11(4), 611–626.

Wei, R., Simon, L., Hu, L., & Michniak-Kohn, B. (2012). Effects of iontophoresis and chemical enhancers on the transport of lidocaine and nicotine across the oral mucosa. *Pharm Res*, 29(4), 961–971.

15

OPTIMIZATION OF TWO- AND THREE-DIMENSIONAL CONTROLLED-RELEASED DRUG-DELIVERY MODELS

15.1 OPTIMUM DESIGN OF CONTROLLED-RELEASED DRUG-DELIVERY SYSTEMS

An important aspect in the manufacturing of controlled-release devices is the estimation of design specifications that meet constraints such as a transdermal drug-delivery rate or the time elapsed before reaching a therapeutic drug concentration. Optimal control of the delivery rate was achieved in a heat-aided transdermal system by manipulating the temperature applied to the skin (Simon, 2007). Parameters, necessary to achieve a desired flux, were computed for an iontophoretic delivery system (Simon et al., 2006).

When analytical expressions are not available, researchers need to conduct several experiments to estimate the drug formulation adequate for a particular release profile. This approach can be expensive and may lead to sub-optimum devices. Berkland et al. (2002) selected adequate mixtures of poly(d,l-lactide-co-glycolide) (PLG) microspheres that yielded constant release of rhodamine and piroxicam for 8 and 14 days, respectively. It is possible to use the experimental data to derive a design equation. In Simon and Fernandes (2004), artificial neural networks (ANNs) were developed to predict the drug permeation rate as a function of the vinyl acetate contents of a copolymer membrane, the membrane thickness and poly(ethylene glycol) compositions. The ANN model was then inverted to compute device properties that would suit end-user requirements.

Two types of problems are encountered in many engineering applications: the **forward** and **inverse problems**. The forward problem consists of predicting the process response given a mathematical model and a set of parameter values. For example, the steady-state flux across a membrane can be estimated from Fick's first law and properties of the system such as the

---

Closed-form Solutions for Drug Transport through Controlled-Release Devices in Two and Three Dimensions, First Edition. Laurent Simon and Juan Ospina.

membrane thickness and the drug concentration in the vehicle. The inverse problem consists of finding the numerical values of the model parameters necessary to produce a particular output. A constrained optimization routine "NMinimize" was implemented in Mathematica to determine the membrane/vehicle properties required to meet a target delivery rate (Simon & Fernandes, 2004). "NMinimize" attempts to find a global minimum of a function F subject to a set of constraints:

$$F(\mathbf{x}) = [\chi_{set} - \chi(\mathbf{x}, \mathbf{p})]^2 \tag{15.1}$$

where $\mathbf{x}$ represents the parameters to be computed, $\mathbf{p}$ is a vector of other constants appearing in the design equation, χ is the output variable and χ_{set} is the desired value of χ (e.g., flux, time constant). The solution to the optimization problem is

$$\mathbf{x}_{\min} = \min_{\mathbf{x}_{low} < \mathbf{x} < \mathbf{x}_{up}} F(\mathbf{x}) \tag{15.2}$$

with $\mathbf{x}_{\min}$ lying between $\mathbf{x}_{low}$ and $\mathbf{x}_{up}$.

15.2 DESIGN OF A 2-D CYLINDRICAL DOSAGE FORM WITH A FINITE MASS TRANSFER COEFFICIENT

The cylindrical model examined in this chapter is similar to the one presented in Section 14.2, the difference being a finite mass transfer coefficient h is now defined at $z = \pm H$. The governing equations are

$$\frac{\partial c}{\partial t} = \eta \left(\frac{1}{r} \frac{\partial}{\partial r} r \frac{\partial c}{\partial r} + \frac{\partial^2 c}{\partial z^2} \right) \quad 0 \le r \le R, -H \le z \le H \tag{15.3}$$

$$c(r, z, 0) = c_0 \tag{15.4}$$

$$c(R, z, t) = c_b \tag{15.5}$$

$$\left. \frac{\partial c\,(r, z, t)}{\partial r} \right|_{r=0} = 0 \tag{15.6}$$

$$\eta \left. \frac{\partial c\,(r, z, t)}{\partial z} \right|_{z=-H} = h[c(r, -H, t) - c_b] \tag{15.7}$$

$$-\eta \left. \frac{\partial c\,(r, z, t)}{\partial z} \right|_{z=H} = h[c(r, H, t) - c_b] \tag{15.8}$$

In this case, c_b is the concentration required to establish equilibrium with the surrounding environment. For this problem, we assumed that the coefficient of mass transfer is still very high at $r = R$. The cumulative amount of drug released is

$$M_t = c_0 \pi R^2 (2H) - 2\pi \int_{-H}^{H} \int_0^R c(r, z, t)\, r dr dz \tag{15.9}$$

The Laplace transform of Eq. (15.3) is given by

$$C(r, z) - c_0 = \frac{\eta \frac{\partial}{\partial r} C(r, z)}{r} + \eta \frac{\partial^2}{\partial r^2} C(r, z) + \eta \frac{\partial^2}{\partial z^2} C(r, z) \tag{15.10}$$

where

$$C(r, z) = \int_0^\infty C(r, z, t)e^{-st}dt \tag{15.11}$$

A solution to Eq. (15.11) is written as $C(r, z) = C(r) + f(r)g(z)$. As a result,

$$sC(r) - c_0 = \frac{\eta \frac{\mathrm{d}}{\mathrm{d}r} C(r)}{r} + \eta \frac{\mathrm{d}^2}{\mathrm{d}r^2} C(r) \tag{15.12}$$

$$\frac{d^2}{dr^2} f(r) = c_1\ f(r) - \frac{\frac{d}{dr} f(r)}{r} \tag{15.13}$$

and

$$\frac{d^2}{dr^2} g(z) = -c_1 g(z) + \frac{sg(z)}{\eta} \tag{15.14}$$

The solution $C(r)$ is

$$C(r) = C_2 J_0\left(r\sqrt{-\frac{s}{\eta}}\right) + C_1 Y_0\left(r\sqrt{-\frac{s}{\eta}}\right) + \frac{c_0}{s} \tag{15.15}$$

After applying the boundary condition (15.6), we find that $C_1 = 0$. Then, Eq.(15.15) becomes

$$C(r) = C_2 J_0\left(r\sqrt{-\frac{s}{\eta}}\right) + \frac{c_0}{s} \tag{15.16}$$

Just as in Chapter 14, the following expression is obtained after using Eq. (15.5):

$$C(r) = \frac{\left(\frac{c_b}{s} - \frac{c_0}{s}\right) J_0\left(r\sqrt{-\frac{s}{\eta}}\right)}{J_0\left(R\sqrt{-\frac{s}{\eta}}\right)} + \frac{c_0}{s} \tag{15.17}$$

Further manipulations lead to (See Chapter 14)

$$C(r, z) = \frac{\left(\frac{c_b}{s} - \frac{c_0}{s}\right) J_0\left(r\sqrt{-\frac{s}{\eta}}\right)}{J_0\left(R\sqrt{-\frac{s}{\eta}}\right)} + \frac{c_0}{s} + \sum_{n=1}^{\infty}\left(iA_n \sinh\left(\frac{z}{R}\sqrt{\frac{sR^2 + {\alpha_{0,n}}^2\eta}{\eta}}\right) + B_n \cosh\left(\frac{z}{R}\sqrt{\frac{sR^2 + {\alpha_{0,n}}^2\eta}{\eta}}\right)\right) J_0\left(\frac{\alpha_{0,n} r}{R}\right) \tag{15.18}$$

The boundary conditions (15.7) and (15.8) give

$$\eta\sum_{n=1}^{\infty}\begin{pmatrix} \frac{iA_n}{R}\sqrt{\frac{sR^2+\alpha_{0,n}{}^2\eta}{\eta}}\cosh\left(\frac{H}{R}\sqrt{\frac{sR^2+\alpha_{0,n}{}^2\eta}{\eta}}\right) \\ -\frac{B_n}{R}\sqrt{\frac{sR^2+\alpha_{0,n}{}^2\eta}{\eta}}\sinh\left(\frac{H}{R}\sqrt{\frac{sR^2+\alpha_{0,n}{}^2\eta}{\eta}}\right) \end{pmatrix} J_0\left(\frac{\alpha_{0,n}r}{R}\right)$$
$$= h\left[\frac{\left(\frac{c_b}{s}-\frac{c_0}{s}\right)J_0\left(r\sqrt{-\frac{s}{\eta}}\right)}{J_0\left(R\sqrt{-\frac{s}{\eta}}\right)} + \sum_{n=0}^{\infty}\begin{pmatrix} iA_n\sinh\left(-\frac{H}{R}\sqrt{\frac{sR^2+\alpha_{0,n}{}^2\eta}{\eta}}\right) + \\ B_n\cosh\left(-\frac{H}{R}\sqrt{\frac{sR^2+\alpha_{0,n}{}^2\eta}{\eta}}\right) \end{pmatrix} J_0\left(\frac{\alpha_{0,n}r}{R}\right)\right] \tag{15.19}$$

and

$$-\eta\sum_{n=1}^{\infty}\begin{pmatrix} \frac{iA_n}{R}\sqrt{\frac{sR^2+\alpha_{0,n}{}^2\eta}{\eta}}\cosh\left(\frac{H}{R}\sqrt{\frac{sR^2+\alpha_{0,n}{}^2\eta}{\eta}}\right) \\ -\frac{B_n}{R}\sqrt{\frac{sR^2+\alpha_{0,n}{}^2\eta}{\eta}}\sinh\left(\frac{H}{R}\sqrt{\frac{sR^2+\alpha_{0,n}{}^2\eta}{\eta}}\right) \end{pmatrix} J_0\left(\frac{\alpha_{0,n}r}{R}\right)$$
$$= h\left[\frac{\left(\frac{c_b}{s}-\frac{c_0}{s}\right)J_0\left(r\sqrt{-\frac{s}{\eta}}\right)}{J_0\left(R\sqrt{-\frac{s}{\eta}}\right)} + \sum_{n=1}^{\infty}\begin{pmatrix} iA_n\sinh\left(\frac{H}{R}\sqrt{\frac{sR^2+\alpha_{0,n}{}^2\eta}{\eta}}\right) + \\ B_n\cosh\left(\frac{H}{R}\sqrt{\frac{sR^2+\alpha_{0,n}{}^2\eta}{\eta}}\right) \end{pmatrix} J_0\left(\frac{\alpha_{0,n}r}{R}\right)\right] \tag{15.20}$$

respectively. After multiplying both sides of Eq. (15.19) by $rJ_0\left(\frac{\alpha_{0,n}r}{R}\right)$ and integrating from $r = 0$ to $r = R$, we have

$$\frac{1}{2}i\sqrt{\eta}A_nR\sqrt{sR^2+\alpha_{0,n}{}^2\eta}(J_0(\alpha_{0,n}))^2\cosh\left(\frac{H}{R}\sqrt{\frac{sR^2+\alpha_{0,n}{}^2\eta}{\eta}}\right)$$
$$-\frac{1}{2}\sqrt{\eta}B_nR\sqrt{sR^2+\alpha_{0,n}{}^2\eta}(J_1(\alpha_{0,n}))^2\sinh\left(\frac{H}{R}\sqrt{\frac{sR^2+\alpha_{0,n}{}^2\eta}{\eta}}\right)$$
$$=\frac{1}{2}\frac{R^2h}{(sR^2+\alpha_{0,n}{}^2\eta)s\alpha_{0,n}J_0\left(R\sqrt{-\frac{s}{\eta}}\right)}$$

$$\times\begin{pmatrix} -iA_n \sinh\left(\frac{H}{R}\sqrt{\frac{sR^2+\alpha_{0,n}{}^2\eta}{\eta}}\right) s^2 J_0\left(R\sqrt{-\frac{s}{\eta}}\right)\alpha_{0,n}R^2(J_1(\alpha_{0,n}))^2 \\ +B_n \cosh\left(\frac{H}{R}\sqrt{\frac{sR^2+\alpha_{0,n}{}^2\eta}{\eta}}\right) s^2 J_0\left(R\sqrt{-\frac{s}{\eta}}\right)\alpha_{0,n}R^2(J_1(\alpha_{0,n}))^2 \\ -2J_0\left(R\sqrt{-\frac{s}{\eta}}\right)c_b R^2 J_1(\alpha_{0,n})s + 2J_0\left(R\sqrt{-\frac{s}{\eta}}\right)c_0 R^2 J_1(\alpha_{0,n})s \\ +B_n \cosh\left(\frac{H}{R}\sqrt{\frac{sR^2+\alpha_{0,n}{}^2\eta}{\eta}}\right) sJ_0\left(R\sqrt{-\frac{s}{\eta}}\right)\alpha_{0,n}{}^3(J_1(\alpha_{0,n}))^2\eta \\ -iA_n \sinh\left(\frac{H}{R}\sqrt{\frac{sR^2+\alpha_{0,n}{}^2\eta}{\eta}}\right) sJ_0\left(R\sqrt{-\frac{s}{\eta}}\right)\alpha_{0,n}{}^3(J_1(\alpha_{0,n}))^2\eta \end{pmatrix} \quad (15.21)$$

Similarly, we multiply both sides of Eq. (15.20) by $rJ_0\left(\frac{\alpha_{0,n}r}{R}\right)$ and integrate from $r = 0$ to $r = R$ and obtain

$$-\frac{1}{2}i\sqrt{\eta}A_n R\sqrt{sR^2+\alpha_{0,n}{}^2\eta}(J_1(\alpha_{0,n}))^2\cosh\left(\frac{H}{R}\sqrt{\frac{sR^2+\alpha_{0,n}{}^2\eta}{\eta}}\right)$$
$$-\frac{1}{2}\sqrt{\eta}B_n R\sqrt{sR^2+\alpha_{0,n}{}^2\eta}(J_1(\alpha_{0,n}))^2\sinh\left(\frac{H}{R}\sqrt{\frac{sR^2+\alpha_{0,n}{}^2\eta}{\eta}}\right)$$
$$=\frac{1}{2}\frac{R^2h}{(sR^2+\alpha_{0,n}{}^2\eta)s\alpha_{0,n}J_0\left(R\sqrt{-\frac{s}{\eta}}\right)}$$
$$\times\begin{pmatrix} -iA_n \sinh\left(\frac{H}{R}\sqrt{\frac{sR^2+\alpha_{0,n}{}^2\eta}{\eta}}\right) s^2 J_0\left(R\sqrt{-\frac{s}{\eta}}\right)\alpha_{0,n}R^2(J_1(\alpha_{0,n}))^2 \\ +B_n \cosh\left(\frac{H}{R}\sqrt{\frac{sR^2+\alpha_{0,n}{}^2\eta}{\eta}}\right) s^2 J_0\left(R\sqrt{-\frac{s}{\eta}}\right)\alpha_{0,n}R^2(J_1(\alpha_{0,n}))^2 \\ -2J_0\left(R\sqrt{-\frac{s}{\eta}}\right)c_b R^2 J_1(\alpha_{0,n})s + 2J_0\left(R\sqrt{-\frac{s}{\eta}}\right)c_0 R^2 J_1(\alpha_{0,n})s \\ +B_n \cosh\left(\frac{H}{R}\sqrt{\frac{sR^2+\alpha_{0,n}{}^2\eta}{\eta}}\right) sJ_0\left(R\sqrt{-\frac{s}{\eta}}\right)\alpha_{0,n}{}^3(J_1(\alpha_{0,n}))^2\eta \\ +iA_n \sinh\left(\frac{H}{R}\sqrt{\frac{sR^2+\alpha_{0,n}{}^2\eta}{\eta}}\right) sJ_0\left(R\sqrt{-\frac{s}{\eta}}\right)\alpha_{0,n}{}^3(J_1(\alpha_{0,n}))^2\eta \end{pmatrix} \quad (15.22)$$

Equations (15.21) and (15.22) are solved for A_n and B_n to give

$$A_n = 0;$$

$$B_n = -4R^3h(c_0 - c_b)e^{\frac{H}{R}\sqrt{\frac{sR^2+\alpha_{0,n}{}^2\eta}{\eta}}}$$

$$\div\,\alpha_{0,n}J_1(\alpha_{0,n})\left(\begin{array}{c} Rh\alpha_{0,n}{}^2\left(e^{\frac{H}{R}\sqrt{\frac{sR^2+\alpha_{0,n}{}^2\eta}{\eta}}}\right)^2+Rh\alpha_{0,n}{}^2\eta-\sqrt{\eta}\sqrt{sR^2+\alpha_{0,n}{}^2\eta}sR^2 \\ -\eta^{3/2}\sqrt{sR^2+\alpha_{0,n}{}^2\eta}\alpha_{0,n}{}^2+\sqrt{\eta}\sqrt{sR^2+\alpha_{0,n}{}^2\eta}sR^2\left(e^{\frac{H}{R}\sqrt{\frac{sR^2+\alpha_{0,n}{}^2\eta}{\eta}}}\right)^2 \\ +R^3hs+\eta^{3/2}\sqrt{sR^2+\alpha_{0,n}{}^2\eta}\alpha_{0,n}{}^2\left(e^{\frac{H}{R}\sqrt{\frac{sR^2+\alpha_{0,n}{}^2\eta}{\eta}}}\right)^2+R^3hs\left(e^{\frac{H}{R}\sqrt{\frac{sR^2+\alpha_{0,n}{}^2\eta}{\eta}}}\right)^2 \end{array}\right) \tag{15.23}$$

Therefore, the concentration is

$$C(r,z)=\frac{\left(\frac{c_b}{s}-\frac{c_0}{s}\right)J_0\left(r\sqrt{-\frac{s}{\eta}}\right)}{J_0\left(R\sqrt{-\frac{s}{\eta}}\right)}+\frac{c_0}{s}+\sum_{n=1}^{\infty}\left(B_n\cosh\left(\frac{z}{R}\sqrt{\frac{sR^2+\alpha_{0,n}{}^2\eta}{\eta}}\right)\right)J_0\left(\frac{\alpha_{0,n}r}{R}\right) \tag{15.24}$$

with B_n defined in Eq. (15.23).

$$\begin{aligned} M(s)=&\frac{2c_0\pi R^2H}{s}-\frac{2\pi RH}{sJ_0\left(\sqrt{-\frac{s}{\eta}}R\right)\sqrt{-\frac{s}{\eta}}} \\ &\times\left(2J_1\left(R\sqrt{-\frac{s}{\eta}}\right)c_b+c_0J_0\left(R\sqrt{-\frac{s}{\eta}}\right)R\sqrt{-\frac{s}{\eta}}-2J_1\left(R\sqrt{-\frac{s}{\eta}}\right)c_0\right) \\ &-2\pi\sum_{n=1}^{\infty}4\frac{R^6h\sqrt{\eta}(c_b-c_0)}{\alpha_{0,n}{}^2}\left(-1+e^{\frac{2H}{R}\sqrt{\frac{sR^2+\alpha_{0,n}{}^2\eta}{\eta}}}\right) \\ &\div\left(\begin{array}{c} \alpha_{0,n}{}^2\sqrt{R^2s+\eta\alpha_{0,n}{}^2}Rh\eta e^{\frac{2H}{R}\sqrt{\frac{sR^2+\alpha_{0,n}{}^2\eta}{\eta}}}+\alpha_{0,n}{}^2\sqrt{R^2s+\eta\alpha_{0,n}{}^2}Rh\eta \\ -2\alpha_{0,n}{}^2\eta^{3/2}sR^2-\sqrt{\eta}s^2R^4-\alpha_{0,n}{}^4\eta^{5/2}+2\alpha_{0,n}{}^2\eta^{3/2}sR^2e^{\frac{2H}{R}\sqrt{\frac{sR^2+\alpha_{0,n}{}^2\eta}{\eta}}} \\ +\sqrt{\eta}s^2R^4e^{\frac{2H}{R}\sqrt{\frac{sR^2+\alpha_{0,n}{}^2\eta}{\eta}}}+\sqrt{R^2s+\eta\alpha_{0,n}{}^2}R^3hs+\alpha_{0,n}{}^4\eta^{5/2}e^{\frac{2H}{R}\sqrt{\frac{sR^2+\alpha_{0,n}{}^2\eta}{\eta}}} \\ +\sqrt{R^2s+\eta\alpha_{0,n}{}^2}R^3hse^{\frac{2H}{R}\sqrt{\frac{sR^2+\alpha_{0,n}{}^2\eta}{\eta}}} \end{array}\right) \end{aligned} \tag{15.25}$$

The inverse Laplace transform of Eq. (15.25) is determined using the **Bromwich integral** and the residue theorem. The fraction of drug released at time t is

$$\frac{M(t)}{M(\infty)}=1-\sum_{n=1}^{\infty}\left(\sum_{m=0}^{\infty}\frac{-8H\sin\left(\beta_m\right)h(-\eta\beta_m+ihH)}{\alpha_{0,n}{}^2\beta_m{}^2(h^2H^2+\eta^2\beta_m{}^2+hH\eta)}e^{i\beta_m-\frac{\eta t\beta_m{}^2}{H^2}-\frac{\eta\alpha_{0,n}{}^2t}{R^2}}\right) \tag{15.26}$$

where

$$e^{2i\beta_m}=\frac{-\eta\beta_m-ihH}{-\eta\beta_m+ihH} \tag{15.27}$$

and

$$M(\infty)=2H\pi R^2(c_0-c_b) \tag{15.28}$$

It is important to design the device so that it releases the drug with a particular time constant. Recall from Chapter 13 that the effective time constant is defined as

$$t_{eff} = \lim_{s\to 0}\left(\frac{1}{s^2} + \frac{d\overline{\psi}(s)}{ds}\right)\left[\lim_{s\to 0}\left(\frac{1}{s} - \overline{\psi}(s)\right)\right]^{-1} \tag{15.29}$$

where

$$\overline{\psi}(s) = \frac{M(s)}{M(\infty)} \tag{15.30}$$

The infinite sum in Eq. (15.25) is truncated to one term to compute a first-order t_{eff}:

$$t_{eff} = \frac{1}{\eta\left[\frac{\beta_0{}^2}{H^2} - \frac{\alpha_{0,1}{}^2}{R^2}\right]} \tag{15.31}$$

or

$$t_{eff} = \frac{0.029R^2\begin{pmatrix} 6473.7H\eta^3 - 12947.3H\eta^3 e^{4.81\frac{H}{R}} - 5383.5H\eta^2 Rh \\ +6473.7H\eta^3 e^{9.62\frac{H}{R}} + 5383.5H\eta^2 e^{9.62\frac{H}{R}} Rh \\ +1119.233358H\eta R^2h^2 e^{9.62\frac{H}{R}} + 4459.532316H\eta R^2 h^2 e^{4.81\frac{H}{R}} \\ +1119.2H\eta R^2h^2 + 692.6h^2R^3\eta + 4442.131200hR^2\eta^2 e^{4.81\frac{H}{R}} \\ -2221.1hR^2\eta^2 - 692.6h^2R^3\eta e^{9.62\frac{H}{R}} - 2221.1hR^2\eta^2 e^{9.62\frac{H}{R}} \end{pmatrix}}{\left[\begin{array}{l} \eta\begin{pmatrix} -193.5H\eta^{3/2} + 193.5H\eta^{3/2}e^{4.81\frac{H}{R}} + 80.4\sqrt{\eta}Rhe^{4.81\frac{H}{R}} \\ 80.4H\sqrt{\eta}Rh + 32hR^2\sqrt{\eta} - 32hR^2\sqrt{\eta}e^{4.81\frac{H}{R}} \end{pmatrix} \\ \times\left(-5.78\eta^{3/2} + 5.78\eta^{3/2}e^{4.81\frac{H}{R}} + 2.40\sqrt{\eta}Rhe^{4.81\frac{H}{R}} + 2.40\sqrt{\eta}Rh\right) \end{array}\right]} \tag{15.32}$$

Equation (15.32) can be applied to fabricate cylindrical devices with a desired time constant. For example, the effective time constant is 45 h for $\eta = 10^{-6}\,\text{cm}^2/\text{s}$,, $R = 1\,\text{cm}$, $H = 5\,\text{cm}$ and $h = 10^{-6}\,\text{cm/s}$. The following command is used to estimate the diffusion coefficient required to obtain $t_{eff} = 300000\,s$.: "*NMinimize*[{(*tau*1[*eta*] − 300000)^2, *eta* > 0}, *eta*]". The output is {0., {*eta* → 5.386 × 10^{-7}}}.

15.3 DESIGN OF A RECTANGULAR PARALLELEPIPED-SHAPED MATRIX WITH A FINITE MASS TRANSFER COEFFICIENT

The polymeric device analyzed in Section 14.3 is re-introduced here (Fig. 14.3). A finite mass transfer coefficient is applied at the surface. The governing equations are

$$\frac{\partial c}{\partial t} = \eta\left(\frac{\partial^2 c}{\partial^2 x^2} + \frac{\partial^2 c}{\partial^2 y^2} + \frac{\partial^2 c}{\partial^2 y^2}\right); -\frac{a}{2} \le x \le \frac{a}{2}, -\frac{b}{2} \le y \le \frac{b}{2}, -\frac{c}{2} \le z \le \frac{c}{2} \tag{15.33}$$

$$c(x, y, z, 0) = c_0 \tag{15.34}$$

$$\eta\left.\frac{\partial c(x, y, z, t)}{\partial x}\right|_{a=-\frac{a}{2}} = h\left(c\left(-\frac{a}{2}, y, z, t\right) - c_b\right) \tag{15.35}$$

$$-\eta \left.\frac{\partial c(x, y, z, t)}{\partial x}\right|_{a=\frac{a}{2}} = h\left(c\left(\frac{a}{2}, y, z, t\right) - c_b\right) \tag{15.36}$$

$$\eta \left.\frac{\partial c(x, y, z, t)}{\partial x}\right|_{y=-\frac{b}{2}} = h\left(c\left(x, -\frac{b}{2}, z, t\right) - c_b\right) \tag{15.37}$$

$$-\eta \left.\frac{\partial c(x, y, z, t)}{\partial x}\right|_{y=\frac{b}{2}} = h\left(c\left(x, \frac{b}{2}, z, t\right) - c_b\right) \tag{15.38}$$

$$\eta \left.\frac{\partial c(x, y, z, t)}{\partial x}\right|_{z=-\frac{c}{2}} = h\left(c\left(x, y, -\frac{c}{2}, t\right) - c_b\right) \tag{15.39}$$

$$-\eta \left.\frac{\partial c(x, y, z, t)}{\partial x}\right|_{z=\frac{c}{2}} = h\left(c\left(x, y, \frac{c}{2}, t\right) - c_b\right) \tag{15.40}$$

The cumulative amount of drug released from the device at time t is

$$M(t) = c_0 abc - \int_{-\frac{a}{2}}^{\frac{a}{2}} \int_{-\frac{b}{2}}^{\frac{b}{2}} \int_{-\frac{c}{2}}^{\frac{c}{2}} c(x, y, z, t)\, dxdydz \tag{15.41}$$

As in Section 14.3, we first solve a one-dimensional problem to get

$$C(x) = C_2 e^{x\sqrt{\frac{s}{\eta}}} + C_1 e^{-x\sqrt{\frac{s}{\eta}}} + \frac{c_0}{s} \tag{15.42}$$

where

$$C(x) = \int_0^\infty c(x, t) e^{-st} dt \tag{15.43}$$

Applying the Laplace transforms of the boundary conditions (15.35) and (15.36) to Eq. (15.42) with $L = a/2$, we obtain

$$\eta\left(C_2\sqrt{\frac{s}{\eta}} e^{-\frac{a}{2}\sqrt{\frac{s}{\eta}}} - C_1\sqrt{\frac{s}{\eta}} e^{\frac{a}{2}\sqrt{\frac{s}{\eta}}}\right) = h\left(C_2 e^{-\frac{a}{2}\sqrt{\frac{s}{\eta}}} + C_1 e^{\frac{a}{2}\sqrt{\frac{s}{\eta}}} + \frac{C_0}{s} - \frac{c_b}{s}\right) \tag{15.44}$$

and

$$-\eta\left(C_2\sqrt{\frac{s}{\eta}} e^{\frac{a}{2}\sqrt{\frac{s}{\eta}}} - C_1\sqrt{\frac{s}{\eta}} e^{-\frac{a}{2}\sqrt{\frac{s}{\eta}}}\right) = h\left(C_2 e^{\frac{a}{2}\sqrt{\frac{s}{\eta}}} + C_1 e^{-\frac{a}{2}\sqrt{\frac{s}{\eta}}} + \frac{C_0}{s} - \frac{c_b}{s}\right) \tag{15.45}$$

We solve for C_1 and C_2:

$$C_1 = C_2 = \frac{h(c_b - c_0)}{s\left(\sqrt{s} e^{-\frac{a}{2}\sqrt{\frac{s}{\eta}}}\sqrt{\eta} - \sqrt{s} e^{\frac{a}{2}\sqrt{\frac{s}{\eta}}}\sqrt{\eta} - e^{-\frac{a}{2}\sqrt{\frac{s}{\eta}}} h - e^{\frac{a}{2}\sqrt{\frac{s}{\eta}}} h\right)} \tag{15.46}$$

Therefore, $C(x)$ becomes

$$C(x) = \frac{h(c_b - c_0)\left(e^{x\sqrt{\frac{s}{\eta}}} + e^{-x\sqrt{\frac{s}{\eta}}}\right)}{s\left(\sqrt{s}e^{-\frac{a}{2}\sqrt{\frac{s}{\eta}}}\sqrt{\eta} - \sqrt{s}e^{\frac{a}{2}\sqrt{\frac{s}{\eta}}}\sqrt{\eta} - e^{-\frac{a}{2}\sqrt{\frac{s}{\eta}}}h - e^{\frac{a}{2}\sqrt{\frac{s}{\eta}}}h\right)} + \frac{c_0}{s} \tag{15.47}$$

The inverse Laplace transform of Eq. (15.47) is computed using the Bromwich integral and the residue theorem:

$$c(x,t) = c_b + \sum_{n=1}^{\infty} \frac{4h(C_0 - c_b)a\eta}{\cos(\beta_n)(a^2h^2 + 4\beta_n^2\eta^2 + 2\eta ah)} \cos\left(\frac{2\beta_n x}{a}\right) e^{-\frac{4\beta_n{}^2\eta t}{a^2}} \tag{15.48}$$

where

$$\sin(\beta_n) = \frac{\cos(\beta_n)ha}{2\beta_n\eta} \tag{15.49}$$

Therefore, the cumulative fraction of drug released is

$$\frac{M(t)}{M(\infty)} = 1 - \sum_{n=1}^{\infty} \frac{2a^2h^2}{\beta_n{}^2(a^2h^2 + 4\beta_n{}^2\eta^2 + 2\eta ah)} e^{-\frac{4\beta_n{}^2\eta t}{a^2}} \tag{15.50}$$

Finally, Eq. (15.50) can be generalized for the three-dimensional problem

$$\frac{M(t)}{M(\infty)} = 1 - \sum_{n=0}^{\infty}\left(\sum_{n=0}^{\infty}\left(\sum_{n=0}^{\infty} g_{mnp} e^{-\left(\frac{4\beta_n{}^2\eta t}{a^2} + \frac{4\delta_n{}^2\eta t}{b^2} + \frac{4\chi_p{}^2\eta t}{c^2}\right)}\right)\right) \tag{15.51}$$

where

$$g_{mnp} = \frac{8h^6a^2b^2c^2}{\beta_n{}^2(a^2h^2 + 4\beta_n{}^2\eta^2 + 2\eta ah)\delta_m{}^2(b^2h^2 + 4\delta_m{}^2\eta^2 + 2\eta bh)\chi_p{}^2(c^2h^2 + 4\chi_p{}^2\eta^2 + 2\eta ch)} \tag{15.52}$$

$$M(\infty) = abc(c_0 - c_b) \tag{15.53}$$

$$\sin(\beta_n) = \frac{\cos(\beta_n)ha}{2\beta_n\eta} \tag{15.54}$$

$$\sin(\delta_m) = \frac{\cos(\delta_m)hb}{2\delta_m\eta} \tag{15.55}$$

and

$$\sin(\chi_p) = \frac{\cos(\chi_p)hc}{2\chi_p\eta} \tag{15.56}$$

A first-order approximation of Eq. (15.51) is

$$\frac{M(t)}{M(\infty)} = 1 - g_{111}e^{-4\eta\left(\frac{\beta_1{}^2}{a^2} + \frac{\delta_1{}^2}{b^2} + \frac{\chi_1{}^2}{c^2}\right)t} \tag{15.57}$$

As a result,

$$t_{eff} = \frac{1}{4\eta\left(\frac{\beta_1^2}{a^2} + \frac{\delta_1^2}{b^2} + \frac{\chi_1^2}{c^2}\right)} \tag{15.58}$$

With a function of the form

$$F(\eta) = [t_{eff\ set} - t_{eff}]^2 \tag{15.59}$$

It is possible to calculate the diffusion coefficient necessary to achieve a time constant of $t_{eff\ set}$.

15.4 SUMMARY

Polymeric devices were designed with a pre-determined time constant t_{eff} for drug release. In the two examples provided, t_{eff} could be manipulated by varying the diffusion coefficient of the drug in the matrix. Closed-formed design equations were developed making it possible to implement routines in Mathematica to carry out the optimization. First-order approximated t_{eff} were obtained for the cylindrical and the rectangular parallelepiped-shaped systems. The percentage of active ingredient exiting the device was expressed as a series solution.

15.5 APPENDIX: MAPLE AND MATHEMATICA CODE LISTINGS

15.5.1 Maple Code: cylind2D.mws (Section 15.2)

```
> restart:with(VectorCalculus):with(inttrans):with(PDETools):
> eq:=diff(C(r,z,t),t)=eta*Laplacian(C(r,z,t),cylindrical[r,theta,z]);
```

$$eq := \frac{\partial}{\partial t}C(r,z,t) = \frac{\eta\left(\frac{\partial}{\partial r}C(r,z,t)+r\left(\frac{\partial^2}{\partial r^2}C(r,z,t)\right)+r\left(\frac{\partial^2}{\partial z^2}C(r,z,t)\right)\right)}{r}$$

```
> eq1:=C(r,z,0)=c[0]:
>
> eq2:=laplace(eq,t,s):
> eq3:=subs(eq1,eq2):
> eq4:=subs(laplace(C(r,z,t),t,s)=C(r,z),eq3):
> eq5:=subs(C(r,z)=C(r),eq4):
> eq6:=dsolve(eq5):
> eq7:=subs(_C1=0,eq6):
> eq8:=subs(r=R,rhs(eq7))=c[b]/s:
> eq9:=isolate(eq8,_C2):
>
> eq10:=subs(eq9,eq7):
> eq11:=subs(c[0]=0,eq4):
> eq12:=pdsolve(eq11,HINT=f(r)*g(z)):
> eq13:=factor(build(eq12)):
> eq14:=subs(_C2=0,eq13):
> eq15:=subs(_C1=1,eq14):
> eq16:=subs(r=R,rhs(eq15))=0:
```

```
> eq17:=BesselJ(0,(-_c[1])^(1/2)*R)=0:
> eq18:=(-_c[1])^(1/2)*R=alpha[0,n]:
> eq19:=isolate(eq18,_c[1]):
> eq20:=simplify(subs(eq19,eq15),power,symbolic):
> eq21:=subs(C=C[n],_C3=A[n],_C4=B[n],eq20):
> eq22:=C(r,z)=rhs(eq10)+Sum(rhs(eq21),n=1..infinity):
> eq23:=subs(z=-H,eta*diff(rhs(eq22),z))=h*(subs(z=-H,rhs(eq22))-
    c[b]/s):
> eq23A:=subs(z=H,-eta*diff(rhs(eq22),z))=h*(subs(z=H,rhs(eq22))-
    c[b]/s):
> eq24:=subs(BesselJ(0,alpha[0,n])=0,int(subs(z=-H,eta*diff(rhs(eq10)+
    rhs(eq21),z))*BesselJ(0,alpha[0,n]/R*r)*r,r=0..R))=
    subs(BesselJ(0,alpha[0,n])=0,int(BesselJ(0,alpha[0,n]/R*r)*r*
    (h*(subs(z=-H,rhs(eq10)+rhs(eq21)-c[b]/s))),r=0..R)):
> eq24A:=subs(BesselJ(0,alpha[0,n])=0,int(subs(z=H,-eta*diff(rhs(eq10)+
    rhs(eq21),z))*BesselJ(0,alpha[0,n]/R*r)*r,r=0..R))=
    subs(BesselJ(0,alpha[0,n])=0,int(BesselJ(0,alpha[0,n]/R*r)*r*
    (h*(subs(z=H,rhs(eq10)+rhs(eq21)-c[b]/s))),r=0..R)):
> eq25:=factor(solve({eq24, eq24A}, {A[n], B[n]})):
> eq26:=subs(eq25,eq22):
> eq27:=Q(s)=2*Pi*int(int(rhs(eq10)*r,r=0..R),z=-H..H)+2*Pi*
    Sum(factor(int(int(subs(eq25,rhs(eq21))*r,r=0..R),z=-H..H)),
    n=1..infinity):
> eq28:=M(s)=c[0]*Pi*R^2*(2*H)/s-Q(s):
> eq29:=subs(eq27,eq28):
> aux1:=2*Pi*int(int(rhs(eq10)*r,r=0..R),z=-H..H):
> eq30:=coeff(convert(series(aux1,s=0,2),polynom),s,-1):
> eq31:=BesselJ(0,(-s/eta)^(1/2)*R)=0:
>
> eq32:=(-s/eta)^(1/2)*R=alpha[0,n]:
> eq33:=isolate(eq32,s):
> eq34:=numer(aux1)*exp(s*t)/diff(denom(aux1),s):
> eq35:=factor(subs(BesselJ(0,alpha[0,n])=0,simplify(subs(eq33,eq34),
    power,symbolic))):
> aux2:=4*R^6*h*eta^(1/2)*(c[b]-c[0])*(-1+exp(2/eta^(1/2)*
    (alpha[0,n]^2*eta+s*R^2)^(1/2)/R*H))/alpha[0,n]^2/(alpha[0,n]^2*
    (alpha[0,n]^2*eta+s*R^2)^(1/2)*R*h*eta*exp(2/eta^(1/2)*
    (alpha[0,n]^2*eta+s*R^2)^(1/2)/R*H)+alpha[0,n]^2*(alpha[0,n]^2*
    eta+s*R^2)^(1/2)*R*h*eta-2*alpha[0,n]^2*eta^(3/2)*s*R^2-
    eta^(1/2)*s^2*R^4-alpha[0,n]^4*eta^(5/2)+2*alpha[0,n]^2*
    eta^(3/2)*s*R^2*exp(2/eta^(1/2)*(alpha[0,n]^2*eta+s*R^2)^(1/2)/
    R*H)+eta^(1/2)*s^2*R^4*exp(2/eta^(1/2)*(alpha[0,n]^2*eta+
    s*R^2)^(1/2)/R*H)+(alpha[0,n]^2*eta+s*R^2)^(1/2)*R^3*h*s+
    alpha[0,n]^4*eta^(5/2)*exp(2/eta^(1/2)*(alpha[0,n]^2*eta+
    s*R^2)^(1/2)/R*H)+(alpha[0,n]^2*eta+s*R^2)^(1/2)*R^3*h*s*
    exp(2/eta^(1/2)*(alpha[0,n]^2*eta+s*R^2)^(1/2)/R*H)):
>
>
>
```

```
> aux4:=-1+exp(2/eta^(1/2)*(alpha[0,n]^2*eta+s*R^2)^(1/2)/R*H)=
    exp(1/eta^(1/2)*(alpha[0,n]^2*eta+s*R^2)^(1/2)/R*H)*
    convert(-exp(-1/eta^(1/2)*(alpha[0,n]^2*eta+s*R^2)^(1/2)/R*H)+
    exp(1/eta^(1/2)*(alpha[0,n]^2*eta+s*R^2)^(1/2)/R*H),sinh):
> aux5:=subs(aux4,aux2):
> eq36:=coeff(convert(series(aux5*exp(s*t),isolate(alpha[0,n]^2*eta+
    s*R^2=0,s),2),polynom),s+alpha[0,n]^2/R^2*eta,-1):
> eq37:=(denom((aux5)))=0:
> eq38:=(1/eta^(1/2)*(alpha[0,n]^2*eta+s*R^2)^(1/2)/R*H)=I*beta[m]:
> eq39:=isolate(eq38,s):
> eq39A:=simplify(subs(eq39,eq37),power,symbolic):
> eq39B:=isolate(eq39A,exp(2*I*beta[m])):
> eq40:=numer(aux5)*exp(s*t)/diff(denom(aux5),s):
> eq41:=simplify(subs(eq39,eq40),power,symbolic):
> eq42:=M(t)=simplify(2*c[0]*Pi*R^2*H-eq30-Sum(eq35,n=1..infinity)-
    Sum(2*Pi*eq36,n=1..infinity)-Sum(Sum(2*Pi*eq41,m=0..infinity),
    n=1..infinity)):
> eq43:=M(infinity)=factor(2*c[0]*Pi*R^2*H-2*Pi*R^2*c[b]*H):
> eq44:=M(t)/M(infinity)=1-Sum(Sum(factor(2*Pi*eq41/(rhs(eq43))),
    m=0..infinity),n=1..infinity):
> eq45:=(beta[m]*H^2*R^2*I-eta*t*beta[m]^2*R^2-eta*t*alpha[0,n]^2*
    H^2)/H^2/R^2=expand((beta[m]*H^2*R^2*I-eta*t*beta[m]^2*R^2-
    eta*t*alpha[0,n]^2*H^2)/H^2/R^2):
> eq46:=subs(eq45,eq44):
>
>
> eq46A:=subs(eq45,simplify(subs(eq39B,eq46)));
```

$$eq46A = \frac{M(t)}{M(\infty)} = 1 - \left(\sum_{n=1}^{\infty}\left(\sum_{m=0}^{\infty}\left(-\frac{8\,h\,e^{I\,\beta_m - \frac{\alpha_{0,n}^2\,\eta\,t}{R^2} - \frac{\eta\,t\,\beta_m^2}{H^2}}\sin\left(\beta_m\right)H(I\,H\,h - \eta\,\beta_m)}{\alpha_{0,n}^2\,\beta_m^2(H^2h^2 + \eta^2\,\beta_m^2 + H\eta\,h)}\right)\right)\right)$$

```
>
```

15.5.2 Maple Code: rectang3D.mws (Section 15.3)

```
> restart:with(inttrans):
> eq:=diff(c(x,t),t)=eta*diff(c(x,t),x,x):
> eq1:=c(x,0)=C[0]:
>
> eq2:=laplace(eq,t,s):
> eq3:=subs(eq1,eq2):
> eq4:=subs(laplace(c(x,t),t,s)=c(x),eq3):
> eq5:=dsolve(eq4):
> eq6:=subs(x=-a/2,eta*diff(rhs(eq5),x))=h*(subs(x=-a/2,rhs(eq5))-
    c[b]/s):
> eq7:=subs(x=a/2,-eta*diff(rhs(eq5),x))=h*(subs(x=a/2,rhs(eq5))-
    c[b]/s):
```

```
> eq8:= {_C1 = -h*(-C[0]+c[b])/(s*(exp(-sqrt(s)*a/(2*sqrt(eta)))*
     sqrt(s)*sqrt(eta)-exp(sqrt(s)*a/(2*sqrt(eta)))*sqrt(s)*sqrt(eta)-
     exp(-sqrt(s)*a/(2*sqrt(eta)))*h-exp(sqrt(s)*a/(2*sqrt(eta)))*h)),
     _C2 = -h*(-C[0]+c[b])/(s*(exp(-sqrt(s)*a/(2*sqrt(eta)))*sqrt(s)*
     sqrt(eta)-exp(sqrt(s)*a/(2*sqrt(eta)))*sqrt(s)*sqrt(eta)-
     exp(-sqrt(s)*a/(2*sqrt(eta)))*h-exp(sqrt(s)*a/(2*sqrt(eta)))*h))}:
>
Eq.8 is the simplified result of eq8 := (solve( {eq6, eq7},
      {_C1, _C2}));
> eq9:=simplify(subs(eq8,eq5)):
> eq10:=c[b]:
> eq11:=denom(rhs(eq9))=0:
> eq12:=1/2/eta^(1/2)*s^(1/2)*a=I*beta[n]:
> eq13:=isolate(eq12,s):
> eq13A:=evalc(simplify(subs(eq13,eq11),power,symbolic))
     assuming eta>0;
```

$$eq13A := -\frac{4\,\beta_n^2\,\eta\left(\frac{4\sin(\beta_n)\,\beta_n\,\eta}{a}-2\cos(\beta_n)\,h\right)}{a^2}=0$$

```
> eq13B:=isolate(eq13A,sin(beta[n])):
>
> eq14:=numer(rhs(eq9))*exp(s*t)/diff(denom(rhs(eq9)),s):
> eq15:=factor(subs(eq13B,evalc(simplify(subs(eq13,eq14),power,
     symbolic) assuming n::integer))):
> eq16:=c(x,t)=eq10+Sum(eq15,n=1..infinity):
> eq17:=M(t)=C[0]*a-c[b]*a-Sum(int(eq15,x=-a/2..a/2),n=1..infinity):
> eq18:=M(t)/M(infinity)=1-Sum(factor(int(eq15/(C[0]*a-c[b]*a),
     x=-a/2..a/2)),n=1..infinity):
> eq18A:=subs(eq13B,eq18):
> aux3:=2*h^2/beta[n]^2*a^2/(2*eta*a*h+4*beta[n]^2*eta^2+a^2*h^2)*
     exp(-4*beta[n]^2*eta/a^2*t):
> aux4:=(subs(a=b,n=m,beta=delta,aux3)):
> aux5:=(subs(a=c,n=p,beta=chi,aux3)):
>
>
> eq20:=M(t)/M(infinity)=1-Sum(Sum(Sum(factor(aux3*aux4*aux5),
     p=0..infinity),m=0..infinity),n=0..infinity);
```

$$eq20 := \frac{M(t)}{M(\infty)} = 1-\left(\sum_{n=0}^{\infty}\left(\sum_{m=0}^{\infty}\left(\sum_{p=0}^{\infty}\left(8\,h^6\,a^2 e^{-\frac{4\,\beta_n^2\,\eta\,t}{a^2}}\,b^2 e^{-\frac{4\,\delta_m^2\,\eta\,t}{b^2}}\,c^2 e^{-\frac{4\,\chi_p^2\,\eta\,t}{c^2}}\right)\Big/\left(\beta_n^2\left(a^2h^2+4\,\eta^2\,\beta_n^2+2\,a\,\eta\,h\right)\delta_m^2(b^2h^2+4\,\eta^2\,\delta_m^2+2\,b\,\eta\,h)\chi_p^2\left(c^2h^2+4\,\eta^2\,\chi_p^2+2\,c\,\eta\,h\right)\right)\right)\right)\right)$$

```
> aux6:=eq13B;
```

>

$aux6 := \sin(\beta_n) = \frac{1}{2}\,\frac{\cos(\beta_n)\,h\,a}{\beta_n\,\eta}$

```
> aux7:=subs(n=m,beta=delta,a=b,aux6);
```

$aux7 := \sin(\delta_m) = \frac{1}{2}\,\frac{\cos(\delta_m)\,h\,b}{\delta_m\,\eta}$

```
> aux8:=subs(n=p,beta=chi,a=c,aux6);
```

$aux8 := \sin(\chi_p) = \frac{1}{2}\,\frac{\cos(\chi_p)\,h\,c}{\chi_p\,\eta}$

>

15.5.3 Mathematica Code: cylind2Dmat.nb (Section 15.2)

---

eq0 = $\partial_t c[r, z, t] = \eta$Laplacian$[c[r, z, t], \{r, \theta, z\},$ "Cylindrical"]// Expand

$c^{(0,0,1)}[r, z, t] = \eta c^{(0,2,0)}[r, z, t] + \frac{\eta c^{(1,0,0)}[r,z,t]}{r} + \eta c^{(2,0,0)}[r, z, t]$

eq0A = $c[r, z, 0] = c[0]$

$c[r, z, 0] = c[0]$

eq = $sc[r, z] - c[r, z, 0] =$
$\frac{1}{r}\left(\eta\left(rc^{(0,2)}[r, z] + c^{(1,0)}[r, z] + rc^{(2,0)}[r, z]\right)\right)$

$sc[r, z] - c[r, z, 0] = \frac{\eta(rc^{(0,2)}[r,z]+c^{(1,0)}[r,z]+rc^{(2,0)}[r,z])}{r}$

eq1 = eq/.eq0A[[1]]−> eq0A[[2]]

$-c[0] + sc[r, z] = \frac{\eta(rc^{(0,2)}[r,z]+c^{(1,0)}[r,z]+rc^{(2,0)}[r,z])}{r}$

eq1A = $sC[r] - c[0] = \frac{\eta(C^{(1)}[r]+rC^{(2)}[r])}{r}$

$-c[0] + sC[r] = \frac{\eta(C'[r]+rC''[r])}{r}$

eq1B = Factor[DSolve[eq1A, $C[r]$, r]]

$\left\{\left\{C[r] \to \frac{c[0]+s\text{BesselJ}\left[0, \frac{ir\sqrt{s}}{\sqrt{\eta}}\right]C[1]+s\text{BesselY}\left[0, -\frac{ir\sqrt{s}}{\sqrt{\eta}}\right]C[2]}{s}\right\}\right\}$

eq1C = ((C[r]/.eq1B)[[1]])/.C[2] → 0

$\frac{c[0]+s\text{BesselJ}\left[0, \frac{ir\sqrt{s}}{\sqrt{\eta}}\right]C[1]}{s}$

eq1D = (eq1C/.r → R) = c[b]/s

$\frac{c[0]+s\text{BesselJ}\left[0, \frac{iR\sqrt{s}}{\sqrt{\eta}}\right]C[1]}{s} = \frac{c[b]}{s}$

eq1E = (Solve[eq1D, C[1]])[[1]]

$\left\{C[1] \to \frac{-c[0]+c[b]}{s\text{BesselJ}\left[0, \frac{iR\sqrt{s}}{\sqrt{\eta}}\right]}\right\}$

eq1F = eq1C/.eq1E

eq8 =

$C[x, t] =$

eq2A+

$$\sum_{n=0}^{\infty}\left(2ae^{-\frac{\beta[n](iax+t\eta\beta[n])}{a^2}}h\left(ie^{\frac{1}{2}i\beta[n]}\left(1+e^{\frac{2ix\beta[n]}{a}}\right)c[b](ah+i\eta\beta[n])+C[0]\left(-ia\left(e^{\frac{1}{2}i\beta[n]}-e^{\frac{ix\beta[n]}{a}}-e^{\frac{i(a+x)\beta[n]}{a}}+e^{\frac{i(a+4x)\beta[n]}{2a}}\right)h+\left(e^{\frac{1}{2}i\beta[n]}+e^{\frac{ix\beta[n]}{a}}-e^{\frac{i(a+x)\beta[n]}{a}}+e^{\frac{i(a+4x)\beta[n]}{2a}}\right)\eta\beta[n]\right)\right)\right)$$

$(\beta[n](ah(ah+2\eta)+\eta^2\beta[n]^2))$

eq2A = (eq1 /. $c[0] \to 0$)

$sc[r, z] = \frac{\eta(rc^{(0,2)}[r,z]+c^{(1,0)}[r,z]+rc^{(2,0)}[r,z])}{r}$

eq2 = eq2A /. $c \to$ `Function`$[\{r, z\}, f[r] * g[z]]$

$sf[r]g[z] = \frac{\eta(g[z]f'[r]+rg[z]f''[r]+rf[r]g''[z])}{r}$

eq3 = `Expand`[eq2[[1]]] / $(f[r] * g[z])$] = `Expand`[eq2[[2]] / $(f[r] * g[z])$]

$s = \frac{\eta f'[r]}{rf[r]} + \frac{\eta f'[r]}{f[r]} + \frac{\eta g''[z]}{g[z]}$

eq3A = $\frac{f'[r]}{rf[r]} + \frac{f'[r]}{f[r]} = c[1]$

$\frac{f'[r]}{rf[r]} + \frac{f'[r]}{f[r]} = c[1]$

eq4 = $s = c[1] * \eta + \frac{\eta g''[z]}{g[z]}$

$s = \eta c[1] + \frac{\eta g'[z]}{g[z]}$

eq5 = `DSolve`[eq3A, $f[r]$, r]

$\{\{f[r] \to$ `BesselJ`$[0, ir\sqrt{c[1]}]C[1] +$ `BesselY`$[0, -ir\sqrt{c[1]}]C[2]\}\}$

eq5A = (`DSolve`[eq4, $g[z]$, z] /. $C[1]->A[1]$) /. $C[2] \to B[1]$

$$\left\{\left\{g[z] \to e^{\frac{z\sqrt{s-\eta c[1]}}{\sqrt{\eta}}}A[1] + e^{-\frac{z\sqrt{s-\eta c[1]}}{\sqrt{\eta}}}B[1]\right\}\right\}$$

eq6 = ($f[r]$ /. eq5)($g[z]$ /. eq5A)[[1]]

$$\left\{\left(e^{\frac{z\sqrt{s-\eta c[1]}}{\sqrt{\eta}}}A[1] + e^{-\frac{z\sqrt{s-\eta c[1]}}{\sqrt{\eta}}}B[1]\right)(\texttt{BesselJ}[0, ir\sqrt{c[1]}]C[1] + \texttt{BesselY}[0, -ir\sqrt{c[1]}]C[2])\right\}$$

eq6A = ((eq6 /. $C[2] \to 0$) /. $C[1] \to 1$)[[1]]

$$\left(e^{\frac{z\sqrt{s-\eta c[1]}}{\sqrt{\eta}}}A[1] + e^{-\frac{z\sqrt{s-\eta c[1]}}{\sqrt{\eta}}}B[1]\right)\texttt{BesselJ}[0, ir\sqrt{c[1]}]$$

eq6C = (eq6A /. $r \to R$) = 0

$$\left(e^{\frac{z\sqrt{s-\eta c[1]}}{\sqrt{\eta}}}A[1] + e^{-\frac{z\sqrt{s-\eta c[1]}}{\sqrt{\eta}}}B[1]\right)\texttt{BesselJ}[0, iR\sqrt{c[1]}] = 0$$

eq6D = `BesselJ`$[0, iR\sqrt{c[1]}] = 0$

`BesselJ`$[0, iR\sqrt{c[1]}] = 0$

eq6E = $iR\sqrt{c[1]} = \alpha[0, n]$

$iR\sqrt{c[1]} = \alpha[0, n]$

eq6F = `Solve`[eq6E, $c[1]$]

$\left\{\left\{c[1] \to -\frac{\alpha[0,n]^2}{R^2}\right\}\right\}$

eq6J = Simplify[eq6A / .eq6F, $\alpha[0, n] > 0 \&\& R > 0$]

$$\left\{e^{-\frac{z\sqrt{s+\frac{\eta\alpha[0,n]^2}{R^2}}}{\sqrt{\eta}}}\left(e^{\frac{2z\sqrt{s+\frac{\eta\alpha[0,n]^2}{R^2}}}{\sqrt{\eta}}}A[1]+B[1]\right)\texttt{BesselJ}\left[0,\ \frac{r\alpha[0,n]}{R}\right]\right\}$$

eq6J1 = (eq6J / .A[1] → A[n]) / .B[1]− > B[n]

$$\left\{e^{-\frac{z\sqrt{s+\frac{\eta\alpha[0,n]^2}{R^2}}}{\sqrt{\eta}}}\left(e^{\frac{2z\sqrt{s+\frac{\eta\alpha[0,n]^2}{R^2}}}{\sqrt{\eta}}}A[n]+B[n]\right)\texttt{BesselJ}\left[0,\ \frac{r\alpha[0,n]}{R}\right]\right\}$$

eq6J1 =

$$\left\{\texttt{BesselJ}\left[0,\ \frac{r\alpha[0,n]}{R}\right]\left(A\,[n]\,\mathrm{Cos}\left[\frac{z\sqrt{-s-\frac{\eta\alpha[0,n]^2}{R^2}}}{\sqrt{\eta}}\right]+B[n]\mathrm{Sin}\left[\frac{z\sqrt{-s-\frac{\eta\alpha[0,n]^2}{R^2}}}{\sqrt{\eta}}\right]\right)\right\}$$

$$\left\{\texttt{BesselJ}\left[0,\ \frac{r\alpha[0,n]}{R}\right]\left(A\,[n]\,\mathrm{Cos}\left[\frac{z\sqrt{-s-\frac{\eta\alpha[0,n]^2}{R^2}}}{\sqrt{\eta}}\right]+B[n]\mathrm{Sin}\left[\frac{z\sqrt{-s-\frac{\eta\alpha[0,n]^2}{R^2}}}{\sqrt{\eta}}\right]\right)\right\}$$

eq6K =
$C[r, z]$ =
eq1F+

$$\sum_{n=1}^{\infty}\texttt{BesselJ}\left[0,\ \frac{r\alpha\,[0,n]}{R}\right]\left(A\,[n]\,\mathrm{Cos}\left[\frac{z\sqrt{-s-\frac{\eta\alpha[0,n]^2}{R^2}}}{\sqrt{\eta}}\right]+B[n]\mathrm{Sin}\left[\frac{z\sqrt{-s-\frac{\eta\alpha[0,n]^2}{R^2}}}{\sqrt{\eta}}\right]\right);$$

$$\text{eq7} = ((\eta(\partial_z(\text{eq6K}[[2]])))\,/.z \to -H) = h(((\text{eq6K}[[2]])\,/.z \to -H) - c[b]/s);$$

$$\text{eq7A} = ((-\eta(\partial_z(\text{eq6K}[[2]])))\,/.z \to H) = h(((\text{eq6K}[[2]])\,/.z \to H) - c[b]/s);$$

eq8 =

$$((\texttt{Integrate}[\eta * (\partial_z((\text{eq1F} + \text{eq6J1}[[1]]))\,/.z \to -H) * r * \texttt{BesselJ}\left[0,\ \tfrac{r\alpha[0,n]}{R}\right], \{r, 0, R\}\Big]\Big)\,/.\texttt{BesselJ}[0, \alpha[0,n]] \to 0) =$$

$$(h * \texttt{Integrate}[r * \texttt{BesselJ}\left[0,\ \tfrac{r\alpha[0,n]}{R}\right] * \left(((\text{eq1F} + \text{eq6J1}\ [[1]])\ /.z \to -H) - \tfrac{c[b]}{s}\right), \{r, 0, R\}\Big]\Big)\ /. \texttt{BesselJ}[0, \alpha[0,n]] \to 0;$$

eq9 =

$$((\texttt{Integrate}[-\eta * (\partial_z((\text{eq1F} + \text{eq6J1}[[1]]))\,/.z \to H) * r * \texttt{BesselJ}\left[0,\ \tfrac{r\alpha[0,n]}{R}\right], \{r, 0, R\}\Big]\Big)\,/.\texttt{BesselJ}[0, \alpha[0,n]] \to 0) =$$

$$(h * \texttt{Integrate}[r * \texttt{BesselJ}\left[0,\ \tfrac{r\alpha[0,n]}{R}\right] * \left(((\text{eq1F} + \text{eq6J1}\ [[1]])\ /.z \to H) - \tfrac{c[b]}{s}\right), \{r, 0, R\}\Big]\Big)\ /. \texttt{BesselJ}[0, \alpha[0,n]] \to 0;$$

eq10 = `Simplify[Solve[`{eq8, eq9}, {$A[n]$, $B[n]$}]]

$$\left\{\left\{A[n]\to -\left(\left(4hR^4\,(c[0]-c[b])\left(h\mathrm{Sin}\left[\frac{H\sqrt{-s-\frac{\eta\alpha[0,n]^2}{R^2}}}{\sqrt{\eta}}\right]+\sqrt{\eta}\mathrm{Cos}\left[\frac{H\sqrt{-s-\frac{\eta\alpha[0,n]^2}{R^2}}}{\sqrt{\eta}}\right]\sqrt{-s-\frac{\eta\alpha[0,n]^2}{R^2}}\right)\right)\Big/\left(\texttt{BesselJ}[1,\alpha[0,n]]\,\alpha[0,n](R^2s+\eta\alpha[0,n]^2)\left(\eta^2\mathrm{Sin}\left[\frac{2H\sqrt{-s-\frac{\eta\alpha[0,n]^2}{R^2}}}{\sqrt{\eta}}\right]\alpha[0,n]^2+R^2\left(\left(h^2+s\eta\right)\mathrm{Sin}\left[\frac{2H\sqrt{-s-\frac{\eta\alpha[0,n]^2}{R^2}}}{\sqrt{\eta}}\right]+2h\sqrt{\eta}\mathrm{Cos}\left[\frac{H\sqrt{-s-\frac{\eta\alpha[0,n]^2}{R^2}}}{\sqrt{\eta}}\right]\sqrt{-s-\frac{\eta\alpha[0,n]^2}{R^2}}\right)\right)\right)\right),\ B[n]\to 0\right\}\right\}$$

eq11 = (eq6K / .eq10)[[1]]

$$C[r,z]=\frac{c[0]+\frac{\texttt{BesselJ}\left[0,\frac{ir\sqrt{s}}{\sqrt{\eta}}\right](-c[0]+c[b])}{\texttt{BesselJ}\left[0,\frac{iR\sqrt{s}}{\sqrt{\eta}}\right]}}{s}+\sum_{n=1}^{\infty}-\left(\left(4hR^4\texttt{BesselJ}\left[0,\frac{r\alpha[0,n]}{R}\right](c[0]-c[b])\mathrm{Cos}\left[\frac{z\sqrt{-s-\frac{\eta\alpha[0,n]^2}{R^2}}}{\sqrt{\eta}}\right]\left(h\mathrm{Sin}\left[\frac{H\sqrt{-s-\frac{\eta\alpha[0,n]^2}{R^2}}}{\sqrt{\eta}}\right]+\sqrt{\eta}\mathrm{Cos}\left[\frac{H\sqrt{-s-\frac{\eta\alpha[0,n]^2}{R^2}}}{\sqrt{\eta}}\right]\sqrt{-s-\frac{n\alpha[0,n]^2}{R^2}}\right)\right)\Big/\left(\texttt{BesselJ}[1,\alpha[0,n]]\,\alpha[0,n](R^2s+\eta\alpha[0,n]^2)\left(\eta^2\mathrm{Sin}\left[\frac{2H\sqrt{-s-\frac{\eta\alpha[0,n]^2}{R^2}}}{\sqrt{\eta}}\right]\alpha[0,n]^2+R^2\left(\left(h^2+s\eta\right)\mathrm{Sin}\left[\frac{2H\sqrt{-s-\frac{\eta\alpha[0,n]^2}{R^2}}}{\sqrt{\eta}}\right]+2h\sqrt{\eta}\mathrm{Cos}\left[\frac{2H\sqrt{-s-\frac{\eta\alpha[0,n]^2}{R^2}}}{\sqrt{\eta}}\right]\sqrt{-s-\frac{n\alpha[0,n]^2}{R^2}}\right)\right)\right)\right)$$

eq12 =
`Simplify[`
`Expand[2 * Pi * Integrate[Integrate[`r * eq1F, {r, 0, R}], {z, $-H$, H}]]] +
(2 * *Pi*) *

$$\sum_{n=1}^{\infty}\left(hR^2\texttt{Integrate}\left[r*\texttt{BesselJ}\left[0,\frac{r\alpha[0,n]}{R}\right],\{r,0,R\}\right](c[0]-c[b])\ \texttt{Integrate}\left[\mathrm{Cos}\left[\frac{z\sqrt{-s-\frac{n\alpha[0,n]^2}{R^2}}}{\sqrt{\eta}}\right],\{z,-H,H\}\right]\ \texttt{Hypergeometric0F1Regularized}\left[2,-\frac{1}{4}\alpha[0,n]^2\right]\right)$$

$$(\texttt{BesselJ}[1,\alpha[0,n]])^2(R^2s+\eta\alpha[0,n]^2)\left(-h\mathrm{Cos}\left[\frac{H\sqrt{-s-\frac{n\alpha[0,n]^2}{R^2}}}{\sqrt{\eta}}\right]+\sqrt{\eta}\mathrm{Sin}\left[\frac{H\sqrt{-s-\frac{n\alpha[0,n]^2}{R^2}}}{\sqrt{\eta}}\right]\sqrt{-s-\frac{\eta\alpha[0,n]^2}{R^2}}\right)\Big)$$

$$\left(2H\pi R^2\left(c[0]\left(\texttt{Hypergeometric0F1Regularized}\left[1,\frac{R^2s}{4\eta}\right]-\texttt{Hypergeometric0F1Regularized}\left[2,\frac{R^2s}{4\eta}\right]\right)+c[b]\texttt{Hypergeometric0F1Regularized}\left[2,\frac{R^2s}{4\eta}\right]\right)\right)/\left(s\texttt{Hypergeometric0F1Regularized}\left[1,\frac{R^2s}{4\eta}\right]\right)+$$

$$2\pi\sum_{n=1}^{\infty}\left(2hR^4\sqrt{\eta}(c[0]-c[b])\ \texttt{Hypergeometric0F1Regularized}\left[2,-\frac{1}{4}\alpha[0,n]^2\right]\mathrm{Sin}\left[\frac{H\sqrt{-s-\frac{\eta\alpha[0,n]^2}{R^2}}}{\sqrt{\eta}}\right]\right)/$$

$$\left(\texttt{BesselJ}[1,\alpha[0,n]]\alpha[0,n](R^2s+\eta\alpha[0,n]^2)\sqrt{-s-\frac{\eta\alpha[0,n]^2}{R^2}}\left(-h\mathrm{Cos}\left[\frac{H\sqrt{-s-\frac{\eta\alpha[0,n]^2}{R^2}}}{\sqrt{\eta}}\right]+\sqrt{\eta}\mathrm{Sin}\left[\frac{H\sqrt{-s-\frac{\eta\alpha[0,n]^2}{R^2}}}{\sqrt{\eta}}\right]\sqrt{-s-\frac{\eta\alpha[0,n]^2}{R^2}}\right)\right)$$

eq13 = $M[s] = 2 * \texttt{Pi} * R\{2 * c[0] * H/s - \mathrm{eq12}$

eq14 =

$$(2H\pi R^2\left(c[0]\left(\texttt{Hypergeometric0F1Regularized}\left[1,\frac{R^2s}{4\eta}\right]-\texttt{Hypergeometric0F1Regularized}\left[2,\frac{R^2s}{4\eta}\right]\right)+c[b]\texttt{Hypergeometric0F1Regularized}\left[2,\frac{R^2s}{4\eta}\right]\right))$$

$$\left(s\texttt{Hypergeometric0F1Regularized}\left[1,\frac{R^2s}{4\eta}\right]\right)$$

$$\left(2H\pi R^2\left(c[0]\left(\texttt{Hypergeometric0F1Regularized}\left[1,\frac{R^2s}{4\eta}\right]-\texttt{Hypergeometric0F1Regularized}\left[2,\frac{R^2s}{4\eta}\right]\right)+c[b]\texttt{Hypergeometric0F1Regularized}\left[2,\frac{R^2s}{4\eta}\right]\right)\right)/\left(s\texttt{Hypergeometric0F1Regularized}\left[1,\frac{R^2s}{4\eta}\right]\right)$$

eq15 = `Residue`[eq14, $\{s, 0\}$]

$2H\pi R^2c[b]$

eq16 =
`Simplify[`

$$\left(2hR^4\sqrt{\eta}(c[0]-c[b])\texttt{Hypergeometric0F1Regularized}\left[2,-\frac{1}{4}\alpha[0,n]^2\right]\mathrm{Sin}\left[\frac{H\sqrt{-s-\frac{\eta\alpha[0,n]^2}{R^2}}}{\sqrt{\eta}}\right]\right)$$

$$\left(\texttt{BesselJ}[1,\alpha[0,n]]\alpha[0,n](R^2s+\eta\alpha[0,n]^2)\sqrt{-s-\frac{\eta\alpha[0,n]^2}{R^2}}\left(-h\mathrm{Cos}\left[\frac{H\sqrt{-s-\frac{\eta\alpha[0,n]^2}{R^2}}}{\sqrt{\eta}}\right]+\sqrt{\eta}\mathrm{Sin}\left[\frac{H\sqrt{-s-\frac{\eta\alpha[0,n]^2}{R^2}}}{\sqrt{\eta}}\right]\sqrt{-s-\frac{\eta\alpha[0,n]^2}{R^2}}\right)\right)\Bigg];$$

$$-h\mathrm{Cos}\left[\frac{H\sqrt{-s-\frac{\eta\alpha[0,n]^2}{R^2}}}{\sqrt{\eta}}\right]+\sqrt{\eta}\mathrm{Sin}\left[\frac{H\sqrt{-s-\frac{\eta\alpha[0,n]^2}{R^2}}}{\sqrt{\eta}}\right]\sqrt{-s-\frac{\eta\alpha[0,n]^2}{R^2}}=0$$

$$\mathrm{eq17}=\frac{H\sqrt{-s-\frac{\eta\alpha[0,n]^2}{R^2}}}{\sqrt{\eta}}=\beta[m]$$

$$\frac{H\sqrt{-s-\frac{\eta\alpha[0,n]^2}{R^2}}}{\sqrt{\eta}} = \beta[m]$$

eq18 = `Solve`[eq17, s]

$$\left\{\left\{s \to \frac{-H^2\eta\alpha[0,n]^2 - R^2\eta\beta[m]^2}{H^2R^2}\right\}\right\}$$

eq18B =
(`Simplify`[

$$\left(-h\text{Cos}\left[\frac{H\sqrt{-s-\frac{\eta\alpha[0,n]^2}{R^2}}}{\sqrt{\eta}}\right] + \sqrt{\eta}\text{Sin}\left[\frac{H\sqrt{-s-\frac{\eta\alpha[0,n]^2}{R^2}}}{\sqrt{\eta}}\right]\sqrt{-s-\frac{\eta\alpha[0,n]^2}{R^2}} = \right.$$

0) / .eq18, $\alpha[0, n] > 0 \&\& \eta > 0 \&\& R > 0 \&\& \beta[m] > 0 \&\& H > 0$])[[1]]

$$hH\text{Cos}[\beta[m]] = \eta\text{Sin}[\beta[m]]\beta[m]$$

eq18C = `Solve`[eq18B, Cos[$\beta[m]$]]

$$\left\{\left\{\text{Cos}[\beta[m]] \to \frac{\eta\text{Sin}[\beta[m]]\beta[m]}{hH}\right\}\right\}$$

eq19 =

$$\left(2hR^4\sqrt{\eta}(c[0]-c[b])\,\texttt{Hypergeometric0F1Regularized}\left[2, -\tfrac{1}{4}\alpha[0,n]^2\right]\text{Sin}\left[\frac{H\sqrt{-s-\frac{\eta\alpha[0,n]^2}{R^2}}}{\sqrt{\eta}}\right]\right) *$$

Exp[$s * t$] /

$$\left(\partial_s\left(\texttt{BesselJ}[1,\alpha[0,n]]\alpha[0,n](R^2s+\eta\alpha[0,n]^2)\sqrt{-s-\frac{\eta\alpha[0,n]^2}{R^2}}\left(-h\text{Cos}\left[\frac{H\sqrt{-s-\frac{\eta\alpha[0,n]^2}{R^2}}}{\sqrt{\eta}}\right] + \sqrt{\eta}\text{Sin}\left[\frac{H\sqrt{-s-\frac{\eta\alpha[0,n]^2}{R^2}}}{\sqrt{\eta}}\right]\sqrt{-s-\frac{\eta\alpha[0,n]^2}{R^2}}\right)\right)\right)$$

$$\left(2e^{st}hR^4\sqrt{\eta}(c[0]-c[b])\,\texttt{Hypergeometric0F1Regularized}\left[2,-\tfrac{1}{4}\alpha[0,n]^2\right]\text{Sin}\left[\frac{H\sqrt{-s-\frac{\eta\alpha[0,n]^2}{R^2}}}{\sqrt{\eta}}\right]\right) /$$

$$\left(\texttt{BesselJ}[1,\alpha[0,n]]\alpha[0,n](R^2s+\eta\alpha[0,n]^2)\sqrt{-s-\frac{\eta\alpha[0,n]^2}{R^2}}\left(-\tfrac{1}{2}H\text{Cos}\left[\frac{H\sqrt{-s-\frac{\eta\alpha[0,n]^2}{R^2}}}{\sqrt{\eta}}\right] - \frac{hH\text{Sin}\left[\frac{H\sqrt{-s-\frac{\eta\alpha[0,n]^2}{R^2}}}{\sqrt{\eta}}\right]}{2\sqrt{\eta}\sqrt{-s-\frac{\eta\alpha[0,n]^2}{R^2}}} - \frac{\sqrt{\eta}\text{Sin}\left[\frac{H\sqrt{-s-\frac{\eta\alpha[0,n]^2}{R^2}}}{\sqrt{\eta}}\right]}{2\sqrt{-s-\frac{\eta\alpha[0,n]^2}{R^2}}}\right) - \right.$$

$$\frac{1}{2\sqrt{-s-\frac{\eta\alpha[0,n]^2}{R^2}}}\texttt{BesselJ}[1,\alpha[0,n]]\alpha[0,n](R^2s+\eta\alpha[0,n]^2)\left(-h\text{Cos}\left[\frac{H\sqrt{-s-\frac{\eta\alpha[0,n]^2}{R^2}}}{\sqrt{\eta}}\right] + \sqrt{\eta}\text{Sin}\left[\frac{H\sqrt{-s-\frac{\eta\alpha[0,n]^2}{R^2}}}{\sqrt{\eta}}\right]\sqrt{-s-\frac{\eta\alpha[0,n]^2}{R^2}}\right) +$$

$$\left.R^2\texttt{BesselJ}[1,\alpha[0,n]]\alpha[0,n]\sqrt{-s-\frac{\eta\alpha[0,n]^2}{R^2}}\left(-h\text{Cos}\left[\frac{H\sqrt{-s-\frac{\eta\alpha[0,n]^2}{R^2}}}{\sqrt{\eta}}\right] + \sqrt{\eta}\text{Sin}\left[\frac{H\sqrt{-s-\frac{\eta\alpha[0,n]^2}{R^2}}}{\sqrt{\eta}}\right]\sqrt{-s-\frac{\eta\alpha[0,n]^2}{R^2}}\right)\right)$$

eq20 = Simplify[eq19 / .eq18,
$\alpha[0, n] > 0 \&\& \eta > 0 \&\& R > 0 \&\& \beta[m] > 0 \&\& H > 0]$

$$\left\{\left(4e^{t\eta\left(-\frac{\alpha[0,n]^2}{R^2}-\frac{\beta[m]^2}{H^2}\right)}hH^2R^2(c[0]-c[b])\,\text{Hypergeometric0F1Regularized}\left[2,-\frac{1}{4}\alpha[0,n]^2\right]\text{Sin}[\beta[m]]\right)\Big/\left(\text{BesselJ}[1,\alpha[0,n]]\alpha[0,n]\beta[m](-3hH\text{Cos}[\beta[m]]+(hH+4\eta)\text{Sin}[\beta[m]]\beta[m]+\eta\text{Cos}[\beta[m]]\beta[m]^2)\right)\right\}$$

eq21 = ((Simplify[(eq20 / .eq18C)])[[1]])[[1]]

$$\left(4e^{t\eta\left(-\frac{\alpha[0,n]^2}{R^2}-\frac{\beta[m]^2}{H^2}\right)}h^2H^3R^2(c[0]-c[b])\,\text{Hypergeometric0F1Regularized}\left[2,-\frac{1}{4}\alpha[0,n]^2\right]\right)\Big/(\text{BesselJ}[1,\alpha[0,n]]\alpha[0,n]\beta[m]^2(hH(hH+\eta)+\eta^2\beta[m]^2))$$

eq22 =

$$M[t]=2H\pi R^2c[0]-\text{eq15}-\sum_{n=1}^{\infty}\sum_{m=0}^{\infty}2*\text{Pi}*\left(4e^{t\eta\left(-\frac{\alpha[0,n]^2}{R^2}-\frac{\beta[m]^2}{H^2}\right)}h^2H^3R^2(c[0]-c[b])\,\text{Hypergeometric0F1Regularized}\left[2,-\frac{1}{4}\alpha[0,n]^2\right]\right)(\text{BesselJ}[1,\alpha[0,n]]\alpha[0,n]\beta[m]^2(hH(hH+\eta)+\eta^2\beta[m]^2))$$

$$M[t]=2H\pi R^2c[0]-2H\pi R^2c[b]-\sum_{n=1}^{\infty}\sum_{m=0}^{\infty}\left(8e^{t\eta\left(-\frac{\alpha[0,n]^2}{R^2}-\frac{\beta[m]^2}{H^2}\right)}h^2H^3\pi R^2(c[0]-c[b])\,\text{Hypergeometric0F1Regularized}\left[2,-\frac{1}{4}\alpha[0,n]^2\right]\right)\Big/(\text{BesselJ}[1,\alpha[0,n]]\alpha[0,n]\beta[m]^2(hH(hH+\eta)+\eta^2\beta[m]^2))$$

eq23 =
Factor[

$$\left(\left(8e^{t\eta\left(-\frac{\alpha[0,n]^2}{R^2}-\frac{\beta[m]^2}{H^2}\right)}h^2H^3\pi R^2(c[0]-c[b])\,\text{Hypergeometric0F1Regularized}\left[2,-\frac{1}{4}\alpha[0,n]^2\right]\right)(\text{BesselJ}[1,\alpha[0,n]]\alpha[0,n]\beta[m]^2(hH(hH+\eta)+\eta^2\beta[m]^2))\right)(2H\pi R^2c[0]-2H\pi R^2c[b])]$$

$$\frac{4e^{t\eta\left(-\frac{\alpha[0,n]^2}{R^2}-\frac{\beta[m]^2}{H^2}\right)}h^2H^2\text{Hypergeometric0F1Regularized}\left[2,-\frac{1}{4}\alpha[0,n]^2\right]}{\text{BesselJ}[1,\alpha[0,n]]\alpha[0,n]\beta[m]^2(h^2H^2+hH\eta+\eta^2\beta[m]^2)}$$

$$\text{eq24}=M[t]/M[\infty]=1-\sum_{n=1}^{\infty}\sum_{m=0}^{\infty}\frac{4e^{t\eta\left(-\frac{\alpha[0,n]^2}{R^2}-\frac{\beta[m]^2}{H^2}\right)}h^2H^2\text{Hypergeometric0F1Regularized}\left[2,-\frac{1}{4}\alpha[0,n]^2\right]}{\text{BesselJ}[1,\alpha[0,n]]\alpha[0,n]\beta[m]^2(h^2H^2+hH\eta+\eta^2\beta[m]^2)}$$

$$\frac{M[t]}{M[\infty]}=1-\sum_{n=1}^{\infty}\sum_{m=0}^{\infty}\frac{4e^{t\eta\left(-\frac{\alpha[0,n]^2}{R^2}-\frac{\beta[m]^2}{H^2}\right)}h^2H^2\text{Hypergeometric0F1Regularized}\left[2,-\frac{1}{4}\alpha[0,n]^2\right]}{\text{BesselJ}[1,\alpha[0,n]]\alpha[0,n]\beta[m]^2(h^2H^2+hH\eta+\eta^2\beta[m]^2)}$$

eq25 = BesselJ[1, α[0, n]] →
(α[0, n]/2) * Hypergeometric0F1Regularized$\left[2, -\frac{1}{4}\alpha[0,n]^2\right]$
BesselJ[1, α[0, n]] → $\frac{1}{2}$Hypergeometric0F1Regularized$\left[2, -\frac{1}{4}\alpha[0,n]^2\right]\alpha[0,n]$
eq26 = eq24/.eq25

$$\frac{M[t]}{M[\infty]} = 1 - \sum_{n=1}^{\infty}\sum_{m=0}^{\infty}\frac{8e^{t\eta\left(-\frac{\alpha[0,n]^2}{R^2}-\frac{\beta[m]^2}{H^2}\right)}h^2H^2}{\alpha[0,n]^2\beta[m]^2(h^2H^2+hH\eta+\eta^2\beta[m]^2)}$$

15.5.4 Mathematica Code: rectang3Dmat.nb (Section 15.3)

eq0 = $\partial_t c[x,t] = \eta * \partial_{x,x}c[x,t]$
$c^{(0,1)}[x,t] = \eta c^{(2,0)}[x,t]$
eq0A = $c[x,0] = C[0]$
$c[x,0] = C[0]$
eq = $s * c[x] - c[x,0] = \eta * \partial_{x,x}c[x]$
$sc[x] - c[x,0] = \eta c''[x]$
eq1 = eq/.eq0A[[1]]- > eq0A[[2]]
$sc[x] - C[0] = \eta c''[x]$
eq1B = (Factor[DSolve[eq1, $c[x]$, x]])[[1]]

$$\left\{c[x] \to \frac{e^{-\frac{\sqrt{sx}}{\sqrt{\eta}}}\left(e^{\frac{\sqrt{sx}}{\sqrt{\eta}}}C[0]+e^{\frac{2\sqrt{sx}}{\sqrt{\eta}}}sC[1]+sC[2]\right)}{s}\right\}$$

eq1C = $\frac{C[0]}{s} + e^{\frac{\sqrt{sx}}{\sqrt{\eta}}}C[1] + e^{-\frac{\sqrt{sx}}{\sqrt{\eta}}}C[2]$

$$\frac{C[0]}{s} + e^{\frac{\sqrt{sx}}{\sqrt{\eta}}}C[1] + e^{-\frac{\sqrt{sx}}{\sqrt{\eta}}}C[2]$$

eq1D = $\eta * ((\partial_x\text{eq1C})/.x \to -a/2) = h * ((\text{eq1C}/.x \to -a/2) - c[b]/s)$

$$\eta\left(\frac{e^{-\frac{a\sqrt{s}}{2\sqrt{\eta}}}\sqrt{s}C[1]}{\sqrt{\eta}} - \frac{e^{\frac{a\sqrt{s}}{2\sqrt{\eta}}}\sqrt{s}C[2]}{\sqrt{\eta}}\right) = h\left(-\frac{c[b]}{s} + \frac{C[0]}{s} + e^{-\frac{a\sqrt{s}}{2\sqrt{\eta}}}C[1] + e^{\frac{a\sqrt{s}}{2\sqrt{\eta}}}C[2]\right)$$

eq1DA = $-\eta * ((\partial_x\text{eq1C})/.x \to a/2) = h * ((\text{eq1C}/.x \to a/2) - c[b]/s)$

$$-\eta\left(\frac{e^{\frac{a\sqrt{s}}{2\sqrt{\eta}}}\sqrt{s}C[1]}{\sqrt{\eta}} - \frac{e^{-\frac{a\sqrt{s}}{2\sqrt{\eta}}}\sqrt{s}C[2]}{\sqrt{\eta}}\right) = h\left(-\frac{c[b]}{s} + \frac{C[0]}{s} + e^{\frac{a\sqrt{s}}{2\sqrt{\eta}}}C[1] + e^{-\frac{a\sqrt{s}}{2\sqrt{\eta}}}C[2]\right)$$

eq1E = Solve[{eq1D, eq1DA}, {$C[1]$, $C[2]$}]

$$\left\{\left\{C[1] \to \frac{e^{\frac{a\sqrt{s}}{2\sqrt{\eta}}}h(c[b]-C[0])}{s\left(h+e^{\frac{a\sqrt{s}}{\sqrt{\eta}}}h-\sqrt{s}\sqrt{\eta}+e^{\frac{a\sqrt{s}}{\sqrt{\eta}}}\sqrt{s}\sqrt{\eta}\right)}, C[2] \to \frac{e^{\frac{a\sqrt{s}}{2\sqrt{\eta}}}h(c[b]-C[0])}{s\left(h+e^{\frac{a\sqrt{s}}{\sqrt{\eta}}}h-\sqrt{s}\sqrt{\eta}+e^{\frac{a\sqrt{s}}{\sqrt{\eta}}}\sqrt{s}\sqrt{\eta}\right)}\right\}\right\}$$

eq1F = (Factor[eq1C/.eq1E])[[1]]

$$\left(e^{-\frac{\sqrt{sx}}{\sqrt{\eta}}}\left(e^{\frac{a\sqrt{s}}{2\sqrt{\eta}}+\frac{2\sqrt{sx}}{\sqrt{\eta}}}hc[b]+e^{\frac{a\sqrt{s}}{2\sqrt{\eta}}}hc[b]+e^{\frac{a\sqrt{s}}{\sqrt{\eta}}+\frac{\sqrt{sx}}{\sqrt{\eta}}}hC[0]-e^{\frac{a\sqrt{s}}{2\sqrt{\eta}}+\frac{2\sqrt{sx}}{\sqrt{\eta}}}hC[0]-e^{\frac{a\sqrt{s}}{2\sqrt{\eta}}}hC[0]+e^{\frac{\sqrt{sx}}{\sqrt{\eta}}}hC[0]+e^{\frac{a\sqrt{s}}{\sqrt{\eta}}+\frac{\sqrt{sx}}{\sqrt{\eta}}}\sqrt{s}\sqrt{\eta}C[0]-e^{\frac{\sqrt{sx}}{\sqrt{\eta}}}\sqrt{s}\sqrt{\eta}C[0]\right)\right)/$$
$$\left(s\left(h+e^{\frac{a\sqrt{s}}{\sqrt{\eta}}}h-\sqrt{s}\sqrt{\eta}+e^{\frac{a\sqrt{s}}{\sqrt{\eta}}}\sqrt{s}\sqrt{\eta}\right)\right)$$

eq2A = Normal[Series[eq1F, $\{s, 0, -1\}$]]$^*s$

$c[b]$

$$\text{eq2} = h+e^{\frac{a\sqrt{s}}{\sqrt{\eta}}}h-\sqrt{s}\sqrt{\eta}+e^{\frac{a\sqrt{s}}{\sqrt{\eta}}}\sqrt{s}\sqrt{\eta}=0$$

$$h+e^{\frac{a\sqrt{s}}{\sqrt{\eta}}}h-\sqrt{s}\sqrt{\eta}+e^{\frac{a\sqrt{s}}{\sqrt{\eta}}}\sqrt{s}\sqrt{\eta}=0$$

$$\text{eq3} = \frac{a\sqrt{s}}{\sqrt{\eta}} = \texttt{Sqrt[-1]} * \beta[n]$$

$$\frac{a\sqrt{s}}{\sqrt{\eta}} = i\beta[n]$$

eq5 = Solve[eq3, s]

$$\left\{\left\{s\to-\frac{\eta\beta[n]^2}{a^2}\right\}\right\}$$

eq5a = Simplify[eq2/.eq5, $\eta > 0$ && $a > 0$ && $\beta[n] > 0$]

$$\{a(1+e^{i\beta[n]})h+i(-1+e^{i\beta[n]})\eta\beta[n]=0\}$$

eq5b = Solve[eq5a/.$e^{i\beta[n]}\to X, X$]

$$\left\{\left\{X\to\frac{-ah+i\eta\beta[n]}{ah+i\eta\beta[n]}\right\}\right\}$$

$$\text{eq5b1} = e^{i\beta[n]}\to\frac{-ah+i\eta\beta[n]}{ah+i\eta\beta[n]}$$

$$e^{i\beta[n]}\to\frac{-ah+i\eta\beta[n]}{ah+i\eta\beta[n]}$$

$$\text{eq5c} = 2 * \text{Cos}[\beta[n]] = \texttt{Simplify}\left[\frac{-ah+i\eta\beta[n]}{ah+i\eta\beta[n]}+\left(\frac{-ah+i\eta\beta[n]}{ah+i\eta\beta[n]}\right)\hat{}\,(-1)\right]$$

$$2\text{Cos}[\beta[n]] = \frac{-2a^2h^2+2\eta^2\beta[n]^2}{a^2h^2+\eta^2\beta[n]^2}$$

eq5A =

$$\left(e^{-\frac{\sqrt{sx}}{\sqrt{\eta}}}\left(e^{\frac{a\sqrt{s}}{2\sqrt{\eta}}+\frac{2\sqrt{sx}}{\sqrt{\eta}}}hc[b]+e^{\frac{a\sqrt{s}}{2\sqrt{\eta}}}hc[b]+e^{\frac{a\sqrt{s}}{\sqrt{\eta}}+\frac{\sqrt{sx}}{\sqrt{\eta}}}hC[0]-e^{\frac{a\sqrt{s}}{2\sqrt{\eta}}+\frac{2\sqrt{sx}}{\sqrt{\eta}}}hC[0]-e^{\frac{a\sqrt{s}}{2\sqrt{\eta}}}hC[0]+e^{\frac{\sqrt{sx}}{\sqrt{\eta}}}hC[0]+e^{\frac{a\sqrt{s}}{\sqrt{\eta}}+\frac{\sqrt{sx}}{\sqrt{\eta}}}\sqrt{s}\sqrt{\eta}C[0]-e^{\frac{\sqrt{sx}}{\sqrt{\eta}}}\sqrt{s}\sqrt{\eta}C[0]\right)\right)*$$

$$\text{Exp}[s*t]\left(\partial_s\left(s\left(h+e^{\frac{a\sqrt{s}}{\sqrt{\eta}}}h-\sqrt{s}\sqrt{\eta}+e^{\frac{a\sqrt{s}}{\sqrt{\eta}}}\sqrt{s}\sqrt{\eta}\right)\right)\right)$$

$$\left(e^{st-\frac{\sqrt{sx}}{\sqrt{\eta}}}\left(e^{\frac{a\sqrt{s}}{2\sqrt{\eta}}+\frac{2\sqrt{sx}}{\sqrt{\eta}}}hc[b]+e^{\frac{a\sqrt{s}}{2\sqrt{\eta}}}hc[b]+e^{\frac{a\sqrt{s}}{\sqrt{\eta}}+\frac{\sqrt{sx}}{\sqrt{\eta}}}hC[0]-e^{\frac{a\sqrt{s}}{2\sqrt{\eta}}+\frac{2\sqrt{sx}}{\sqrt{\eta}}}hC[0]-e^{\frac{a\sqrt{s}}{2\sqrt{\eta}}}hC[0]+e^{\frac{\sqrt{sx}}{\sqrt{\eta}}}hC[0]+e^{\frac{a\sqrt{s}}{\sqrt{\eta}}+\frac{\sqrt{sx}}{\sqrt{\eta}}}\sqrt{s}\sqrt{\eta}C[0]-e^{\frac{\sqrt{sx}}{\sqrt{\eta}}}\sqrt{s}\sqrt{\eta}C[0]\right)\right)/$$
$$\left(h+e^{\frac{a\sqrt{s}}{\sqrt{\eta}}}h+s\left(\frac{1}{2}ae^{\frac{a\sqrt{s}}{\sqrt{\eta}}}+\frac{ae^{\frac{a\sqrt{s}}{\sqrt{\eta}}}}{2\sqrt{s}}\frac{h}{\sqrt{\eta}}-\frac{\sqrt{\eta}}{2\sqrt{s}}+\frac{e^{\frac{a\sqrt{s}}{\sqrt{\eta}}}\sqrt{\eta}}{2\sqrt{s}}\right)-\sqrt{s}\sqrt{\eta}+e^{\frac{a\sqrt{s}}{\sqrt{\eta}}}\sqrt{s}\sqrt{\eta}\right)$$

eq6 = Simplify[eq5A/.eq5, $\eta > 0$ && $a > 0$ && $\beta[n] > 0$]

$$\left\{\left(e^{-\frac{\beta[n](iax+i\eta\beta[n])}{a^2}}\left(-2iae^{\frac{1}{2}i\beta[n]}\left(1+e^{\frac{2ix\beta[n]}{a}}\right)hc[b]+2C[0]\left(ia\left(e^{\frac{1}{2}i\beta[n]}-e^{\frac{ix\beta[n]}{a}}-e^{\frac{i(a+x)\beta[n]}{a}}+e^{\frac{i(a+4x)\beta[n]}{2a}}\right)h+e^{\frac{ix\beta[n]}{a}}(-1+e^{i\beta[n]})\eta\beta[n]\right)\right)\right)/\right.$$
$$\left.(-2ia(1+e^{i\beta[n]})h+(ae^{i\beta[n]}h+3(-1+e^{i\beta[n]})\eta)\beta[n]+ie^{i\beta[n]}\eta\beta[n]^2)\right\}$$

eq7 = (`Simplify`[(eq6/.eq5b1)])[[1]]

$$\left(2ae^{-\frac{\beta[n](iax+i\eta\beta[n])}{a^2}}h\left(ie^{\frac{1}{2}i\beta[n]}\left(1+e^{\frac{2ix\beta[n]}{a}}\right)c[b](ah+i\eta\beta[n])+C[0]\left(-ia\left(e^{\frac{1}{2}i\beta[n]}-e^{\frac{ix\beta[n]}{a}}-e^{\frac{i(a+x)\beta[n]}{a}}+e^{\frac{i(a+4x)\beta[n]}{2a}}\right)h+\left(e^{\frac{1}{2}i\beta[n]}+e^{\frac{ix\beta[n]}{a}}-e^{\frac{i(a+x)\beta[n]}{a}}+e^{\frac{i(a+4x)\beta[n]}{2a}}\right)\eta\beta[n]\right)\right)\right)/$$
$$(\beta[n](ah(ah+2\eta)+\eta^2\beta[n]^2))$$

eq8 =
$C[x,t]=$
eq2A+

$$\sum_{n=0}^{\infty}\left(2ae^{-\frac{\beta[n](iax+t\eta\beta[n])}{a^2}}h\right.$$
$$\left(ie^{\frac{1}{2}i\beta[n]}\left(1+e^{\frac{2ix\beta[n]}{a}}\right)c[b](ah+i\eta\beta[n])+\right.$$
$$C[0]\left(-ia\left(e^{\frac{1}{2}i\beta[n]}-e^{\frac{ix\beta[n]}{a}}-e^{\frac{i(a+x)\,\beta[n]}{a}}+e^{\frac{i(a+4x)\,\beta[n]}{2a}}\right)h+\right.$$
$$\left.\left.\left.\left(e^{\frac{1}{2}i\beta[n]}+e^{\frac{ix\beta[n]}{a}}-e^{\frac{i(a+x)\beta[n]}{a}}+e^{\frac{i(a+4x)\beta[n]}{2a}}\right)\eta\beta[n]\right)\right)\right)$$
$$(\beta[n](ah(ah+2\eta)+\eta^2\beta[n]^2))$$

$$C[x,t]=c[b]+\sum_{n=0}^{\infty}\left(2ae^{-\frac{\beta[n](iax+i\eta\beta[n])}{a^2}}h\left(ie^{\frac{1}{2}i\beta[n]}\left(1+e^{\frac{2ix\beta[n]}{a}}\right)c[b](ah+i\eta\beta[n])+C[0]\left(-ia\left(e^{\frac{1}{2}i\beta[n]}-e^{\frac{ix\beta[n]}{a}}-e^{\frac{i(a+x)\beta[n]}{a}}+e^{\frac{i(a+4x)\beta[n]}{2a}}\right)h+\left(e^{\frac{1}{2}i\beta[n]}+e^{\frac{ix\beta[n]}{a}}-e^{\frac{i(a+x)\beta[n]}{a}}+e^{\frac{i(a+4x)\beta[n]}{2a}}\right)\eta\beta[n]\right)\right)\right)/$$
$$(\beta[n](ah(ah+2\eta)+\eta^2\beta[n]^2))$$

eq9 =
$M[t]=2*(a/2)*c[0]-2*(a/2)*c[b]-$

$$\sum_{n=0}^{\infty}\texttt{Integrate[}$$
$$\left(\left(2ae^{-\frac{\beta[n](iax+t\eta\beta[n])}{a^2}}h\right.\right.$$
$$\left(ie^{\frac{1}{2}i\beta[n]}\left(1+e^{\frac{2ix\beta[n]}{a}}\right)c[b](ah+i\eta\beta[n])+\right.$$
$$\left.\left.\left(e^{\frac{1}{2}i\beta[n]}+e^{\frac{ix\beta[n]}{a}}-e^{\frac{i(a+x)\beta[n]}{a}}+e^{\frac{i(a+4x)\beta[n]}{2a}}\right)\eta\beta[n]\right)\right)$$
$$(\beta[n](ah(ah+2\eta)+\eta^2\beta[n]^2))),\{x,-a/2,a/2\}]$$

$$M[t]=ac[0]-ac[b]-\sum_{n=0}^{\infty}\left(2a^2e^{-\frac{t\eta\beta[n]^2}{a^2}}h\left(2\left(-1+e^{i\beta[n]}\right)c[b](ah+i\eta\beta[n])+C[0]\left(2ah+\texttt{i}ah\beta[n]+2i\eta\beta[n]+\eta\beta[n]^2-e^{i\beta[n]}(2i+\beta[n])(-iah+\eta\beta[n])\right)\right)\right)/$$
$$(\beta[n]^2((ah(ah+2\eta)+\eta^2\beta[n]^2))$$

eq10 =

$$\left(\left(\left(\left(2a^2e^{-\frac{t\eta\beta[n]^2}{a^2}}\ h(2(-1+e^{i\beta[n]})c[b](ah+i\eta\beta[n])+\right.\right.\right.\right.$$
$$C[0](2ah+iah\beta[n]+2i\eta\beta[n]+\eta\beta[n]^2-$$
$$e^{i\beta[n]}(2i+\beta[n])(-iah+\eta\beta[n]))))$$
$$(\beta[n]^2(ah(ah+2\eta)+\eta^2\beta[n]^2))/.\mathrm{eq5b1}))(aC[0]-ac[b]),$$
$$\eta>0\&\&a>0\&\&\beta[n]>0\&\&h>0]$$

$$\frac{8a^2 e^{-\frac{t\eta\beta[n]^2}{a^2}} h^2}{\beta[n]^2(ah(ah+2\eta)+\eta^2\beta[n]^2)}$$

$$\text{eq11} = M[t]/M[\infty] = 1 - \sum_{n=0}^{\infty} \frac{8a^2 e^{-\frac{t\eta\beta[n]^2}{a^2}} h^2}{\beta[n]^2(ah(ah+2\eta)+\eta^2\beta[n]^2)}$$

$$\frac{M[t]}{M[\infty]} = 1 - \sum_{n=0}^{\infty} \frac{8a^2 e^{-\frac{t\eta\beta[n]^2}{a^2}} h^2}{\beta[n]^2(ah(ah+2\eta)+\eta^2\beta[n]^2)}$$

$$\text{eq13} = \frac{M[t]}{M[\infty]} = 1 - \left(\sum_{n=0}^{\infty} \frac{8a^2 e^{-\frac{t\eta\beta[n]^2}{a^2}} h^2}{\beta[n]^2(ah(ah+2\eta)+\eta^2\beta[n]^2)}\right) * \left(\sum_{m=0}^{\infty} \frac{8b^2 e^{-\frac{t\eta\delta[m]^2}{b^2}} h^2}{\delta[m]^2(bh(bh+2\eta)+\eta^2\delta[m]^2)}\right) * \left(\sum_{k=0}^{\infty} \frac{8c^2 e^{-\frac{t\eta\lambda[k]^2}{c^2}} h^2}{\lambda[k]^2(ch(ch+2\eta)+\eta^2\lambda[k]^2)}\right)$$

$$\frac{M[t]}{M[\infty]} = 1 - \left(\sum_{n=0}^{\infty} \frac{8a^2 e^{-\frac{t\eta\beta[n]^2}{a^2}} h^2}{\beta[n](ah(ah+2\eta)+\eta^2\beta[n]^2)}\right) \left(\sum_{m=0}^{\infty} \frac{8b^2 e^{-\frac{t\eta\delta[m]^2}{b^2}} h^2}{\delta[m]^2(bh(bh+2\eta)+\eta^2\delta[m]^2)}\right) \left(\sum_{k=0}^{\infty} \frac{8c^2 e^{-\frac{t\eta\lambda[k]^2}{c^2}} h^2}{\lambda[k]^2(ch(ch+2\eta)+\eta^2\lambda[k]^2)}\right)$$

PROBLEMS

15.1. Consider the cylindrical device defined in Section 15.2. Plot the effective time constant (in hours) as a function of the diffusion coefficient (range: $[5 \times 10^{-7}\,\text{cm}^2/\text{s}, 2 \times 10^{-6}\,\text{cm}^2/\text{s}]$). Select, from the graph, the diffusion coefficient that corresponds to a time constant of 45 hrs. Use a first-order approximation and the following specifications: $R = 1\,\text{cm}$, $H = 5\,\text{cm}$ and $h = 10^{-6}\,\text{cm}/\text{s}$.

15.2. Consider the cylindrical device defined in Section 15.2. Find the height H necessary to achieve a time constant of 30 hrs. Use a first-order approximation and the following specifications: $R = 1\,\text{cm}$, $\eta = 1.5 \times 10^{-6}\,\text{cm}^2/s$ and $h = 10^{-6}\,\text{cm}/\text{s}$.

15.3. Consider the cylindrical device defined in Section 15.2. Plot the effective time constant (in hours) as a function of the diffusion coefficient (range: $[5 \times 10^{-7}, 2 \times 10^{-6}]$). Select, from the graph, the diffusion coefficient that corresponds to a time constant of 45 hrs. Use a second-order approximation and the following specifications: $R = 1\,\text{cm}$, $H = 5\,\text{cm}$ and $h = 10^{-6}\,\text{cm}/\text{s}$.

15.4. Given the rectangular parallelepiped-shaped matrix in Section 15.3, find the relationship between the effective time constant and the diffusion coefficient. Use a first-order approximation.

15.5. Given the rectangular parallelepiped-shaped matrix in Section 15.3, plot the effective time constant (in hours) as a function of the diffusion coefficient (range: $[50\,\mu\text{m}^2/\text{h}, 500\,\mu\text{m}^2/\text{h}]$). Use a first-order approximation and the following dimension: $a = 75\,\mu\text{m}$, $b = 180\,\mu\text{m}$, $c = 380\,\mu\text{m}$ and $h = 180\,\mu\text{m}/\text{h}$.

REFERENCES

Berkland, C, King, M, Cox, A, Kim, K, & Pack, D. W. (2002). Precise control of PLG microsphere size provides enhanced control of drug release rate. *J Control Rel* 82: 137–147.

Simon, L. (2007). Analysis of heat-aided membrane-controlled drug release from a process control perspective. *Int J Heat Mass Transf*, 50(11–12), 2425–2433.

Simon, L, & Fernandes, M. (2004). Neural network-based prediction and optimization of estradiol release from ethylene–vinyl acetate membranes. *Comput Chem Eng*, 28(11), 2407–2419.

Simon, L, Weltner, A. N., Wang, Y, & Michniak, B. (2006). A parametric study of iontophoretic transdermal drug-delivery systems. *J Memb Sci*, 278(1–2), 124–132.

INDEX

Closed-form Solutions for Drug Transport through Controlled-Release Devices in Two and Three Dimensions, First Edition. Laurent Simon and Juan Ospina.